Nanoenergy

Flavio L Souza · Edson R Leite
Editors

Nanoenergy

Nanotechnology Applied for Energy Production

Second Edition

Springer

Editors
Flavio L Souza
Universidade Federal do ABC
Santo André
Brazil

Edson R Leite
Federal University of São Carlos
São Carlos, São Paulo
Brazil

ISBN 978-3-319-62799-1 ISBN 978-3-319-62800-4 (eBook)
https://doi.org/10.1007/978-3-319-62800-4

Library of Congress Control Number: 2017946666

Printed on acid-free paper

This Springer imprint is published by Springer Nature
The registered company is Springer International Publishing AG
The registered company address is: Gewerbestrasse 11, 6330 Cham, Switzerland

Preface

The worldwide energy demand has been currently addressed by hundreds of scientific published articles, books, newspapers, symposiums, scientific and technological conferences in the field. Accordingly, to these sources considerable enhancement on the worldwide energy consumption was projected for reaching the double and triple in 2050 and 2100 compared with 13 TW (Terawatts) needed nowadays. The increased global energy consumption with an uncontrollable CO_2 emission from fossil energy resource that supply the carbon-based technology is the challenge facing the humanity in this century. Free-carbon technology is an alternative way to produce clean and sustainable source of energy to prevent the urban air pollution and provide an environmental friendly. To reach this scenario a scalable and cost-effective methods to design new materials for engineering highly efficiency free-carbon technologies and fuels must be developed. In this sense, the future energy demand and CO_2 emission control has been part of the governmental program worldwide with huge financial support to the academy and industrial to create and developing technologies to store and produce free-carbon fuels. Early of 2012, several books on energy topic addressing the main achievements in Europa, USA, Asia by the leading experts were published. In the year after (2013), the first edition of the Nanoenergy book edited by springer was launched given an overview of the main developments in Brazil by leading experts on the energy field. In this context, motivated by the great interest of the scientific community this second edition of the Nanoenergy book brings new contributions and updated version of the first edition chapters regarding of recent progress in renewable, sustainable and clean energy production and storage. Indeed, the second edition of Nanoenergy book summarizes the latest advances in material, scalable-cost-effective manufacturing methods and technology applications by researcher's experts in major fields of physical, chemistry and material science in Brazil related to the energy production with a special chapter contribution on clean energy storage from a group in France. This second edition consists in 10 chapters, which six from the first edition were revised or completely changed to cover the current state-of-art in the area with three additional new contribution on the hot topic in energy field related to the solar fuel production, hydrogen fuel storage and wastewater conversion into chemical fuel. The fourth initial chapter is focused on the solar energy conversion into electricity via organic-inorganic or dye sensitizer cells; or hydrogen solar fuel via photo-chemical and electrochemical process. The followed chapters will address achievements in materials and devices application on biofuels, development of electrocatysts in nanoscale for oxygen reduction and ethanol oxidation reactions, lithium-ion, metal-hydrated and neutralization battery and hydrogen storage using metal-alloys. The books begin with an exciting critical overview of the classical bulk hetero-junction device focused on the replacement of the traditional use of fullerene derivative by inorganic and metal nanoparticles incorporation. The following contribution is focused on the latest achievements on dye-sensitizers which have allowed to modulate their photochemical and photophysical properties to rise the solar devices efficiency. The fundamental characteristics, synthetic process, charge transfer and other aspects are also reviewed. The next two chapter will address the state-of the-art on direct solar conversion into oxygen and hydrogen gas by using Earth abundant elements in electrode and suspended form. The experimental advances, chemical surface and defects, water oxidation reaction (OER) mechanism, charge transfer models, optical properties, structural effects and morphology design will be highlighted given an overview of the last decades followed by a critical view of remaining material and device drawbacks. To keep the multidisciplinary focus of this book, the next chapter gives an exciting overview of the biocatalytic phenomena when the enzymes are attached onto a solid platform describing the influence of this interface on charger transfer, which is the main drawback in this field that need to be overcome for improve the device efficiency. It will be also reviewed the fundamentals of thermodynamic and kinetics of this bioelectrodes till the most important challenges in miniaturizing this biodevices and developing bio-inspired technological systems. The development of electrocatalysts designed in nanoscale will be discussed in Chap. Developments in Electrocatalystsfor Oxygen Reduction and Ethanol Oxidation focusing on the main challenges and progress on oxygen reduction reaction (ORR) and ethanol oxidation reaction for low-temperature fuel cell application. The recent achievements on performance, charge capacity, most of the fundamental aspect such as charge transport models, electro-to-chemical reaction and mechanism on lithium-ion, metal-hydride and neutralization battery will also be covered on this second edition by followed chaps. Nanocomposites from V_2O_5 and Lithium-Ion Batteries to Neutralization Batteries. Assuring that most of the aspects on energy field will be addressed, the last chapter brings the latest advances on hydrogen fuel storage by using new ternary intermetallic based magnesium materials introducing the fishing approach, as powerful tool to discover new magnesium-rich phases materials. This ternary class of material crystalizes within a new 2D structure leading an original physical property, which are discussed in this chapter in terms of mechanical, electrical, magnetic and hydrogen sorption.

The editors truly appreciate the efforts and commitment of the current and new authors to provide revised and extended manuscripts to this second edition of Nanoenegy book. We hope that this book provides information and guidance to the students, scientists, engineering, governments and companies interested in developing zero-emission coal technologies. In other words, all the contributions summarize an exciting overview and understanding on nanoscience and nanotechnology leading an essential comprehension about the urgency to consolidate both knowledges to reach the desired technology.

Santo André, Brazil
May 2017

Prof. Dr. Flavio L Souza
Prof. Dr. Edson R Leite

The original version of the book was revised:
Incorrect editor names have been updated.
The erratum to the book is available at
https://doi.org/10.1007/978-3-319-62800-4_11

Contents

Hybrid Solar Cells: Effects of the Incorporation of Inorganic Nanoparticles into Bulk Heterojunction Organic Solar Cells

Jilian Nei de Freitas, João Paulo de Carvalho Alves and Ana Flávia Nogueira

1 Introduction

There is a continuously growing demand for clean and renewable energy, impelled by the need of bringing electricity to remote areas and due to an increase of the population, which requires more (and safer) energy, at the same time minimizing the impacts on Earth and nature. Solar energy is considered a promising alternative to fulfill these aims.

For many decades the photovoltaic industry has been dominated by solid-state devices based mainly on silicon [1]. The energy conversion efficiency of the best monocrystalline Si photovoltaic cells is ~25% [2, 3], which is very close to its theoretical limit of 31% [4]. However, the manufacturing of Si-based devices is very expensive due to the requirements for high purity crystalline semiconductor substrates [5]. Such drawbacks results in the high costs associated with solar energy exploration [6]. In order to increase the market share of photovoltaic technology, considering, for example, the application in low-scale consumer goods, such as cells phones, laptops, energetic bags and clothes, etc., the development of low-cost devices is extremely necessary.

In this scenario, organic solar cells (OSC) appear as very interesting candidates. Since these devices are usually assembled with organic semiconductors, either small molecules or polymers, they show great promise due to the synthetic variability of organic materials, their low-temperature processing (similar to that

---

J.N. de Freitas (✉)
Centre for Information Technology Renato Archer (CTI), Rod. D. Pedro I, Km 143,6, Campinas, SP 13069-901, Brazil
e-mail: jilian.freitas@cti.gov.br

J.N. de Freitas · J.P. de Carvalho Alves · A.F. Nogueira (✉)
Laboratory of Nanotechnology and Solar Energy, Chemistry Institute, University of Campinas (UNICAMP), P.O. Box 6154, 13083-970 Campinas, SP, Brazil
e-mail: anaflavia@iqm.unicamp.br

F.L. de Souza and E. Leite (eds.), *Nanoenergy*,
DOI 10.1007/978-3-319-62800-4_1

applied to common plastics), and the possibility of producing lightweight, flexible, easily manufactured, and inexpensive solar cells. Moreover, the high optical absorption coefficients of conducting polymers, in comparison to silicon, provide the possibility of preparation very thin (100–200 nm) solar cells.

Recent progress in the field of polymer-based OSC has led to devices with PCE of $\sim 10\%$ [7, 8]. To further enhance the competitiveness of polymeric OSC, the efficiency and long-term stability remain crucial issues. The photocurrent in these solar cells is limited by the light-harvesting capability of the polymers used in the active layer of the device. The synthesis of new small band gap polymers has been intensively investigated for the purpose of overcoming such drawback [9]. Nevertheless, it remains a complicated matter, since changing the band gap energy usually changes the energetic value of the highest occupied molecular orbital (HOMO), which have unfavorable implications on the open-circuit voltage of the cells. Morphology is also important in this context since it directly affects the charge transport, and an intimate contact between donor and acceptor materials on a nanoscale range may be difficult to achieve due to phase separation. A better understanding of the processes at the nanoscale level, in layer-to-layer interfaces, is necessary and the exact role of phase separation remains subject of active research.

To overcome some of the drawbacks, different types of acceptor materials have been applied in the photoactive layer of OSC. When one of the components of the device is replaced by an inorganic counterpart, these devices are referred to as hybrid solar cells (HSC). Figure 1 shows the structure and dimensions of nanomaterials typically used in OSC and HSC.

The use of inorganic nanoparticles in optoelectronic devices has some advantages, mainly related to the versatility of these materials, which can be synthesized in a great variety of sizes and shapes, according to the desired properties. Usually, the so-called inorganic "nanoparticles" are structures that present at least one dimension with size between 1 and 100 nm. Since these materials are very small, their properties such as absorption, emission, electron affinity, etc., depend on the size (diameter) of the nanoparticle. For example, as the diameter decreases, the absorption maximum is blue-shifted, as a result of a change in the band gap level due to the quantum confinement effect. Besides the quantum size effect, inorganic nanoparticles offer advantage because they can be can be synthesized in a great variety of shapes, such as spheres, prisms, rods, wires, and even larger and more complex structures, such as tetrapods or hyperbranched nanocrystals. For these materials, not only the optical properties but also the solubility can be controlled by varying size or shape of the nanostructures. Figure 2 shows examples of metal nanoparticles with various shapes and sizes, and the absorption characteristics of colloidal nanoparticle solutions [10].

In this chapter, recent progress on the application of inorganic nanoparticles in OSC is discussed. The chapter is divided into three sections: the first contains basic concepts of the assembly and principle of operation of classical OSC; the second

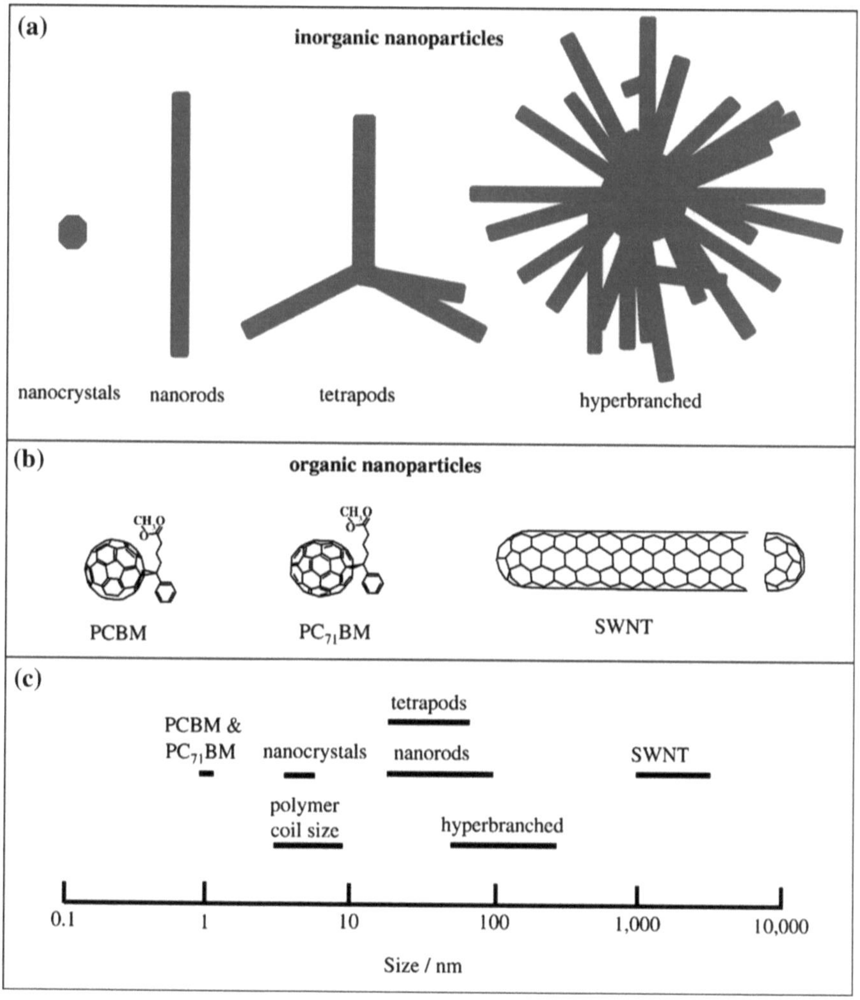

Fig. 1 Structural depictions and approximate dimensions of nanomaterials used in polymer photovoltaic cells. The structures shown in **a** are for CdSe nanoparticles. The structures shown in **b** are for carbon-based nanomaterials used in OSC. The size ranges shown in **c** are estimates based on literature reports for these materials. Reprinted from Ref. [24]. Copyright 2008, with permission from Elsevier

and third parts review recent results on OSC containing inorganic semiconductor nanoparticles and metal nanoparticles, respectively. Reviews on the synthesis and properties of semiconductor nanoparticles [11–16] and metal nanoparticles [17–22], as well as other reviews on the application of inorganic nanoparticles in opto-electronic devices [23–25] can be found elsewhere.

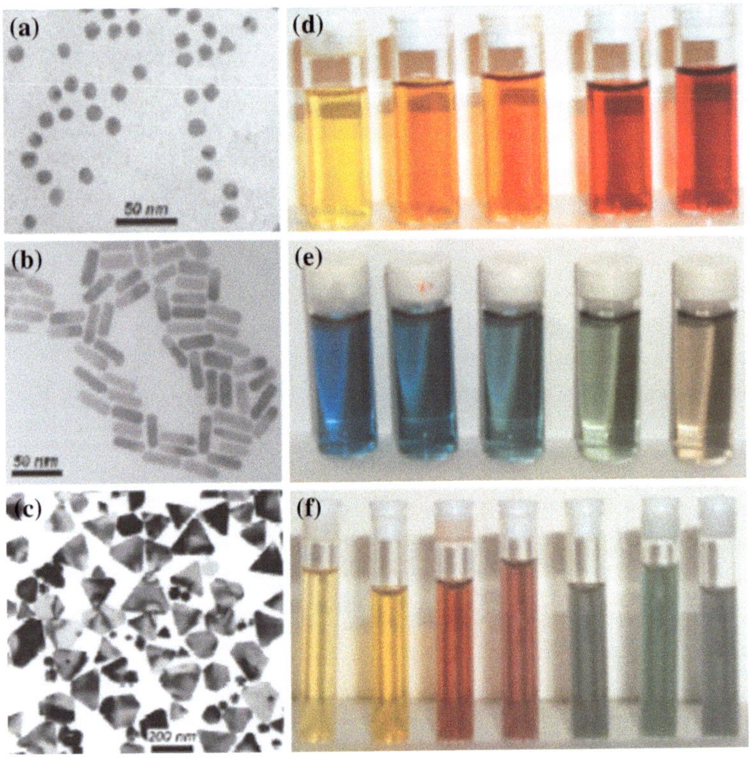

Fig. 2 (*Left*) Transmission electron micrographs of Au nanospheres and nanorods (**a**, **b**) and Ag nanoprisms (**c**). (*Right*) Photographs of colloidal dispersions of Au–Ag alloy nanoparticles with increasing Au concentration (**d**), Au nanorods of increasing aspect ratio (**e**) and Ag nanoprisms with increasing lateral side (**f**). Reprinted from Ref. [10]. Copyright 2004, with permission from Elsevier

2 Organic Solar Cells

The first OSC was developed by Tang [26] in 1986. The device consisted of a thin, two-layer film fabricated from copper phthalocyanine and a perylene tetracarboxylic derivative deposited between two electrodes and presented ~1% efficiency. Since then, different types of organic materials have been applied in OSC, such as antracene, perylene, porphyrins, phthalocyanines, fullerenes, carbon nanotubes, graphene, oligomers, and conjugated polymers.

Organic semiconductors differ from inorganic semiconductors in many aspects. The charge carriers generated in a photoactive organic material are not spontaneously dissociated but form a bound electron–hole pair (exciton) due to relatively high-binding energies (of the order 400 meV [27]), in comparison to a few meV observed for inorganic semiconductors. This is a consequence of the low electric permittivity and localized electron and hole wave functions in organic semiconductors, enhancing the Coulombic attraction between electron and hole [28]. Also the noncovalent electronic interactions between organic molecules are weak compared to the strong interatomic electronic interactions of covalently bonded inorganic semiconductor materials like silicon, so the wave function of the electron is spatially restricted, allowing it to be localized in the potential well of its conjugate hole (and vice versa). As result, a tightly bound electron–hole pair (Frenkel exciton or mobile excited state) is the usual product of light absorption in organic semiconductors [29]. Therefore, in conjugated polymers at room temperature only approximately 10% of the photoexcitations are spontaneously dissociated into free charge carriers [30]. The typical lifetime of excitons is hundreds of picoseconds [31], after which they recombine radiatively or non-radiatively. In consequence, the power conversion efficiency (PCE) of pure conjugated polymer-based solar cells is typically low, ca. 1–2% [32]. It is necessary to combine polymers with a suitable electron acceptor material, to provide an efficient dissociation of excitons and separation of the charge carriers and, therefore, enhance PCE. The use of an intermixed layer based on two materials with different electron affinity also allows the transport of the electrons and holes in the separate materials, with reduced probability of recombination.

Classical OSC are based on the combination of an electron donor/hole transporting material, usually conducting polymers or small organic molecules, and an electron acceptor/electron-transporting material, usually fullerene (C_{60}) and its derivatives. Since light absorption results in the formation of excitons, these devices are also known as excitonic solar cells [33]. The small exciton diffusion length observed in organic semiconductors (i.e., 10–20 nm [34]), leads to the fact that only the excitons generated near the interface can effectively split. The energetic differences in the electron affinity and ionization potential of the two organic materials give rise to the driving force for exciton splitting at the interface. On the other hand, it is important to consider that organic films of the order of $\sim$100–200 nm or more are required to absorb a significant fraction of the incident light.

The overall processes occurring in polymer/fullerene OSC are represented in Fig. 3, and may be described as follows:

- Absorption of photons;
- Generation of exciton pairs in the photoactive material;
- Diffusion of excitons in the photoactive material toward the donor/acceptor interface;
- Dissociation of exciton and separation of the charge carriers at the boundary between donor and acceptor materials;
- Transport of the holes and electrons to the electrodes;
- Collection of the holes and electrons by the electrodes.

Due to the excitonic nature of these devices, bilayer OSC, where two layers of different materials are co-deposited on top of each other, usually show lower efficiency due to the small interface for exciton dissociation [35]. Bulk heterojunction (BHJ) OSC, where the two materials are organized in an interpenetrated network,

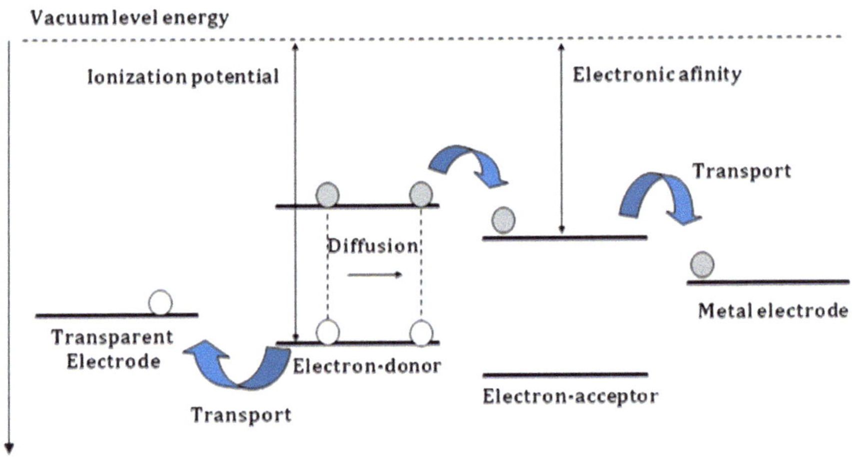

Fig. 3 Simplified energy diagram and the main steps of the photovoltaic process in a typical OSC. The incident photon creates an electron–hole pair (exciton). The exciton diffuses to the interface between the donor and acceptor materials, where it is separated into free charges. These charges need to be transported and collected on the respective electrodes to produce a photocurrent

present an elegant form of minimizing the problems of exciton dissociation. In 1995, it was shown that using this approach the performance of OSC could be significantly improved [36, 37]. The nanoscale mixture of two components allows the formation an interface throughout all the film extension and thus, the photogenerated excitons rapidly split before recombination.

The development of the BHJ architecture and the discovery of photoinduced electron transfer from polymers to C_{60} [38] were the breakthrough for the development of more efficient polymer-based OSC. The photoinduced charge transfer in these materials is irreversible and very fast (~45 fs), with efficiencies of ~100% [39], while the recombination process is very slow (on the order of μs-ms) [40].

In a classical BHJ OSC device, indium tin oxide (ITO)-coated glass is used as conductive, transparent substrate. ITO has a high work function and optical transparency. This substrate is usually covered with a layer of a commercially available highly conducting polymer blend of poly(3,4-ethylene-dioxy)-thiophene: polystyrene-sulfonic acid (PDEOT:PSS). This material increases the work function of the electrode (the work functions are ~4.7 eV [41] and ~5 eV [42] for ITO and PEDOT:PSS, respectively) and smoothen the surface of the substrate for deposition of the active layer. In contrast, the top electrode should be a metal of low work function, such as Al, Mg, or Ca, to allow transport of electrons and injection into the metal. A thin layer of LiF is usually employed to improve the electron injecting process [43]. The selected metal should also provide an ohmic contact with the semiconductor.

A crucial point for charge separation is the alignment of the energy levels of the materials at the polymer/fullerene interface. It is energetically favorable when the energies of the HOMO and the lowest unoccupied molecular orbital (LUMO) of the conducting polymer lie at higher values than the HOMO and LUMO of fullerene, respectively. This ensures that electrons from the polymer are accepted by the fullerene, whereas holes remain in the polymer. Appropriate alignment of the energy levels is also important at the interfaces between conducting polymer and the hole collecting electrode, and between fullerene and the metal electrode.

The driving force for selective transport of electrons and holes to the opposite electrodes is attributed to a built-in potential [44]. As in the metal–insulator–metal model described by Parker [45], the difference between the work functions of the collecting electrodes induces the formation of an electric field, responsible for a selective charge transport. The diffusion concept was also used to explain selective charge transport. This concept considers that charge carrier transport is induced by concentration gradients, associated with the use of selective electrodes [46]. Both concepts have in common the fact that charge transport is directed by the asymmetry of the electrodes. Gregg and Hanna [47] proposed a model, where charge transport is assigned to the chemical potential gradient, formed by charge generation at the interface.

Morphology also plays a crucial role in BHJ OSC [48–52]. An efficient generation of charges depends on the formation of an interface in the exciton diffusion length, which is favored by a homogenous mixture of the components, while efficient charge transport to the electrodes depends on the formation of percolation pathways from that interface to the respective electrode, which is favored by a certain degree of phase separation. Furthermore, to achieve efficient charge transport, the concentration of the acceptor material (i.e., fullerene derivative) must be high in most cases.

Since both charge carriers exist in the heterojunction, there is a possibility that they might recombine before reaching the electrodes. There are two possible paths for recombination: (i) bimolecular recombination, where the electron and hole originated from the dissociation of a singlet exciton recombine with each other; and (ii) the recombination of electrons or holes deriving from the dissociation of different excitons. Although a large interface is good for exciton splitting, it also enhances the probability of recombination of the charges. Recombination is also strongly influenced by unbalanced hole and electron mobility. For the materials routinely employed in OSC, the electron mobility exceeds hole mobility by two or more orders of magnitude. Therefore, the improvement of hole transport mobility in polymers is an important parameter to improve the photovoltaic performance [53]. The recombination is also enhanced if the charge transport towards the electrodes is hindered by the domain boundaries between the electron donating and accepting materials [54], leading to a strong morphology-dependent behavior.

In excitonic solar cells, the dark current is usually governed by a mechanism different from the photocurrent. The dark current is limited by charge carrier injection at the electrodes, while the photocurrent derives from interfacial exciton dissociation. In organic semicondutors with low degrees of impurities, the dark current can be many orders of magnitude lower than the photocurrent under forward bias [48].

Photocurrent (or short-circuit current, Jsc) is considered to be proportional to light-harvesting, since broader light absorption should give rise to increased exciton generation. Most materials used in OSC strongly absorb light in the spectral range below 600 nm. Thus, a way to improve the photocurrent and, in consequence, the efficiency, is to develop new materials with improved light-harvesting. Photocurrent also depends on the film morphology (a balance between charge separation, transport and recombination), as discussed before.

The open circuit potential (Voc) in OSC is directly related to the energy difference between the HOMO level of the donor and the LUMO level of the acceptor [55–57]. Scharber et al. [58] proposed a relationship between Voc and the values of the HOMO (E_{HOMO}) energy level of the conducting polymer and the LUMO (E_{LUMO}) energy level of the fullerene derivative (Eq. (1), where *e* is the elemental charge).

$$V_{oc} = (1/e)(E_{HOMO}(\text{polymer}) - E_{LUMO}(\text{fullerene})) - 0.3\text{V} \quad (1)$$

Gadisa et al. [59] observed a linear relation of Voc with the HOMO energy level of the conducting polymer employed in the heterojunction. On the other hand, Yamanari et al. [60] did not observe any direct relation of Voc and HOMO values. These reports suggest that Voc in OSC is mainly related to the electronic structure of the acceptor, which may be tuned by molecular engineering. The value of Voc is also influenced by an interfacial factor associated with different morphologies of composite films [61].

The influence of the metal electrode on Voc and on overall device performance has been investigated by many authors [62–65]. The insensitivity of voltage behavior on the electrode characteristics reported by some authors was ascribed to the Fermi level pinning between metal electrodes and fullerenes via charged interfacial states [56]. However, other authors have shown that Voc depends on the work function of the electrodes for the cases where a Schottky junction is formed, and is independent of this parameter when an ohmic contact is established [63]. Ramsdale et al. [64] reported a linear relation of Voc with the metal electrode work function in bilayer devices. The dependency of Voc on the work function of a PEDOT:PSS based electrode was also shown, where the voltage values could be manipulated by doping of the electrode at different levels [66].

In 2001, Shaheen et al. [67] reported, for the first time, a PCE of 2.5% for a BHJ OSC assembled with [6,6]-phenyl C_{61}-butyric acid methyl ester (PCBM) and poly [2-methoxy-5-(3,7-dimethyloctyloxy)]-1,4-phenylenevinylene (MDMO-PPV). By changing the film deposition conditions and the concentration of PCBM the PCE was increased to 2.9% [68]. When MDMO-PPV was substituted with poly (3-hexylthiophene) (P3HT), over 3% of PCE was obtained [69, 70]. The further increase in efficiency using P3HT was related to the improved light-harvesting and charge transport of this material in comparison to MDMO-PPV or other poly (p-phenylene vinylene) derivatives used before [54, 71–73]. A better control of the nanomorphology of the P3HT/PCBM film, obtained by changing the solvent and the introduction of post-production treatment (annealing), further improved the efficiency of such devices [74, 75]. Annealing is considered to contribute to increase the photocurrent in the following ways: (i) crystallization of PCBM and P3HT [76–80], which improves the light absorption and charge transport, and (ii) increase in the yield of dissociated charges, correlated with a decrease of the ionization potential of P3HT [81].

The addition of small alkyl thiol molecules or other additives [82, 83], the use of low band gap conjugated polymers [9, 84] in combination with $PC_{71}BM$, and optimization of device design (i.e., tandem solar cells) [85, 86] further improved the PCE of these devices to ~7% [75, 87–89]. Table 1 summarizes the record PCEs achieved in the years between 2010 and 2016. There was an expressive enhancement of PCE in the period, from 7.4 to 10.8%, achieved with the use of a low band gap polymer and an inverted solar cell architecture, and by a control of the aggregation and morphology of the bulk heterojunction.

3 Semiconductor Nanoparticles in Organic Solar Cells

Considering the application in solar cells, inorganic semiconductor nanoparticles have interesting characteristics: (i) tunable optical properties; (ii) high extinction coefficients; (iii) high intrinsic dipole moments, which may contribute to the

Table 1 Highest PCEs achieved for BHJ OSC in the years between 2010 and 2016

BHJ	Device configuration	PCE (%)	Reference
PTB7:$PC_{71}BM$	ITO/PEDOT:PSS/PTB7:$PC_{71}BM$/Ca/Al	7.4	[88]
PTB7:$PC_{71}BM$	ITO/PEDOT:PSS/PTB7:$PC_{71}BM$/PFN/Al	8.39	[90]
PTB7:$PC_{71}BM$	ITO/PFN/PTB7:$PC_{71}BM$/MoO_3/Al	9.2	[91]
PBDTTT-EFT: $PC_{71}BM$	ITO/PEDOT:PSS/PBSTTT-EFT: $PC_{71}BM$/Ca/Al	9.0	[92]
PffBT4T-2OD: $PC_{71}BM$	ITO/ZnO/PffBT4T-2OD:$PC_{71}BM$/MoO_3/ Al	10.8	[7]
PTB7-Th:$PC_{71}BM$	ITO/PFN/PTB7-Th:$PC_{71}BM$/MoO_3/Al	10.61	[8]
PTB7-Th:$PC_{71}BM$	ITO/ZnO/PEOz/PTB7-Th: $PC_{71}BM$/MoO_3/Ag	9.57	[93]

PTB7 poly({4,8-bis[(2-ethylhexyl)oxy]benzo[1,2-b:4,5-b′]dithiophene-2,6-diyl}{3-fluoro-2-[(2-ethylhexyl)carbonyl]thieno[3,4-b]thiophenediyl}), *$PC_{71}BM$* phenyl-C_{71}-butyric acid methyl ester, *ITO* indium tin oxide, *PEDOT:PSS* poly(3,4-ethylenedioxythiophene):polystyrene sulfonate, *PFN* poly[(9,9-bis(3′-(*N,N*-dimethylamino)propyl)-2,7-fluorene)-alt-2,7-(9,9–dioctyl-fluorene)], *PBDTTT-EFT* poly[4,8-bis(5-(2-ethylhexyl)thiophen-2-yl)benzo[1,2-b;4,5-b′]dithiophene-2,6-diyl-alt-(4-(2-ethylhexyl)-3-fluorothieno[3,4-b]thiophene-)-2-carboxylate-2-6-diyl)], *PffBT4T-2OD* poly[(5,6-difluoro-2,1,3-benzothiadiazol-4,7-diyl)-alt-(3,3‴-di(2-octyldodecyl)-2,2′;5′,2″;5″,2‴-quaterthiophen-5,5‴-diyl)], *PTB7-Th* poly[4,8-bis(5-(2-ethylhexyl)thiophen-2-yl)benzo[1,2-b;4,5-b′]dithiophene-2,6-diyl-alt-(4-(2-ethylhexyl)-3-fluorothieno[3,4-b]thiophene-)-2-carboxylate-2-6-diyl)], *PEOz* poly(2-ethyl-2-oxazoline), *PDEPB* poly(*N*-dodecyl-2-ethynylpyridinium bro mide)

separation of charges; and (iv) multiple exciton generation, i.e., the absorption of one single photon may lead to the generation of more than one exciton [94–99]. Compact films of inorganic semiconductor nanoparticles also present very high charge transport characteristics [100].

Hybrid solar cells assembled with the substitution of fullerenes by a vast variety of inorganic semiconductor nanoparticles, such as $CuInS_2$ [101–110], $CuInSe_2$ [111], FeS_2 [112–114], Sb_2S_3 [115, 116], Bi_2S_3 [117] SnS [118], Ge [119], InP [120] and even Si [121], have been investigated as promising alternatives. Probably, the most used materials in these hybrid solar cells are the CdE and PbE nanostructures (where E = S, Se or Te) and, therefore, this section will focus on the results obtained upon the introduction of these materials in polymer-based OSC.

3.1 Cd-Based Inorganic Nanoparticles

In 2002, Huynh et al. [122] combined CdSe spherical nanoparticles of 7 nm and CdSe nanorods (7 nm $\times$ 60 nm) with P3HT and observed that the maximum incident photon-to-current efficiency (IPCE) value was increased from $\sim$20 to $\sim$55% when the nanospheres were substituted by the nanorods. The best device presented Jsc of 5.7 mA cm^{-2}, Voc of 0.7 V and FF of 40%, with a PCE of 1.7%. Later, it was found that, by optimizing the solvent mixture used during film deposition, even higher IPCE values could be obtained [123, 124]. It was also shown that the use of nanorods instead of spherical particles allows the formation of a highly connected network of particles homogeneously distributed through the polymer film, i.e., nanorod/polymer films were found to be more spatially homogeneous than the spherical nanoparticles/polymer films [125].

Other CdSe structures, such as tetrapods [126, 127] and hyperbranched nanocrystals [128] have also been used in combination with P3HT or MDMO-PPV, resulting in devices with PCE around 1–2%. Sun et al. [128] compared the performance of solar cells assembled with CdSe nanorods or tetrapods and observed that the branched nanoparticles lead to more efficient devices, because the elongated and branched particles provided extended electric pathways. For example, charge separation was found to be efficient if the exciton is generated in a CdSe branch in direct contact with P3HT. [129] Furthermore, it has been argued that the use of tetrapods with longer arms aids in a better spatial connectivity, leading to better charge transport [130]. Dayal et al. [131] used microwave conductivity measurements to show that the increased aspect ratio from dots to rods to tetrapods provides conduits for electrons to move away from the dissociation sites, thus improving the generation of long-lived charge carriers.

In 2005, Sun et al. [132] used CdSe tetrapods in combination with P3HT and films spin-cast from 1,2,4-trichlorobenzene solutions resulted in devices with PCE of 2.8%, a much higher value than that obtained using films spin-cast from chloroform. The solvent used during film deposition significantly affects the morphology, and its manipulation is a key parameter in hybrid systems. In 2003, Huynh et al. [124] had already demonstrated that binary solvent mixtures, such as pyridine/chloroform, could be used to control the dispersion of the nanoparticles in the polymer matrix. By varying the ratio between the solvents in the mixture, where pyridine act as a ligand for the nanoparticles, the phase separation could be tuned from the micrometer scale to the nanometer scale.

In 2010, a PCE of 3.2% was reported by Dayal et al. [133] for a bulk heterojunction solar cell assembled with CdSe tetrapods and a low band gap polymer. The success of this device was attributed to a combination of the advantages of tetrapods with the improved light-harvesting properties of the polymer PCPDTBT (poly[2,6-(4,4-bis-(2-ethylhexyl)-4H-cyclopenta[2,1-b;3,4-b0]dithiophene)-alt-4,7-(2,1,3-benzothiadiazole)]). In 2011, Kuo et al. [134] investigated mixtures of CdSe tetrapods and the low band gap polymer PDTTTPD, comprising 2,5-di (thiophen-2-yl)thieno[3,2-b]thiophene and thieno[3,4-c]pyrrole4,6-dione units. Upon annealing at 130 °C for 20 min, the PCE increased from 1 to 2.9%. The removal of surface pyridine molecules led to a reduction in the inter-particle distance, increasing the CdSe packing density, which was presumed to be the responsible for improving the electron transport. Other authors reported similar effects using thermal annealing [135–137].

In 2012, an even higher PCE of 3.42% was reported by Celik et al. [138] for a combination of the polymer PCPDTBT with CdSe nanorods, instead of the tetrapods. The improvement was achieved by performing a series of washing steps with methanol and n-hexane, followed by ligand exchange with pyridine. This procedure led to the removal of excess free ligands from the surface of the nanoparticles.

Chemical vapor annealing is another post-film deposition treatment that have been used to tune the surface of the nanoparticles and the film morphology. Wu et al. [139] used 1,3-benzenedithiol (BDT) to treat CdSe nanorods/P3HT hybrid films and the devices achieved a PCE of 2.65%. The BDT molecules diffuse into the film and react with the CdSe surface, replacing bulky alkylphosphonic acid residual molecules, thus changing the aggregation of CdSe nanorods in the film.

Despite the several reports showing reasonable PCE for devices containing elongated or branched nanoparticles, there are several attempts to make hybrid solar cells using spherical CdSe nanoparticles that resulted in low efficiency devices (PCE < 1%). In 2006, Choi et al. [140] reported hybrid solar cells based on a mixture of CdSe nanoparticles of 5 nm with P3HT or MEH-PPV (poly (1-methoxy-4-(2-ethylhexyloxy-2,5-phenylenevinylene)) and obtained Jsc of 2.16 μA cm^{-2}, Voc of 1.0 V, FF of 20% and PCE of 0.05% for the best sample. In 2007, Tang et al. [141] used spherical nanoparticles covered with 2-mercaptoacetic acid in solar cells in combination with MEH-PPV. The devices delivered Jsc of only 2.6 μA cm^{-2}, Voc of 0.58 V and FF of 28%. Han et al. [142] used spherical CdSe nanoparticles crystallized in *zinc blend* structure, with 4.5 nm of diameter, covered with 1-octadecene and oleic acid (OA), and combined with the polymer MEH-PPV in solar cells. After annealing (post-production thermal treatment), the devices delivered Jsc of 2.0 mA cm^{-2}, Voc of 0.90 V, FF of 47% and PCE of 0.85%. In 2010, Jiang et al. reported a PCE of 0.25% with hybrid films of CdS/P3HT by the addition of an aromatic acid interface modifier [143]. In 2011, a PCE of 0.34% was reported by Chen et al. [144] for a device containing CdS nanoarrays and MEH-PPV. In 2015, Mohamed et al. [145] reported a device containing 30 wt % of CdS nanoparticles capped with thiophenol and MEH-PPV, that delivered Jsc of 0.053 mA cm^{-2}, Voc of 0.34 V, FF of 28% and PCE of 0.01% under 50 mW cm^{-2} of irradiation. It was considered that the capping agents may not passivate completely the surface of nanoparticles during the synthesis and, therefore, a tendency for aggregation may occur in the films (the surface roughness increased from 1 nm for a MEH-PPV film, to 27 nm for the composite material).

An expressive PCE of 2% was reached using spherical CdSe nanoparticles in combination with P3HT in 2010, as reported by Zhou et al. [146]. The authors achieved that efficiency by treating the nanoparticles with hexanoic acid, which removed the excess of surfactants accumulated around the quantum dot surfaces. The acid-treated nanoparticles are expected to exhibit higher efficiencies than non-treated nanoparticles because the connectivity of the interpenetrating nanoparticle network is possibly improved by partial elimination of the insulating ligand shell.

In 2011, Zhou et al. [147] applied the acid-assisted washing procedure to spherical CdSe nanoparticles (mean diameter of 4.7 nm) capped with different ligands, in combination with the low band gap polymer PCPDTBT. A PCE of 2.7% was obtained after optimization of the nanoparticle:polymer ratio, film thickness, thermal annealing and use of different cathode materials. The same group reported optimized devices using hexanoic acid-treated CdSe nanoparticles (mean diameter of 7.1 nm) and PCPDTBT, exhibiting a further improvement of the PCE up to 3.1% [148]. The size of the nanoparticles affected the efficiency because the larger nanoparticles might improve the connectivity of the inorganic network. The authors also showed that mixing spherical nanoparticles with nanorods could be beneficial for device performance, because the spherical particles reduce the nanorod–nanorod horizontal aggregation, improving the electron transport in the vertical direction [149]. In 2012, Fu et al. [149] reported a similar value of PCE (3.09%) for P3HT: CdSe devices, that was achieved by using n-butanethiol to remove the OA chains on the surface of CdSe nanoparticles. In 2013, the same group used acetic acid instead of n-butanethiol for the post-deposition treatment of P3HT:CdSe films, but found that higher efficiencies were obtained using the previously reported treatment (with n-butanethiol), because the acetic acid is not strong enough to replace all of the original ligands [150].

Ren et al. [151] reported a record PCE of 4.1% in 2011, for a hybrid device using spherical CdS nanoparticles with 4.0 nm of diameter, capped with n-butylamine, and combined with P3HT nanowires. In what was called grafting process, polymeric nanowire structures were formed by first dissolving P3HT and CdS in 1,2-dichlorobenzene and octane, respectively, and then mixing the two solutions together. The best device fabricated using this method, followed by a post-production ligand exchange processes using 1,2-ethanedithiol, delivered Jsc of 10.9 mA cm^{-2}, Voc of 1.1 V, FF of 0.35, and PCE of 4.1%. It was argued that, with this method, phase segregation could be minimized, while the P3HT/CdS interface is maximized. The formation of bicontinuous donor–acceptor phases and a well-defined interface in the hybrid photoactive layer could enhance the efficiency of charge separation and transport.

A different approach to control the interface and intermixing between the polymer and the nanoparticles was reported by Dixit et al. [152] in 2012. The authors obtained BHJ devices by combining P3HT with CdSe(ZnS) core–shell nanoparticles, starting from a bilayer structure and induced the mixing with thermal inter diffusion. The optimized device was thermally treated for at 120 °C for 7 min, delivering Jsc of 1.48 mA cm^{-2}, Voc of 0.62 V, FF of 0.45 and PCE of 5.1% under 8.2 mW cm^{-2} of irradiation.

A comparison between the reports from different authors reveal that other parameters, rather than the shape of the nanoparticles (i.e., dot, rod, tetrapod, hyperbranched, etc.), might be more determinant to the achievement of hybrid devices with higher efficiencies. Amongst these factors, the surface chemistry of the nanoparticle, film morphology and the control of the interface between the polymer and the inorganic material at a molecular level possibly play a more fundamental rule in these hybrid cells.

Figure 4 shows the structures of P3HT, MDMO-PPV, a conjugated polymer containing fluorene and thiophene units (PFT) and CdSe nanoparticles with 4.0 nm of preferential diameter, covered with trioctylphosphine oxide (TOPO). The absorption characteristics of films of these materials and solutions of CdSe

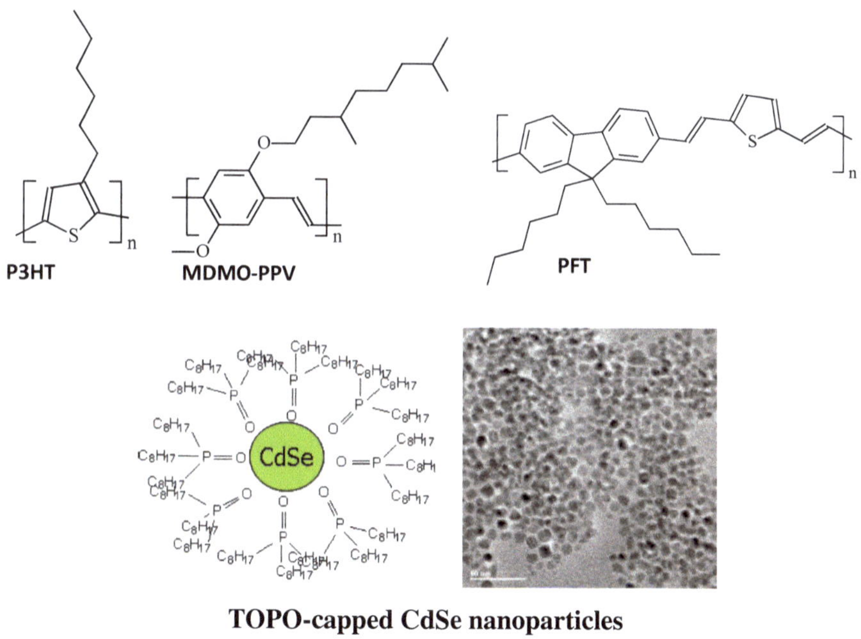

Fig. 4 Structures of the polymers P3HT. MDMO-PPV and PFT, and high-resolution transmission electron microscopy image (scale bar = 50 nm) of CdSe nanoparticles covered with trioctylphosphine oxide (TOPO). More information about PFT characteristics can be found elsewhere [153]

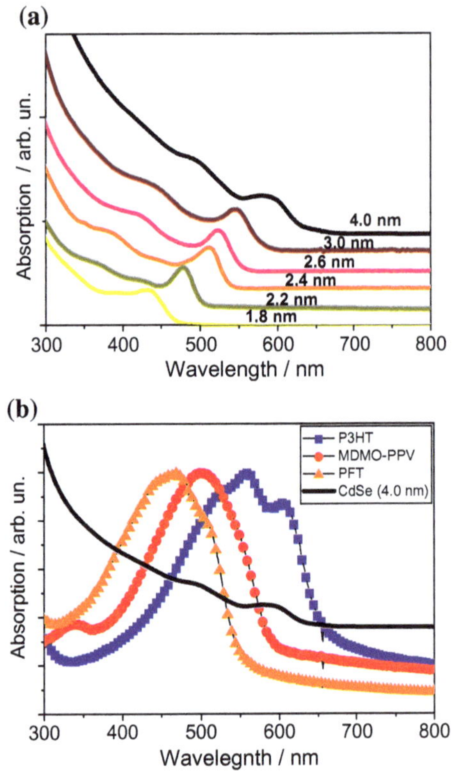

Fig. 5 Absorption spectra of **a** toluene solutions of CdSe nanoparticles with different sizes and **b** CdSe nanoparticles with 4.0 nm of preferential diameter and P3HT, MDMO-PPV, or PFT films. More information about PFT characteristics can be found elsewhere [154]

nanoparticles with different sizes are shown in Fig. 5. It can be seen that the absorption of MDMO-PPV or PFT and CdSe are complementary. Thus, it would be expected that, for hybrid cells assembled with the combination of these materials, both the polymer and the inorganic nanoparticles might contribute to increase light-harvesting.

Solar cells assembled with the mixture of TOPO-capped CdSe nanoparticles and PFT delivered low values of photocurrent and fill factor (Jsc ~150 μA cm^{-2}, FF ~0.25) [154]. The low Jsc and FF indicated a poor diode behavior for these systems, probably caused by poor electrical contacts at the interfaces and losses by recombination during charge transport. Such condition could be caused by a poor interaction between the polymer and CdSe, with the formation of phase separation or heterogeneous morphologies, or by a poor charge transfer process and/or poor charge transport. On the other hand, the devices showed very high open circuit voltage values (Voc ~0.6–0.8 V). As discussed in the previous section, for this kind of device Voc depends directly on the energetic difference between the HOMO of the polymer (−5.4 eV for PFT) and the LUMO of the electron acceptor material. Therefore, the high Voc observed in devices assembled with CdSe were attributed to a more favorable energy of the LUMO level of CdSe in comparison to the LUMO of PCBM [155]. Furthermore, Huynh et al. [155] found that, similar to OSC cells, the Voc of hybrid solar cells is not directly dependent on the difference of the work functions of the collecting electrodes because of the Fermi level pinning of metal electrode to the surface states close to the lowest unoccupied energy level of the semiconductor.

Some reports have shown a quenching of the polymer emission by interaction with CdSe nanoparticles that may be considered as a first indication of the existence of charge transfer between these materials, either by the injection of electrons (from the polymer to CdSe) or by injection of holes (from CdSe to the polymer) [156–160]. Nevetheless, it is generally accepted that the presence of TOPO hinders efficient charge transfer and might also prevent charge transport through hopping in the nanoparticle phase (inter-particle charge transport) [161, 162], leading to devices with low Jsc, FF and PCE. Considering that, many authors have focused on modifications of the nanoparticles surface, in order to obtain more favorable interactions of the nanoparticles with the polymers, while trying to minimize the traps and defects in the surface of the nanoparticle and maximize the dispersion of the inorganic materials in polymeric matrixes.

Ligand exchange, a routinely used process, is a powerful tool to change the properties of the nanoparticles. For example, the exchange of TOPO/dodecylamine with 3-mercaptopropionic acid (MPA) quenches the band edge emission and enhances the deep trap emission [163], while ligand exchange with EDT or 1,2-ethanediamine significantly improved the exciton dissociation yield and/or charge carrier mobility in CdSe films [164].

Pyridine has been widely used as capping ligand. These molecules have great affinity for the CdSe surface and, therefore, easily replace the bulky surfactants usually employed during the synthesis, such as TOPO, OA and others. The advantage of pyridine is that the smaller radius of these molecules allow a better interaction between nanoparticles and polymers, therefore facilitating charge transfer, which has led to devices with higher efficiencies [23, 159]. On the other hand, some authors reported that the use of pyridine also leads to the formation of agglomerates of CdSe, which might damage film morphology [141]. Lokteva et al. [165] reported the impact of a single or multiple pyridine treatments on the properties of OA-capped CdSe nanoparticles. Interestingly, repeated pyridine treatment was found to lead to more complete ligand exchange, which in turn enabled more efficient charge transfer, but at the same time, the performance of the solar cell was found to be reduced. The authors correlated that fact with the increased aggregation tendency of repeatedly pyridine-treated particles, which negatively influenced the morphology, as well as with a larger amount of surface defects in particles stabilized by the weak pyridine ligand shell.

Several surfactant-free strategies have also been investigated, aiming at the formation of more favorable interfaces for charge transfer.

Dasgupta et al. [166] investigated the incorporation of nanorods containing a p–n junction ($Cu_2S|CdS$ heterostructure) into a P3HT matrix to form hybrid BHJ devices. The epitaxial attachment between Cu_2S and CdS in the nanorods without any surfactant at the interface was expected to allow the formation of a depletion region at the junction between p- and n-sections in each nanorod. Thus, upon illumination, charge separation or exciton dissociation would occur at the nanorods themselves in addition to the conventional polymer:nanorod interfaces. It was observed that the parameters of solar cells depended on the relative length of Cu_2S in the $Cu_2S|CdS$ p–n junctions. In these devices, three different interfaces existed for dissociation of charge carriers: (1) P3HT:Cu_2S, (2) P3HT:CdS, and (3) $Cu_2S|$ CdS. While the former two junctions are excitonic in the nature, the p–n junction is expected to be the most efficient one in separating charge carriers. On the other hand, the efficiency of carrier transport through Cu_2S and CdS depends on the type of semiconductor: with an increase in the length of p-section in a nanorod, the moieties for electron transport is reduced, leading to a decrease in Jsc. Therefore, it was stated that the relative length of Cu_2S in $Cu_2S|CdS$ p–n junctions affected: (i) the relative surface area of P3HT:Cu_2S and P3HT:CdS interfaces, (ii) carrier transport through Cu_2S and CdS nanorods, and (iii) the width of depletion region at the $Cu_2S|CdS$ junction. An optimized device performance was achieved with the $Cu_2S|CdS$ (40:60) nanorod junction, with a PCE of 3.7%. This value is higher than the PCE observed for devices based on P3HT:CdS nanorods (2.4%) or P3HT:Cu_2S nanorods (0.3%). In the optimized device, the increased charge separation process probably superseded the detrimental effect of decreased electron-transporting pathways, reaching a balance and improved overall performance.

Another explored approach consists in the formation of Cd-chalcogenide nanoparticles in situ, by promoting the decomposition of a suitable precursor previously dispersed in the polymer matrix. Cappel et al. [168] used transient absorption spectroscopy (TAS) to investigate CdS:P3HT films prepared by in situ decomposition of a xanthate precursor of CdS and found that the charge generation via hole transfer from CdS to P3HT is more efficient than charge generation from P3HT to CdS. The polymer structure (especially concerning the presence of bulky ramifications) and the nature of the surfactant molecules on the nanoparticle surfaces seem to play a major role in the effectiveness of the photoinduced charge transfer processes in these hybrid systems [160]. TAS has been a tool used in several reports to prove that efficient charge injection processes indeed occur between inorganic nanoparticles and conjugated polymers [157, 158, 167].

Wood et al. [168] investigated the evolution of the film morphology during thermal annealing using in situ resonant Raman spectroscopy, aiming at a simultaneous monitoring of the formation of CdS nanoparticles via chemical decomposition of a Cd xanthate precursor and the changes in the molecular order of a P3HT matrix. The formation of CdS is fast, reaching completion within 2 min of heating to 160 °C, while the P3HT shows a corresponding disruption in its molecular order. This behavior is opposed to the increase in molecular order observed for thermal annealing of neat P3HT. It is considered that the polymer could act as a capping agent for the CdS nanoparticles and such interaction would result in disruption of the P3HT molecular order, as the polymer chains conform to the CdS surface. It would also lead to increased solubility of the inorganic nanoparticles in the polymer matrix. Alternatively, the decomposition process itself may cause localized damage, as some volatile products evolve and pass through the polymer film. In either case some degree of polymer disorder should be expected, and was considered intrinsic to the method.

Some authors reported the preparation of surfactant-free nanoparticles and their application in solar cells by depositing the hybrid photoactive layer by spray coating methods [169, 170]. Using this approach, Kumar et al. [171] reported that the inorganic nanoparticles have great tendency to aggregate, i.e., the average particle size was found to be 30 nm in their work. Such large aggregates affect the morphology and PCE of 0.6 and 1.02% was obtained for P3HT:CdS and PCPDTBT:CdS hybrid devices, respectively.

From a morphology perspective, phase separation is a phenomenon frequently observed in polymer–nanoparticle hybrid systems, and may lead to the formation of morphologies where the nanoparticles are organized in "islands" dispersed in a polymer "sea". This might cause poor charge transport in the inorganic phase. The crystallization, interchain interaction and phase separation of active layers depend on the concentration of the composite mixture and its loading amount [104, 105].

For polymer/CdSe mixtures, optimized photovoltaic responses have been obtained when 60–90 wt% of nanoparticles were incorporated into the polymer matrix [129, 171, 172, 223]. The need for a high concentration of nanoparticles is related to the nature of charge transport in the CdSe phase (high loading is a necessary condition for achieving a percolation network in the inorganic phase for charge transport through hopping). Furthermore, the extent of phase separation and aggregation of nanoparticles increase with the amount of CdSe loading [173]. Another aspect to consider is that, keeping the same concentration of CdSe nanoparticles dispersed in a polymer matrix, a percolation network is more easily formed when larger nanoparticles are used [162].

Dowland et al. [174] showed that, as the CdS concentration increases, the size of P3HT domains is reduced, favoring the migration of excitons within the polymer phase to an interface, thus enhancing the charge generation yield. A higher concentration of the nanoparticles may also result in a more crystalline character and a higher degree of interconnectivity within the CdS domains. On the other hand, the hole transport is favored by larger and more crystalline P3HT domains. Therefore, it was inferred that films with higher CdS concentrations having smaller, less crystalline P3HT domains, lead to higher charge generation, but lower photocurrent extraction. It was also shown that charge generation from P3HT excitons is heavily dependent on the nanomorphology, while charge generation through hole transfer from CdS was mostly unaffected by the morphology.

Two elegant approaches have been used in pursuit of CdSe-polymer composites with more controlled morphologies. The first consists in modification of the nanoparticle surface via ligand exchange with oligomers [175–177], which should facilitate the dispersion of the nanoparticles into a polymeric matrix. Sih et al. [178] synthesized a series of oligothiophenes containing thiol, which were used to functionalize the surface of CdSe nanorods previously functionalized with tetradecylphosphonic acid via ligand exchange. The authors observed that the emission characteristics of the nanorods depended on the number of oligothiophenes attached to the surface. Oligothiophene molecules containing aniline also have been used to passivate CdSe nanoparticles previously capped with thiol or carbonyl groups via ligand exchange [179].

The second approach consists of directly grafting the polymers to the nanoparticle, or grafting of a monomer followed by polymerization [180–183]. To this end, Zhang et al. [184] passivated CdSe nanorods with molecules containing thiol groups and aryl bromides. Afterwards, P3HT derivatives containing vinylic groups in the chains were reacted with the aryl bromides via Heck coupling. The nanocomposite obtained was compared to the one obtained by a simple physical mixture of P3HT and CdSe, revealing a better dispersion of the inorganic nanoparticles in the grafted materials, as well as an improved charge transfer behavior. Xu et al. [185] performed a similar approach, attaching P3HT to CdSe nanoparticles via Heck coupling and observed similar improvements in the composite morphology and charge transfer processes. In another example, poly(*N*-vinylcarbazole) (PVK) was grafted onto CdSe nanoparticles (CdSe-PVK) via the atom transfer radical polymerization of *N*-vinylcarbazole with an OH-capped-CdSe previously reacted with 2-bromoisobutyryl bromide [186]. The photovoltaic properties of devices assembled with P3HT were improved by using PVK-CdSe in comparison to OH-capped CdSe, due to enhanced compatibility between P3HT and PKV-CdSe, as seen with atomic force microscopy (AFM) images.

Besides the combination of Cd-chalcogenide nanoparticles with polymers, these materials have also been widely used as sensitizer [187–193] or co-sensitizer [194, 195] for TiO_2 in dye-sensitized solar cells (DSSC). The combination of CdSe with other nanoparticles, such as CdS [196–198], CdTe [199], ZnS [200] and SiO_2 [201], as sensitizers for DSSC has also been shown. The charge injection from CdSe into the conduction band of TiO_2 is a fast, efficient process. Similar behavior has been reported for CdSe (or PbSe) and ZnO [202, 203]. Efficient charge (or energy) transfer from CdSe to a series of organic dyes has also been shown [204–206].

Charge transfer from CdSe to multi-walled carbon nanotubes [207, 208] and from CdTe to single-walled carbon nanotubes [209], and energy transfer from CdSe/ZnS nanocrystals to graphene sheets [210] have also been reported. The ability of TOPO-capped CdSe nanoparticles to absorb light and inject electrons in C_{60} was shown by Biebersdorf et al. [211]. Advantage might be taken from these types of interactions, with application of inorganic nanoparticles in ternary systems containing bulk heterojuctions of polymer, PCBM and CdSe nanoparticles, for example [155, 212]. In these ternary systems it is expected that, upon the incidence of light, both the polymer and CdSe nanoparticles absorb photons and generate excitons, which could then lead to electron transfer from the polymer to the inorganic nanoparticle or PCBM, and from the nanoparticle to PCBM, or hole injection from CdSe to the polymer. Hole transport is accomplished by the polymer, whereas the electrons are transported in the PCBM phase.

Figure 6 shows the current–voltage (*J–V*) characteristics and IPCE curves obtained for solar cells assembled with ternary mixtures of PFT/CdSe/PCBM, using nanoparticles with different sizes (2.0, 3.0 or 4.0 nm), keeping the CdSe:PCBM ratio constant at 1:1 wt%. The results obtained for devices assembled with the

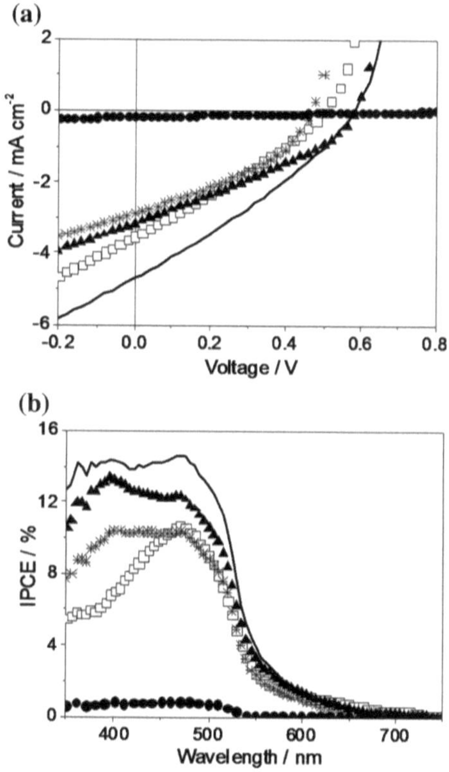

Fig. 6 *J–V* curves at 100 mW cm^{-2} (**a**) and IPCE spectra (**b**) of solar cells (active area ~0.1 cm^2) assembled with (*square*) PFT/PCBM, (*bullet*) PFT/CdSe (size = 4.0 nm), or ternary mixtures PFT/PCBM/CdSe containing nanoparticles with preferential diameters (*star*) 2.0 nm, (*black triangle*) 3.0 nm, or (*thick line*) 4.0 nm: For the ternary mixtures, the CdSe:PCBM ratio was kept constant at 1:1 wt%. For all samples, the total amount of PCBM and/or CdSe was kept constant at 80 wt% (20 wt% of polymer). Reprinted from Ref. [214]. Copyright 2010, Society of Photo Optical Instrumentation Engineers

binary mixtures PFT/PCBM and PFT/CdSe (4.0 nm) are also presented for comparison. The overall FF values observed for the ternary systems, although higher than that observed for the two-component based solar cells, are still low (~30%). This might be related to the poor light absorption and limited charge transport observed for the polymer used in the cell (absorption maximum at 430 nm, and hole mobility ~10^{-6} cm^2 V^{-1} s^{-1} for PFT [154].

In general, it was observed that Voc increases when CdSe nanoparticles are incorporated into the PFT/PCBM system, because of the more favorable energy of the LUMO level of CdSe in comparison to the LUMO of PCBM. In a previous work, we observed that the concentration of CdSe in the ternary mixtures also affects Voc: the higher the concentration of CdSe, the higher is Voc [155]. The fact that Voc is sensitive to the presence of CdSe nanoparticles indicates that the new interface created after addition of this material is effective, and probably contributes somehow to charge generation processes.

The photocurrent is also dependent on the concentration of CdSe. There is an optimum condition (CdSe:PCBM 1:1 wt% ratio) at which Jsc is much higher [155]. Considering that the TOPO-capped CdSe has a poor charge transport characteristic, the improvement of Jsc for a specific concentration of materials was explained as follows: an increase in the CdSe content leads to an increase in light-absorption (more excitons are generated and split, since CdSe is expected to inject charges into PCBM); on the other hand, the decrease in PCBM concentration is detrimental the electron transport to the aluminum electrode. Therefore, there is a compromise between the generation of more charges and their effective transport, reflected by the optimum concentration of CdSe and PCBM in the active layer.

Photocurrent is strongly dependent on the CdSe size as well, as shown in Fig. 6a. A decrease in Jsc is observed when smaller nanoparticles are used (preferential diameters of 2.0 or 3.0 nm). When larger nanoparticles are used (preferential diameter of 4.0 nm) an increase in Jsc was observed compared to the device assembled with PFT/PCBM only, leading to a significant enhancement of PCE from 0.5 to 0.8%.

In these ternary systems, CdSe nanoparticles may contribute to current generation in two different ways: (i) increasing the light-harvesting, since this material strongly absorbs visible light; or (ii) changing the film morphology. The contribution of CdSe for light-harvesting can be seen in the IPCE curves (Fig. 6b). It is evident that the addition of CdSe increases the IPCE values, and also changes the curve profile in the higher energy region (below 450 nm). However, if this contribution in light-harvesting was the main effect, the addition of CdSe would lead to an increase in the Jsc for all the devices, even when using the smaller nanoparticles. Thus, a morphologic effect is more likely to be the crucial parameter determining the efficiency of these devices.

The morphology can be seen in Fig. 7, where the AFM images obtained for films of PFT/PCBM, PFT/CdSe, and ternary mixtures PFT/CdSe/PCBM (1:1 wt% CdSe:PCBM ratio) with nanoparticles of different sizes are displayed. The polymer concentration was kept constant in all samples. The images show that the morphology of the PFT/PCBM film is very smooth, while for the sample composed of PFT/CdSe it is much rougher. For each different nanoparticle size the morphology changes drastically despite the fact that the same material concentrations are used. The surface becomes rougher as the size of CdSe nanoparticles increases. In principle, rougher surfaces could be related to the formation of larger aggregates of CdSe, or phase separation with the formation of a CdSe layer in the surface. The AFM image obtained for films of bare CdSe (4.0 nm of preferential diameter, for example) show that the nanoparticle aggregates are smaller and the surface roughness is lower compared to the characteristics of the film of PFT/PCBM/CdSe (4.0 nm), suggesting that the surface of the ternary system is not mainly composed of bare CdSe. This is an indication that the nanoparticles are distributed throughout the polymer/PCBM phase and do not suffer extended phase separation at the surface in the ternary systems.

The morphologic effect in these ternary systems was also observed with high-resolution transmission electron microscopy (HRTEM) images and optical microscopy images [155, 213]. The increase of CdSe nanoparticle concentration or size increases the agglomeration of CdSe, at the same time decreasing the agglomeration and the crystallite sizes of PCBM. The morphology changes observed are clearly related to the improvement of efficiency after the incorporation

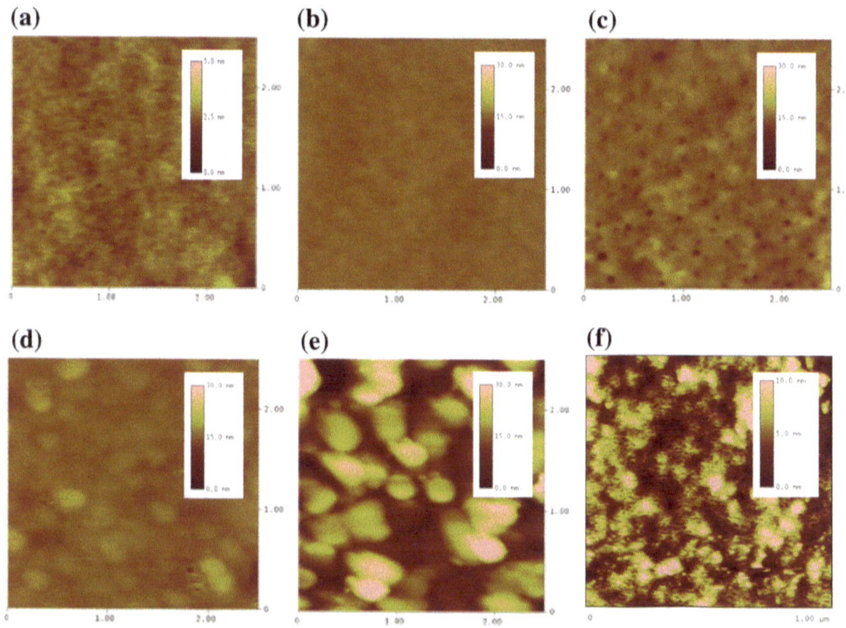

Fig. 7 AFM images obtained in the tapping mode for films of PFT containing 80 wt% of **a** PCBM, **b** PCBM:CdSe (2.0 nm), **c** PCBM:CdSe (3.0 nm), **d** PCBM:CdSe (4.0 nm), **e** CdSe (4.0 nm) and films of **f** pure CdSe film (4.0 nm), deposited onto PEDOT:PSS covered ITO-glass substrates. For the ternary component mixtures, the PCBM:CdSe ratio was kept constant at 1:1 wt%. The scan area is (2.5 μm × 2.5 μm). The scale indicates **a** 5 nm, **b**, **c**, **d**, **e** 30 nm, or **f** 10 nm. Reprinted from Ref. [214]. Copyright 2010, Society of Photo Optical Instrumentation Engineers

of CdSe nanoparticles in the PFT/PCBM mixture, especially when larger nanoparticles are used.

The effects of CdSe incorporation on the morphology of P3HT/PCBM mixtures were also investigated. Figure 8 displays AFM images for films of P3HT/PCBM (50 wt% of PCBM) and P3HT/PCBM/CdSe (size = 4.0 nm, PCBM:CdSe 1:1 wt%, total amount of 50 wt%) with and without post-production treatment: annealing was carried out in air, for 10 min at 110 °C.

After annealing, the RMS surface roughness of P3HT/PCBM film increased from 1.6 to 4.4 nm. Most reports show that surface roughness of P3HT/PCBM films increase after annealing, due to demixing and formation of P3HT crystalline regions and PCBM aggregates [81, 214, 215]. On the other hand, a few papers observed the opposite trend [216, 217]. These differences are attributed to different conditions of film preparation (solvent used, concentration of PCBM, fast or slow cooling/drying, time and temperature of annealing, etc.) [75, 217, 218]. After introduction of CdSe the RMS surface roughness is considerably increased to 10.6 nm. Annealing of the ternary system leads to the formation of larger agglomerates (RMS ~15.9 nm), probably related to a stronger phase separation of CdSe [213].

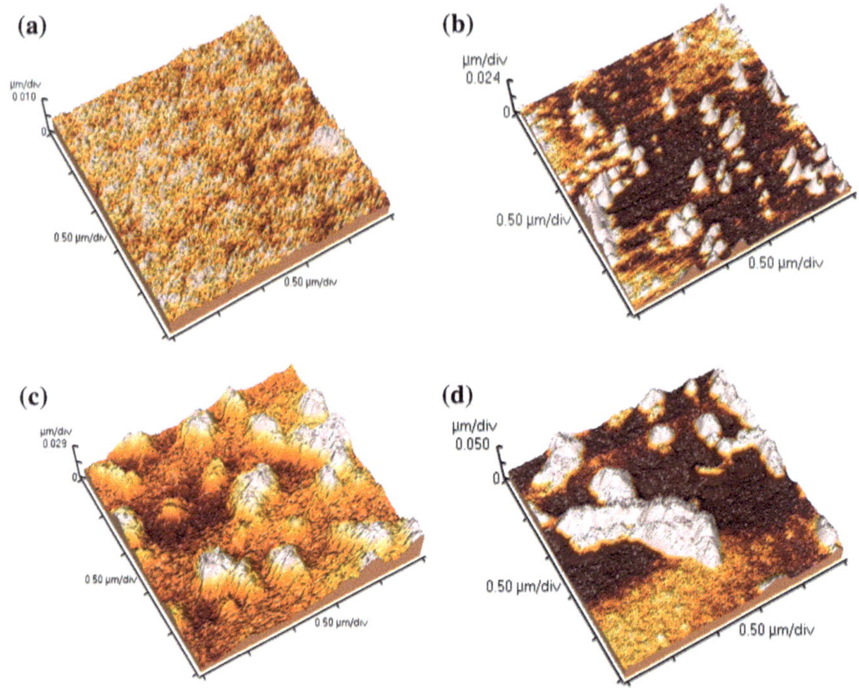

Fig. 8 AFM images obtained in the tapping mode for films of **a**, **b** P3HT/PCBM and **c**, **d** P3HT/PCBM/CdSe (4.0 nm), deposited onto PEDOT:PSS covered ITO-glass substrates, **a**, **c** before and **b**, **d** after annealing. The scan area is (2.5 μm × 2.5 μm)

The absorption characteristics of these films were also investigated. As expected, annealing induces the reorganization and crystallization of P3HT, resulting in a redshift of the absorption maxima and in the appearance of shoulders related to vibronic transitions. The addition of CdSe causes a similar effect in the absorption characteristics, probably related to the change in morphology induced by the presence of the inorganic nanoparticles. The annealing of the P3HT/PCBM/CdSe film does not lead to further significant changes in the absorption characteristics of the polymer in this system.

Interestingly, the incorporation of CdSe into P3HT/PCBM deteriorated the performance of the solar cells, differently from what was observed for the PFT/PCBM/CdSe system. Photophysical measurements revealed that the difference observed for the devices with polymers P3HT and PFT could be associated with different polymer–nanoparticle interactions [219]. The higher content of thiophene units in the P3HT polymer contributes significantly to the formation of a charge transfer complex between this polymer and CdSe, and to the deactivation of electron transfer process between the polymer and PCBM. In this type of complex, electrons can be transferred from P3HT to the CdSe and, since the TOPO-capped CdSe nanoparticles are not good charge carriers, electrons can be trapped on their surface, not being collected by the electrodes. From this perspective, only the P3HT segments not complexed with CdSe would be free to perform an effective electron transfer to PCBM, as the CdSe may act as a trap for most of the electrons from P3HT. For PFT systems, since this polymer has fluorene units in its segments, the possibility of formation of a charge transfer complex with CdSe is reduced. In other words, in PFT/PCBM/CdSe devices, both the polymer and CdSe nanoparticles are more available to interact with PCBM, which is responsible for the transport of electrons. In P3HT/PCBM/CdSe devices, the main interaction is between the polymer and CdSe, and PCBM molecules remains isolated, not being able to carry on the electron transport [220].

On the other hand, other reports found an enhancement of the photovoltaic characteristics of P3HT/PCBM solar cells after incorporation of CdS nanocrystals. The CdS quantum dots were synthesized with nearly monodispersed size, with mean diameter around ~5.0 nm and capped with OA, and were introduced into the BHJ OSC based on P3HT/PCBM spin-cast from dichlorobenzene, leading to a PCE increase from 2.03 to 2.52%, due to the enhancements observed in Jsc and FF [220]. The improvement of Jsc was correlated to a light absorption enhancement in the range of 350–550 nm, originated from the intrinsic absorption of CdS and the light scattering at the interfaces between CdS and P3HT:PCBMd. The authors also found that, by increasing the CdS concentration, the Jsc decreased. This effect was attributed to fact that, at high concentration, the agglomeration of CdS nanoparticles is more expressive, increasing the probability of recombination of the electrons by traps at the nanoparticles surface.

In 2016, Sharma et al. [221] investigated the incorporation of CdS nanoparticles synthesized with microwave synthesis into a PTB7:PCBM BHJ OSC device with an inverted architecture (ITO | ZnO | PTB7:PCBM:CdS | MoO_3 | Ag). The length and diameter of the nanoparticles ranged from 10 to 20 nm and 6 to10 nm respectively, characterizing low aspect ratio nanorods, rather than dots. The authors observed an improvement of the Jsc and PCE of the device at low CdS concentration (2–4 wt%), while at higher concentrations (6.7 or 10 wt%) the Jsc decreased significantly. The PCE was enhanced from 6.44% (without CdS) to 7.14% upon addition of 4 wt% of the CdS nanoparticles in the active layer. The trend observed in Jsc upon incorporation of CdS was similar to the trend observed for the series resistance. The initial decrease in the series resistance was correlated to a facilitate charge transfer from PTB7 to PCBM due to the presence of CdS, by a cascade mechanism. Further photoluminescence quenching was observed for the samples containing 2 and 4 wt% CdS in comparison to the sample containing only PTB7: PCBM, and was considered an evidence of a more efficient exciton dissociation in the presence of CdS. With UV–Vis absorption spectroscopy was confirmed that CdS has a poorer light absorption in comparison to PCBM and, therefore, it was ruled out that the contribution of CdS incorporation to Jsc increase could be mainly related to an increased light-harvesting. It was also considered by the authors that the CdS nanoparticles could provide additional interfacial area for charge transfer and/or charge transport. In such scenario, the addition of higher concentrations of CdS to the BHJ OSC might lead to agglomeration, which decreases the interfacial area, thus being detrimental to the device performance. With AFM was demonstrated that the RMS surface roughness increased with the increasing CdS concentration.

Table 2 summarizes the data reported for hybrid BHJ solar cells by different authors. As a general trend, the reported photocurrent and efficiency values vary significantly in quantum dots-based BHJ solar cells. This arises from problems inherent to the nanoparticle/polymer systems, such as poor dispersion properties of nanoparticles in solution and the phase separation frequently observed when films are deposited. The presence of bulky capping molecules on the nanoparticle surface, used to prevent aggregation during the synthesis, usually increases the solubility of these nanoparticles and their physical interaction with the polymer matrix, but also hinders charge transfer and the charge transport processes. All these parameters are crucial because they affect both the operation and the reproducibility of the solar cells.

3.2 Pb-Based Inorganic Nanoparticles

PbS and PbSe nanoparticles have shown great prominence in the development of photovoltaic devices. As is the case for the Cd-chalcogenide nanoparticles, the quantum confinement effect in Pb-chalcogenide nanoparticles allows the band gap energy values to be tuned with nanoparticle size modification, allowing adjustments in the positions of the valence and conduction bands of the semiconductor. In bulk PbS, the Bohr radius for the exciton is 20 nm, while for PbSe it is 46 nm, and the confinement effect begins to appear in nanoparticles with radii smaller than those values. When the radius of the nanoparticle is significantly smaller than the Bohr radius, the quantum confinement effect is intense, so that the energy value of the band gap becomes highly susceptible to the size of the nanoparticle [225–227]. For example, the band gap of PbS nanoparticles can be tuned over all near-infrared spectral range due to their large exciton Borh radius, and a bulk band gap of 0.41 eV [228]. Because of these features, PbS and PbSe show the possibility to extend light-harvesting in photovoltaic devices, and these materials have been used in hybrid and all-inorganic devices. The use of PbS and PbSe nanoparticles is also facilitated by the fact that the synthesis processes are often reproducible, with a fairly efficient size control. Another aspect is the possibility of obtaining nanoparticles dispersions with high stability. Classical synthesis methods to prepare Pb-chalcogenide nanoparticles generally use OA as organic stabilizer [229]. However, these oleate ligands are bulky and limit the charge transport and transfer processes when the inorganic nanoparticles are applied in solar cells.

In 2005, Watt et al. [230] reported the combination between the polymer MEH-PPV and PbS nanoparticles, obtaining devices with 0.7% of PCE under white light. In 2006, Cui et al. [231] demonstrated that the absorption of PbSe nanoparticles in the infrared region could contribute significantly to the photocurrent generation. In that work, PbSe nanoparticles with 6 nm of diameter, stabilized

Table 2 Performance parameters for a series of BHJ hybrid solar cells assembled with CdE (E = S, Se, or Te) and polymers, reported by different authors

Active layer	Nanoparticle structure	Solvent	Jsc (mA cm^{-2})	Voc (V)	FF (%)	PCE (%)	IPCE (max) (%)	Reference
CdS:PTB7:PCBM	Low aspect ratio nanorods (6–10 × 10–20 nm)	CB	14.64	0.74	65.6	7.14	~70	[222]
CdSe/ZnS:P3HT		CB/Toluene	1.48	0.62	45	5.1 (under 8 mW cm^{-2})	–	[153]
CdS:P3HT	Quantum dots (4 nm)	DCB/octane	10.9	1.1	35	4.1	–	[152]
CdSe:PCPDTBT	Nanorods	$CHCl_3$/py/TCB	9.90	0.66	50	3.23	~50	[139]
CdSe:PCPDTBT	Tetrapods	$CHCl_3$/py/TCB	9.02	0.674	51.5	3.13	55	[134]
CdSe:PCPDTBT	Nanorods/Quantum dots (4.7 nm)	CB	8.70	0.63	56	3.1	~45	[149]
CdSe:P3HT	Quantum dots (6.7 nm)	CB/py	7.91	0.68	56	3.0	~70	[150]
CdSe:PCPDTBT	Quantum dots (4.7 nm)	CB/TCB	8.16	0.59	55.3	2.65	55	[148]
CdSe:P3HT	Nanorods (15 × 30–100 nm)	$CHCl_3$/TCB	9.70	0.55	49.4	2.65	–	[140]
CdSe:P3HT	Nanorods (5 × 65 nm)	$CHCl_3$/py/TCB	8.79	0.62	50	2.6	60	[125]
CdSe:MDMO-PPV	Tetrapods	$CHCl_3$/py/TCB	6.42	0.76	44	2.4	52	[133]
CdSe:APFO-3	Branched	Xylene	7.23	0.95	38	2.4	44	[158]
CdSe:P3HT	Hyperbranched	$CHCl_3$/py	7.00	0.60	50	2.2	23	[129]

(continued)

Table 2 (continued)

Active layer	Nanoparticle structure	Solvent	Jsc (mA cm^{-2})	Voc (V)	FF (%)	PCE (%)	IPCE (max) (%)	Reference
CdSe:P3HT	Quantum dots (5.5 nm)	DBC	5.80	0.623	56	2.0	50	[147]
CdSe:P3HT	Quantum dots (~6 nm)	CB/py	5.40	0.65	56	1.95	~55	[151]
CdSe: MDMO-PPV	Tetrapods	$CHCl_3$/ py/DCB	7.30	0.65	35	1.8	45	[128]
CdSe:P3HT	Nanorods (7 × 60 nm)	$CHCl_3$/py	5.70	0.70	40	1.7	54	[123]
CdSe: P3HT-NH_2	Nanorods (7 × 30 m)	$CHCl_3$/py	5.80	0.57	40	1.4	–	[172]
CdSe:PF-PFT: PCBM	Quantum dots (4 nm)	Toluene	5.00	0.68	39	1.3	–	[155]
CdSe: MEH-PPV	Tetrapods	$CHCl_3$/py/CB	2.86	0.69	46	1.13	47	[127]
CdS:P3HT	Quantum dots	CB	3.54	0.61	33.3	0.72	36.5	[222]
CdSe: MEH-PPV	Quantum dots (3–5 nm)	CB/py	1.64	0.95	41.6	0.647	–	[143]
CdS:PBT	Nanorods	–	0.52	0.84	31	0.38	–	[223]
CdSe:P3HT	Quantum dots (5.5 nm)	$CHCl_3$/py	0.002	1.00	20	0.05	~0.1	[141]
CdSe-PVK: P3HT	Quantum dots	DCB	0.385	0.166	31.30	0.020	–	[187]
CdSe:PA_2	Nanorods	$CHCl_3$/py	0.147	0.264	28.4	0.011	–	[224]
CdSe: MEH-PPV	Quantum dots (5 nm)	$CHCl_3$	0.05	0.50	26	0.6 (at 514 nm)	12	[162]

(continued)

Table 2 (continued)

Active layer	Nanoparticle structure	Solvent	Jsc (mA cm^{-2})	Voc (V)	FF (%)	PCE (%)	IPCE (max) (%)	Reference
CdSe:P3HT	Nanorods (7 × 60 nm)	$CHCl_3$/py	–	–	–	–	59	[124]
CdSe: MEH-PPV	Quantum dots	$CHCl_3$	0.003	0.58	28	–	–	[142]
CdS:PCPDTBT	Quantum dots (~30 nm)	CB	5.68	0.584	30.8	1.02	–	[171]
CdS:P3HT: PCBM	Quantum dots (5.2 nm)	DCB	7.23	0.61	57	2.52	–	[221]

CB chlorobenzene; *DCB* 1,2-dichlorobenzene; *TCB* 1,2,4-trichlorobenzene; *py* pyridine; *PCBM* [6,6]-phenyl C_{61}-butyric acid methyl ester; *APFO-3* poly(2,7-(9,9-dioctyl-fluorene)-*alt*-5,5-(4′,7′-di-2-thienyl-2′,1′,3′-benzothiadiazole)); *MDMO-PPV* poly[2-methoxy-5-(3,7-dimethyloctyloxy)]-1,4-phenylenevinylene; *MEH-PPV* poly(1-methoxy-4-(2-ethylhexyloxy-2,5-phenylenevinylene)); *P3HT* poly(3-hexylthiophene); *PA_2* poly(2-(2-(4-octylthiophen2-yl)thiophen-3-yl) acetic acid); *PBT* Polybithiophene; *PCPDTBT* poly[2,6-(4,4-bis-(2-ethylhexyl)-4*H*-cyclopenta[2,1-*b*;3,4-*b′*]dithiophene)-*alt*-4,7-(2,1,3-benzothiadiazole)]; *PDTPQx* poly(2,3-didecyl-quinoxaline-5,8-diyl-*alt*-*N*-octyldithieno[3,2-*b*:2′,3′-*d*]pyrrole); *PF-PFT* poly[(9,9-dihexyl-9H-fluorene-2,7-diyl)-1,2-ethenediyl-1,4-phenylene-1,2-ethenediyl]-co-[(9,9-dihexyl-9H-fluorene-2,7-diyl)-1,2-ethenediyl-2,5-thiophene-1,2-ethenediyl]; *PVK* poly(*N*-vinylcarbazole); *PTB7* poly [(4,8-bis-(2-ethylhexyloxy)-benzo(1,2-b:4,5-b0) dithiophene)-2,6-diyl-alt-(4-(2-ethylhexyl)-3-fluorothieno[3,4-b] thiophene -)-2-carboxylate-2-6-diyl)] with oleate ligands, were associated with P3HT and the hybrid solar cells delivered a PCE of 0.14%. In 2007, other works investigated P3HT:PbSe and MEH-PPV: PbSe devices, but these hybrid solar cells still presented low efficiencies [232, 233]. Other works reported the use of amines as capping ligands for PbS, but the hybrid devices also showed low efficiencies [234, 235].

In 2009, Seo et al. [236] developed a ligand exchange method with acetic acid, a short-chain ligand, for treatment of P3HT:PbS films. OA-capped PbS nanoparticles were mixed with P3HT for film deposition and then the film was immersed into a acetic acid solution. FTIR spectroscopy was used to demonstrate that the OA molecules were replaced with acetic acid. Despite the low efficiencies obtained (0.01%), a two orders of magnitude enhancement of Jsc was observed for the devices treated with acetic, in comparison to the untreated devices.

It is noted that, for some time, the hybrid devices based on polymer:Pb-chalcogenide mixtures presented much lower efficiencies that the hybrid devices containing polymer: Cd-chalcogenide mixtures. Both type of nanoparticles show light-harvesting features that could contribute to the photocurrent generation when applied in the active layer of the solar cells. In fact, the PbS and PbSe nanoparticles show an even more extended harvesting of the spectrum. Nevertheless, the performance of devices containing these nanoparticles was significantly lower, revealing that other crucial factors had to be considered when designing hybrid cells with such materials.

The energy levels are one of the primordial factors for the development of efficient hybrid devices and are even more sensitive in the case of PbS and PbSe structures, because of the smaller band gap of these materials. The alignment of the energy levels at the interface between polymers and inorganic nanoparticles can facilitate or hinder the electron and hole injection processes. The combination of conjugated polymers and nanoparticles can form type I or type II heterojunction structures (Fig. 9) at the hybrid interface, depending on the optical characteristics of the polymer and composition and size of inorganic materials [237].

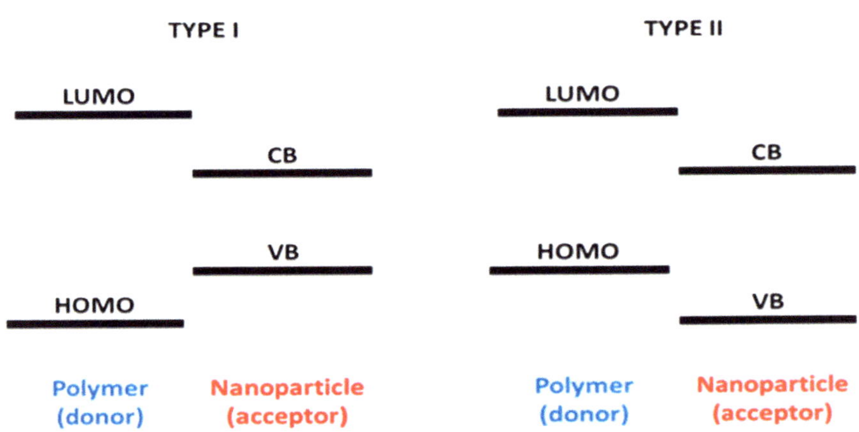

Fig. 9 Illustration of the donor–acceptor energy-level alignment in hybrid interfaces: HOMO and LUMO of polymer donor materials and valence band (VB) and conduction band (CB) of inorganic nanoparticle acceptors, forming type I or type II heterojunction structures

Better results in hybrid devices containing PbS or PbSe nanoparticles were obtained when these materials were combined with low band gap polymers, due to the formation of favorable type II structures, which promotes more efficient charge separation and transport processes. The combination between Pb-chalcogenide nanoparticles and low band gap polymers has provided a better energy-level alignment, a suitable morphology and a better electron–hole transport balance due to the high electron and hole mobilities of the materials.

Noone et al. [238] reported hybrid polymer-PbS devices based on three new donor–acceptor conjugated polymers combined with butyl amine-capped PbS nanoparticles with 3 nm of diameter. The polymers used were PDTPQx (poly(2,3- didecyl-quinoxaline-5,8-diyl-alt-*N*-octyldithieno[3,2-b:20,30-d]pyrrole)), PDTPPPz (poly (2,3-didecyl-pyrido[3,4-b]pyrazine-5,8- diyl-alt-*N*-dodecyldithieno[3,2-b:20,30-d] pyrrole)), and PDTPBT (poly(2,6-bis(3-*n*-dodecylthiophen-2-yl)-alt-*N*-dodecyldith ieno[3,2-b:20,30-d]pyrrole)). Photoinduced absorption spectroscopy (PIA) indicated the formation of long-lived polarons in the PDTPQx/PbS films and the external quantum efficiency (EQE) of the PDTPQx/PbS device was approximately two orders of magnitude larger than the devices containing the other two polymers. Nevertheless, a modest PCE of 0.55% was obtained when the PDTPQx:PbS ratio was 1:9.

The development of photovoltaic devices based on inorganic nanoparticles with higher efficiencies occurred when ligand exchange treatments using thiol molecules as linkers were carried out after the deposition of the all-inorganic films (post-deposition treatments) [239, 240]. Inorganic nanoparticles dispersed in a solvent may easily form aggregates when treated with thiols molecules, and post-deposition exchange processes can solve this problem, while also promoting the removal of the bulky original ligands and morphology reorganization. Ligand exchange processes might enhance polymer–nanoparticle and nanoparticle–nanoparticle interfacial coupling and improve surface passivation [241].

It is interesting to note that, not only the use of low band gap polymers, but also a change of the surface characteristics of the inorganic nanoparticles were necessary to reach appreciable photovoltaic responses in the hybrid devices.

In 2011, Seo et al. [242] used a post-deposition ligand exchange process by casting EDT on top of hybrid films based on mixtures of PDTPBT with OA-capped PbS nanoparticles. The device with PDTPBT:PbS delivered a PCE of 2.07%. An even higher performance was obtained with the insertion of a TiO_2 layer on top of the hybrid film, to promote a more efficient electron transfer and extraction process, and using PbS nanoparticles with first excitonic peak at 795 nm, reaching a PCE of 3.78%.

In 2012, Zhang et al. [243] investigated a planar junction configuration in an inverted device structure (ITO/ZnO/PbS/polymer/MoOx/Ag) and demonstrated, for the first time, that the quenching of the PbS photoluminescence could be attributed to hole transfer from PbS to polymers. PbS nanoparticles with first excitonic peaks at 810 or 1020 nm were used and the original oleate ligands were replaced with BDT by a post-deposition process. PCDTBT and PCPDTTBTT (poly[2,6-(4,4-bis (2-ethylhexyl)-4H-cyclopenta-[2,1-b:3,4-b′]dithiophene-alt-4,7-bis(thiophen-2-yl) benzo-2,1,3-thiadiazole]) were combined with the PbS nanoparticles, forming a type I heterojunction for PCDTBT:PbS and a type II for PCPDTTBTT-PbS. Devices without polymer layer presented an average PCE of 1.89%, while the insertion of a 15 nm-thick PCPDTTBTT layer increased the PCE to 4.22%. In contrast to PCPDTTBTT, a 21 nm-thick PCDTBT layer reduced significantly the efficiency values due to a formation of type I heterojunction, which blocked the hole transport to the electrodes.

Piliego et al. [244] combined PbS nanocrystals and the low band gap polymer PDPPTPT, poly[{2,5-bis(2-hexyldecyl)-2,3,5,6-tetrahydro-3,6-dioxopyrrolo[3,4-c] pyrrole-1,4-diyl}-alt-{[2,20-(1,4-phenylene)bis-thiophene]-5,50-diyl}]. The optimized device (PDPPTPT:PbS ratio 1:9, with post-deposition exchange process with 1,4-benzenedithiol) delivered a PCE of 2.9%.

In 2013, Liu et al. [245] fabricated hybrid solar cells based on the low band gap polymer PDTPBT and PbS_xSe_{1-x} nanoparticles and high efficiencies were obtained. A donor–(donor:acceptor)–acceptor structure (D-D:A-A) was used and an extra pure nanoparticle layer was deposited on top of the hybrid film. This strategy can significantly reduce charge recombination, promoting efficient charge separation, and transport processes. The best devices delivered a Jsc of 14.66 mA cm^{-2} and PCE of 5.50%, the highest efficiency achieved for solution-processed hybrid polymer–nanoparticle devices. These results were attributed to the high performance of the materials, optimization of the polymer:nanoparticle ratio, and mainly to the vertical D-D:A-A structure.

Nam et al. [246] investigated hybrid devices containing mixtures of PSBTBT (poly[2,6-4,4′-bis(2-ethylhexyl)dithienol[3,2-b:2′,3′-d]silole)-alt-4,7(2,1,3-benzothi adiazole)]) with PbS_xSe_{1-x} nanoparticles and nanorods. The spherical nanoparticles were synthesized with ~4 nm of diameter and the nanorods with diameters of 4 nm and length of 30–40 nm. A PCE of 3.4% was obtained and was related to the broad-range absorption of light and to the efficient transport of electrons via aligned pathways within nanorods interconnected by nanoparticles.

Colbert et al. [247] prepared devices based on a mixture between PbS and the polymer PDTPQx-HD (poly(2,3-bis(2-hexyldecyl)quinoxaline-5,8-diyl-alt-N-(2-hexyldecyl)dithieno[3,2-b:20,30-d]pyrrole)) and showed a spectroscopic evidence that the hole transfer produces long-lived charges in hybrid films. At first, PbS nanoparticles were treated by tetrabutylammonium iodide, and then the PDTPQx-HD:PbS films were treated with a solution of MPA. The devices delivered an EQE of 15%. It was observed that both polymer and nanoparticles could contribute to light-harvesting. However, the contribution of absorbed photons with energy in the visible region is higher than those with energy in the infrared region.

In 2014, Nam et al. [248] reported hybrid solar cells based on a mixture of PSBTBT and PbS nanoparticles and obtained values of efficiency of 3.48%. This photovoltaic performance was correlated with the diameter of the PbS nanoparticles, which allowed the formation of favorable type II heterojunction with donor material.

In 2015, a PCE of 4.23% was obtained for polymer-PbS devices with post-deposition treatment with thiol ligands, as reported by Yuan et al. [249].

The authors investigated a series of new organic donor–acceptor polymers with varying optical band gap and treated the hybrid films with BDT. For the cell assembly, a thin layer of PbS nanoparticles was deposited on top of the polymer/PbS hybrid film, which can act as electron transport layer, reducing charge recombination, and cover defects on the film surface. The best results were obtained with the polymer PDBT (poly(dibenzothiophene)) (optical gap of 1.44 eV) and the polymer:PbS ratio was 1:19 w/w. This work showed the importance of the architecture optimization for improvement of the device performance.

P3HT is one of the most used polymers in polymer/fullerene and polymer/Cd-chalcogenide solar cells. However, for some time, appreciable values of efficiency reported for polymer-PbS based photovoltaic devices were obtained when PbS nanoparticles were combined with low band gap polymers, whereas wide band gap polymers commonly showed low efficiencies when combined with Pb-chalcogenide nanoparticles. In 2014, Firdaus et al. [250] investigated solution-phase and post-deposition ligand exchange methods for surface modification of PbS nanoparticles and their application in P3HT:PbS (2.4 nm of diameter) heterojunctions. The best performance (PCE of 1%) was obtained for post-deposition ligand exchange using 1,4-benzenedithiol and small-sized PbS nanoparticles. In 2015, by optimization of the architecture of the device, the average efficiency was increased to 2.17% [251]. In the same year, Borrielo et al. [252] applied a post-deposition treatment with EDT on P3HT:PbS (2 nm of diameter) films and obtained devices with 0.75% of efficiency. An increase of Jsc and PCE was observed compared to hybrid films containing PbS coated with bulky oleate ligands. Photoluminescence measurements showed that the incorporation of nanoparticles promoted a fluorescence quenching of P3HT and, after the EDT treatment, the quenching was even more intense. The results showed that EDT treatment promoted a more effective exciton dissociation and transfer to PbS nanoparticles.

Giansante et al. [253] reported devices based on hybrid films of P3HT-PbS, using a post-deposition treatment with MPA, reaching PCE values of 3%. The authors demonstrated that organic ligands influence the interactions between P3HT e PbS, affecting the morphology of hybrid films prepared from solution. Oleate-capped PbS showed a uniform distribution in polymer matrix due to the interaction between the hexyl side chains of P3HT and the aliphatic chains of oleate-capped PbS, whereas arenethiolate-capped PbS presented a cubic close-packing, resulting from strong ligand–ligand interactions. Thus, it was concluded that non-covalent interactions between polymers and nanoparticles influence the morphology of the films and the efficiency of solution-processed hybrid solar cells. Meanwhile, electron microscopy images showed that MPA post-deposition treatment did not significantly change the morphology of P3HT/arenethiolate-capped PbS composites.

Contrasting to that work, it was found that post-deposition treatment with EDT strongly affects the morphology of hybrid polymer:PbS films, as recently reported by Alves [254] and illustrated in Fig. 10. The AFM images of hybrid films of OA-capped PbS combined with the polymers P3HT and PSBTBT, before and after the application of post-production treatment with EDT, show that the RMS values increased significantly after the treatment. The RMS surface roughness values estimated for P3HT/PbS and PSBTBT/PbS films were 11 and 5 nm, respectively. After EDT treatment, the RMS increased to 24 and 9 nm, respectively. This fact can be attributed to a morphological reorganization of hybrid films when the bulky OA molecules are replaced with EDT, a short-chain ligand, which probably reduces the inter-particle distance, leading to stronger agglomeration of the PbS nanoparticles.

Besides affecting the morphology, ligand exchange processes also significantly influence the photoluminescence dynamics of inorganic and hybrid films. Kahmann et al. [255] observed that the lifetime of PCPDTBT mixed with OA-capped PbS is similar to that of pristine PCPDTBT, and transient absorption measurements showed a similar response in the PCPDTBT spectra for both samples. On the other hand, for PCPDTBT mixed with BDT-capped PbS there was a significant reduction of photoluminescence lifetime, and a transient absorption signal of long-lived species was observed. These features were attributed to a more efficient exciton dissociation at the PCPDTBT/BDT-capped PbS interface and the formation of hole

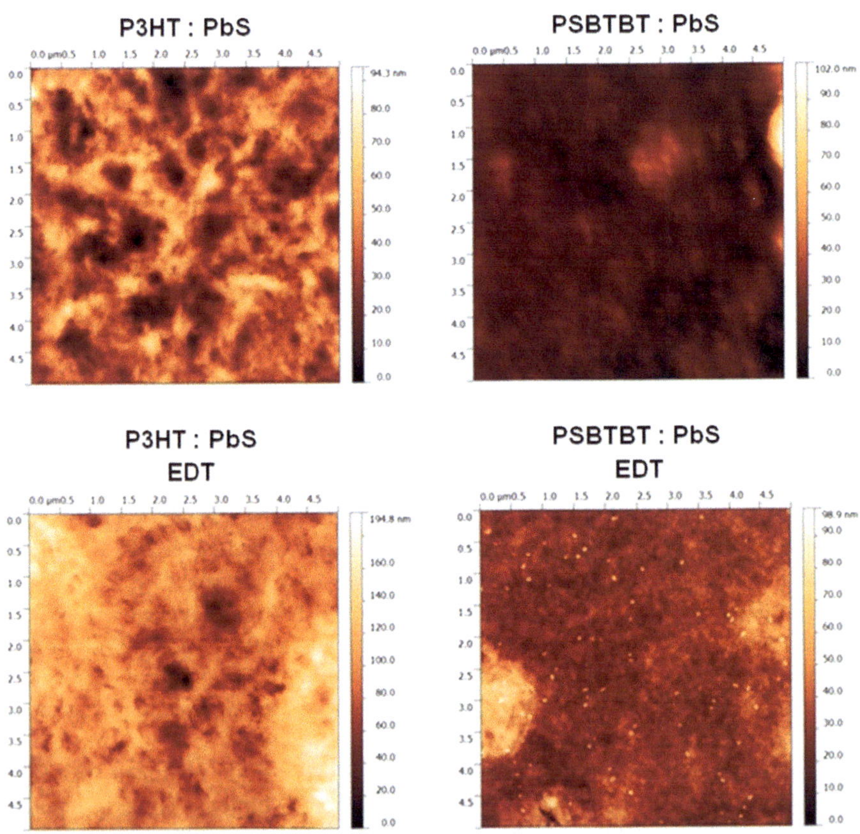

Fig. 10 AFM images obtained for P3HT/PbS and PSBTBT/PbS films, before and after EDT post-deposition treatment

polarons on the polymer and electron transfer to PbS. Choi et al. [256] investigated the relation between exciton dissociation and inter-nanoparticle coupling in films of PbS nanoparticles. The ligand exchange process changes the inter-nanoparticle separation distance, as observed with grazing incident small-angle X-ray scattering measurements, affecting the exciton dissociation dynamics. OA-capped PbS nanoparticles films presented an average inter-nanoparticle distance of 4.5 nm and the resonant energy transfer is the dominant process, whereas in BDT-capped PbS nanoparticles film the exciton dissociation between neighboring-nanoparticles occurs via tunneling mechanism.

The role of the polymer is also an important aspect in hybrid devices. Mastria et al. [257] investigated polymer:PbS devices with the BDT post-deposition treatment, using mixtures of PbS nanoparticles (2.8 nm of diameter) and polymers with different band gaps: MEH-PPV, P3HT, and TQ1, with absorption spectra spanning the visible region, and LBPP-4 e PDPP3T, with absorption in near-infrared region. Despite the fact that the low band gap polymers shows extended light-harvesting capacities, the best results, with improved photocurrent densities, were obtained with the wide band gap polymers. The IPCE spectra showed that both materials, polymer and PbS nanoparticles, could contribute to the photocurrent generation, for all the hybrid devices studied. Nevertheless, the samples containing wide band gap polymers presented higher photoluminescence quenching of the polymer signal, as well as larger PIA signals, suggesting that these polymers could promote more efficient exciton and charge transfer processes when compared to the low band gap polymers investigated. For devices based on these low band gap polymers, low photocurrent densities were obtained due to the less efficient energy and charge transfer processes, and promotion of the population of midgap trap states. Therefore, it was demonstrated that the engineering of the polymer energy levels also has a crucial role in all the process associated to photocurrent generation in hybrid devices.

The effects of different ligands on the performance of hybrid solar cells was also investigated, as reported by Colbert et al. [258]. Films of PbS nanoparticles and the polymer PTB1 (poly[(4,8-bis(octyloxy)benzo(1,2-b:4,5-b′)-dithiophene-2,6-diyl) (2-((dodecyloxy)carbonyl)thieno(3,4-b)thiophenediyl))) were prepared and submitted to ligand exchange process with EDT, MPA, malonic acid, or inorganic halide ions from tetrabutylammonium iodide (TBAI). MPA-treated devices delivered an average PCE of 2.59%, while the other ligands delivered lower efficiencies (less than 1.7%). The best results observed for MPA-treated devices were associated with a substantially higher Voc value delivered by these devices. Transient photovoltage and charge extraction techniques showed that the ligand exchange process with MPA promoted longer carrier recombination lifetimes at open-circuit conditions. These lifetimes were 1–2 orders of magnitude longer than the ones found for EDT-treated and TBAI-treated devices.

Although the state of the art for hybrid devices is based on devices with post-deposition ligand exchange of hybrid films, recently Lu et al. [259] reported the use of PbI_2 as a capping ligand for PbS nanoparticles, by exchanging the ligand prior to the mixture with the polymer and deposition of the hybrid film.

These nanoparticles exhibited good stability when dispersed in a mixture of dichlorobenzene and butylamine, solvents in which the polymer Si-PCPDTBT (poly[(4,4′-bis(2-ethylhexyl)dithieno[3,2-b:2′,3′-d]silole)-2,6-diyl-alt-(2,1,3-benzothiadiazole)-4,7-diyl]) was solubilized. The device achieved a PCE of approximately 4.78%, which is amongst the highest values reported for hybrid devices. Time-resolved photoluminescence spectroscopy indicated that the energy or charge transfer processes are efficient when PbI_2-capped PbS nanoparticles are used in combination with Si-PCPDTBT.

Table 3 summarizes the PCE values reported for polymer/Pb-chalcogenide devices by different authors. The best results are presented by low band gap polymers as donor specie, due to a more appropriate energy-level alignment at the interface with the quantum dots, and use of very high nanoparticle concentration, condition required for the percolation networks formation. The problems associated with these devices are mainly related to the interfaces between the different organic and inorganic materials and the limitation of charge transfer and transport processes involving the inorganic nanoparticles. The crystallinity of donor and acceptor materials at heterojunction interface and traps on the surface of nanoparticles also are important factors [260]. Due to poor nanoparticle dispersion and aggregation process in organic solvents, capping ligands on the nanoparticle surface are required. Ligand exchange process are used to improve polymer–nanoparticle interactions and nanoparticles surface passivation. Post-deposition exchange processes promote better results in devices than solution-phase process. Generally, solution-phase exchange processes limit the nanoparticle dispersion in organic solvents, and agglomeration is favored, whereas post-deposition exchange processes improve the charge transfer and charge transport process, with morphology reorganization. In spite of the improvement of hybrid photovoltaic devices in the last years, the efficiencies achieved remain much lower than those reported for all-inorganic devices (highest PCE of 11% approximately [261]). The easier morphology control, lower phase separation and deposition of multilayers of nanoparticle can explain the best results obtained for all-inorganic devices.

4 Metal Nanoparticles in Organic Solar Cells

Similar to the case of inorganic semiconductor nanoparticles, metal nanoparticles also have optical properties strongly dependent on size and shape. For example, bulk gold looks yellow in reflected light, while thin Au films look blue in transmission, and this color changes as the nanoparticle size is decreased. These characteristics of metal nanoparticles can be used to change the properties of their surrounding media. An interesting example is the Lycurgus Cup [262], which possesses the unique feature of changing color depending upon the light in which it is viewed: it looks green in reflected light and looks red when a light is shone from inside and is transmitted through the glass. This effect is attributed to the presence of Au and Ag nanoparticles in the glass.

Table 3 Performance of a series of BHJ hybrid solar cells assembled with PbE (E = S and/or Se) and polymers, reported by different authors

Active layer	Device configuration	Polymer:nanoparticle ratio	Ligand exchange	PCE (%)	Reference
PDTPBT:PbS_xSe_{1-x}	ITO/PEDOT:PSS/PDTPBT/PDTPBT:PbS_xSe_{1-x}/PbS_xSe_{1-x}/LiF/Al	1:15	BDT	5.50	[246]
Si-PCPDTBT:PbS	ITO/PEDOT:PSS/Si-PCPDTBT:PbS/ZnO/Al	1:15	PbI_2	4.78	[260]
PDBT:PbS	ITO/PEDOT:PSS/PDBT:PbS/PbS/LiF/Al	1:19	BDT	4.23	[250]
PCPDTTBTT:PbS	ITO/ZnO/PbS/PCPDTTBTT/MoOx/Ag	–	BDT	4.22	[244]
PDBF:PbS	ITO/PEDOT:PSS/PDBF:PbS/PbS/LiF/Al	1:19	BDT	3.85	[250]
PDTPBT:PbS	ITO/PEDOT:PSS/PDTPBT:PbS/TiO_2/LiF/Al	1:9	EDT	3.78	[243]
PDTD:PbS	ITO/PEDOT:PSS/PDTD:PbS/PbS/LiF/Al	1:19	BDT	3.65	[250]
PSBTBT:PbS	ITO/PEDOT:PSS/PSBTBT:PbS/BCP/Al	1.5:8.5	EDT	3.48	[249]
PSBTBT:PbS_xSe_{1-x}	ITO/PEDOT:PSS/PSBTBT:PbS_xSe_{1-x}/ZnO/Al	1:9	EDT	3.40	[247]
PDTT:PbS	ITO/PEDOT:PSS/PDTT:PbS/PbS/LiF/Al	1:19	BDT	3.01	[250]
P3HT:PbS	ITO/PEDOT:PSS/P3HT:PbS/LiF/Al	1:9	p-methyl-benzenethiol/MPA	3.00	[254]
PDPPTPT:PbS	ITO/PEDOT:PSS/PDPPTPT:PbS/LiF/Al	1:9	BDT	2.90	[245]
P3HT:PbS	ITO/PEDOT:PSS/P3HT:PbS/PbS/LiF/Al	1:19	BDT	2.65	[250]
PTB1:PbS	ITO/TiO_2/PTB1:PbS/MoOx/Ag	1:9	MPA	2.59	[259]
PDFT:PbS	ITO/PEDOT:PSS/PDFT:PbS/PbS/LiF/Al	1:19	BDT	2.55	[250]
P3HT:PbS	ITO/PEDOT:PSS/P3HT/P3HT:PbS/Ca/Ag	1:9	BDT	2.17	[252]
P3HT:PbS	ITO/PEDOT:PSS/P3HT:PbS/LiF/Al	1:9	BDT	1.98	[258]
TQ1:PbS	ITO/PEDOT:PSS/TQ1:PbS/LiF/Al	1:9	BDT	1.93	[258]

(continued)

Table 3 (continued)

Active layer	Device configuration	Polymer:nanoparticle ratio	Ligand exchange	PCE (%)	Reference
MEH-PPV: PbS	ITO/PEDOT:PSS/MEH-PPV:PbS/LiF/Al	1:9	BDT	1.70	[258]
LBPP-4:PbS	ITO/PEDOT:PSS/LBPP-4:PbS/LiF/Al	1:9	BDT	1.38	[258]
P3HT:PbS	ITO/PEDOT:PSS/P3HT/P3HT:PbS/Ca/Ag	1:9	BDT	1.00	[251]
PDPP3T:PbS	ITO/PEDOT:PSS/PDPP3T:PbS/LiF/Al	1:9	BDT	0.85	[258]
P3HT:PbS	ITO/PEDOT:PSS/P3HT:PbS/Al	1:9	EDT	0.75	[253]
MEH-PPV: PbS	ITO/PEDOT:PSS/MEH-PPV:PbS/Al	1:1	–	0.70	[231]
PDTPQx:PbS	ITO/PEDOT:PSS/PDTPQx:PbS/Al	1:9	–	0.55	[239]
P3HT:PbSe	ITO/PEDOT:PSS/P3HT:PbSe/Al	1:4	–	0.14	[232]
P3HT:PbSe	ITO/PEDOT:PSS/P3HT:PbSe/Al	1:6	–	0.04	[233]
P3HT:PbS	ITO/PEDOT:PSS/P3HT:PbS/Al	1:1	Acetic acid	0.01	[237]

ITO indium tin oxide, *PEDOT:PSS* poly(3,4-ethylenedioxythiophene) :polystyrene sulfonate, *MEH-PPV* poly[2-methoxy-5-(2-ethylhexyloxy)-1,4-phenylenevinylene], *P3HT* poly(3-hexylthiophene)-2,5-diyl, *PDTPQx* poly(2,3-didecyl-quinoxaline-5,8-diyl-alt-N-octyldithieno[3,2-b:2′,3′-d]pyrrole), *PDTPBT* poly(2,6-bis(3-*n*-dodecylthiophen-2-yl)-alt-*N*-dodecyldithieno[3,2-b:2′,3′-d]pyrrole), *PCPDTTBTT* poly[2,6-(4,4-bis(2-ethylhexyl)-4H-cyclopenta-[2,1-b:3,4-b’]dithiophene-alt-4,7-bis(thiophen-2-yl)benzo-2,1,3-thiadiazole], *PSBTBT* poly[(4,4′-bis(2-ethylhexyl)dithieno[3,2-b:2′,3′-d]silole)-2,6-diyl-alt-(2,1,3-benzothiadiazole)-4,7-diyl], *TQ1* poly[2,3-bis(3-octyloxyphenyl)quinoxaline-5,8-diyl-alt-thiophene-2,5-diyl]), *PDPP3T* poly(diketopyrrolopyrrole-terthiophene), *PTB1* poly((4,8-bis(octyloxy)benzo(1,2-b:4,5-b’)dithiophene-2,6-diyl)(2-((dodecyloxy)carbonyl)thieno(3,4-b)thiophenediyl)), *PDBF* poly(2-(6-fluoro-7-methyl-benzo[c][1,2,5]thiadiazol-4-yl)-4-(heptadecan-9-yl)-6-methyl-4H-dithieno[3,2-b:2′,3′-d]pyrrole), *PDTD* poly(2,5-bis(2-ethylhexyl)-3-(5-(4-(heptadecan-9-yl)-6-methyl-4H-dithieno[3,2-b:2′,3′-d]pyrrol-2-yl)thiophen-2-yl)-6-(5-methylthiophen-2-yl)-2,5-dihydropyrrolo[3,4-c]pyrrole-1,4-dione), *PDTT* poly(4-(heptadecan-9-yl)-2-methyl-6-(5-(7-(5-methylfuran-2-yl)benzo[c]-[1,2,5]thiadiazol-4-yl)furan-2-yl)-4H-dithieno[3,2-b:2′,3′-d]pyrrole), *PDFT* poly(4-(heptadecan-9-yl)-2-methyl-6-(5-(7-(5-methylthiophen-2-yl)benzo[c]-[1,2,5]thiadiazol-4-yl)thiophen-2-yl)-4H-dithieno[3,2-b:2′,3′-d]-pyrrole), *Si-PCPDTBT* poly[(4,4′-bis(2-ethylhexyl)dithieno[3,2-b:2′,3′-d]silole)-2,6-diyl-alt-(2,1,3-benzothiadiazole)-4,7-diyl], *BDT* 1,3-benzenedithiol, *EDT* 1,2-ethanedithiol, MPA = 3-mercaptopropionic acid

The optical properties of small metal nanoparticles are actually dominated by collective oscillation of the conduction electrons, resulting from the interaction with the electromagnetic field. A restoring force in the nanoparticles tries to compensate for this, resulting in a unique resonance wavelength [263]. These effects are called surface plasmon resonance [264], and correspond to the frequency of oscillation of conduction electrons in response to the alternating electric field of an incident electromagnetic radiation. Figure 11 illustrates this phenomenon. The oscillation wavelength depends on many factors, including the particle size and shape, and the local dielectric environment [265–268]. In addition, when nanoparticles are sufficiently close together, interactions between neighboring particles arise. For elongated particles (1D systems), the resonance wavelength depends on the orientation of the electric field, giving two oscillations, transverse and longitudinal, as illustrated in Fig. 11. The longitudinal oscillation is very sensitive to the aspect ratio of the particles [269], leading to color changes.

Metal nanoparticles can exhibit strong surface plasmon resonances localized at UV, visible, and near-infrared wavelengths. However, only a few metals, such as Au, Ag, and Cu, possess plasmon resonances in the visible spectrum, which give rise to intense colors [265, 270, 271].

Surface plasmons decay radiatively or non-radiatively, giving rise to scattering or absorption of light. Light scattering from metal nanoparticles near their local surface plasmon excitation is a promising way to increase light absorption in solar cells. The strong interaction of photons with metal nanoparticles induces the formation of an electromagnetic field in the regions near these particles. If a semiconductor is located in the surroundings, for example, the absorption of light by this

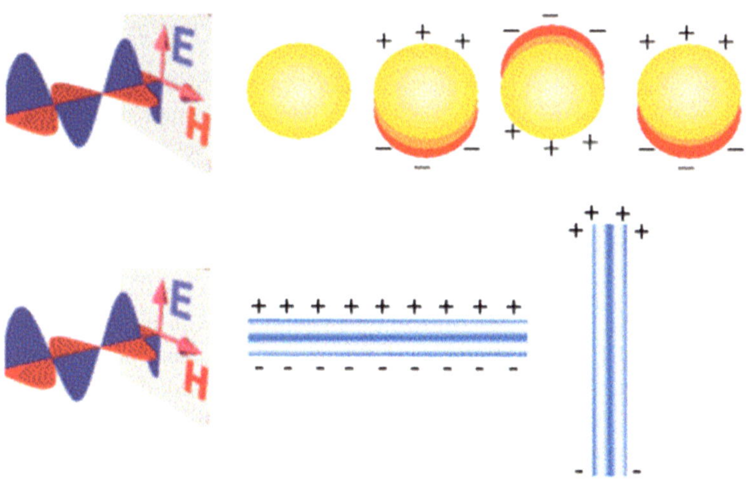

Fig. 11 (*Top*) Schematic drawing of the interaction of the electromagnetic radiation with a metal nanosphere. A dipole is induced, which oscillates in phase with the electric field of the incoming light. (*Bottom*) Transverse and longitudinal oscillation of electrons in a metal nanorod. Reprinted from Ref. [10]. Copyright 2004, with permission from Elsevier

material may be increased, resulting in increased exciton generation. Furthermore, the excitons may split at the interface semiconductor/metal nanoparticle due to the electromagnetic field. These effects are expected to improve the photocurrent in solar cells.

Metal nanoparticles have been shown to improve the performance of silicon-based solar cells [272–282], GaAs solar cells [283], and quantum well solar cells [284]. For these inorganic devices, the principal method of enhancement is scattering of incident light, which increases light-trapping and, therefore, might reduce reflection. For metal nanoparticles situated on the front surface of a device, the nanoparticles must scatter light into the device (i.e., in the forward direction) to reduce reflection, and must scatter at oblique angles to improve light-trapping. Any light absorbed by the nanoparticle is lost as heat, so absorption should be minimized. If the nanoparticles are strongly absorbing and/or backscattering, they will decrease the device efficiency.

The incorporation of Au or Ag nanoparticles in TiO_2 [285–292] or ZnO [293–295] nanostructures, particularly aiming at the application in DSSC, has also been demonstrated. The contact between bulk gold and bulk TiO_2 can form a Schottky barrier of ~ 1.0 eV [273, 296], which can be difficult for electrons to overcome, thus making interfacial charge separation unfavorable. Nevertheless, electron transfer from Au nanoparticles to the conduction band of TiO_2 has been observed with transient absorption spectroscopy [288, 297]. Photoexcited electrons can also be transferred from Au nanoparticles to the ZnO conduction band. The Schottky barrier at the ZnO/Au interface blocks the electron transfer back from ZnO to the dye and electrolyte, and thus increases the electron density of the ZnO conduction band [295]. For TiO_2 based DSSC, there are a few different mechanisms, which may be considered as candidates for the plasmonic contribution to the photocurrent [286]:

- Internal plasmon decay into energetic electron–hole pairs in the metal nanostructure, followed by subsequent charge carrier injection over, or tunneling through, the Schottky barrier between the metal and semiconductor substrate;
- Direct near-field coupling to electronic transitions in the dye molecules;
- Near-field coupling to TiO_2 band gap transitions;
- Far-field scattering leading to prolonged optical path and, in particular, coupling to wave guided modes in the TiO_2 film.

The interactions of Ag and Au metal nanoparticles with many different molecules and materials have been observed, which could account for the development of new types of plasmonic solar cells. For example, Schottky diode junctions were formed between CdS nanowires and Au nanowires and delivered 0.92 mA cm^{-2} of current under one sun illumination [298]. The incorporation of Au nanowires within nanocrystalline CdSe increases the extraction of photogenerated carriers, and the photoresponse of CdSe/Au hybrid materials can be controlled by changing the conductivity of the Au nanowires [299]. Electron transfer was also observed in Au–CdS core-shell nanocrystals. The as-synthesized Au–CdS nanocrystals exhibited superior photocatalytic performance under visible-light illumination compared to other relevant commercial materials, demonstrating their potential as an effective visible-light-driven photocatalysts [300].

Remarkable enhancements in the photocurrent action and fluorescence excitation spectra were observed for porphyrin and phthalocyanine when considerable amounts of Ag nanoparticles were deposited onto the ITO electrode. Raman scattering measurements suggested the effects of enhanced electric fields resulting from localized surface plasmon resonance and light scattering on the photocurrent enhancement [301].

The addition of Au nanoclusters to P3HT resulted in an increased photoluminescence, explained by polymer chain separation induced by the presence of the nanoparticles [302]. Addition of Au to a blue light emitting polymer also resulted in enhanced luminescent stability [303]. Parfennov et al. [304] reported that Ag nanoparticles increased the photoluminescence efficiency by increasing the exciton decay time of the surrounded fluorophore, while Sarathy et al. [305] observed fluorescence enhancement of poly(3-octylthiophene) (P3OT) that was near the Ag nanoparticles by means of a near-field scanning optical microscope experiment. The fluorescence of Au/MEH-PPV nanocomposites was changed by over an order of magnitude by controlling the Au particle size, the spatial distribution of nanoparticles throughout the MEH-PPV host, and the ligand chain layer thickness. Smaller particles were more efficient at quenching. As the ligand chain length increased, the quenching became less efficient [306].

Considering the application in OSC, metal nanoparticles have been explored in different approaches: (i) in the metal electrode for collection of electrons (i) as a buffer layer, (ii) as an interfacial layer, or (iii) in the BHJ active layer.

Au and Ag evaporated films have been widely used as metal-electrodes for charge collection in OSC [307, 308]. Usually, the use of these metals is associated with decreased Voc in comparison to Al cathodes. However, Au/LiF/Al layered cathodes showed 20–30% improved Jsc and η over devices without Au. The introduction of a nanotextured Au thin film was observed to increase the absorption of a P3HT/PCBM thin film through plasmon-assisted localization of the electromagnetic field of the incident light [309].

Electron transfer from metal to C_{60} was reported [310–312] and the interaction between Ag and C_{60} is considered stronger than that between Au and C_{60}. On the one hand, this interaction allows strategies such as the use of a Ag or Au layer before the metal cathode [310]. On the other hand, some authors consider that this effect may be responsible for a Voc drop observed when large amounts of nanoparticles are used, since they might reduce the electron affinity of C_{60} [328].

The incorporation of metal nanoparticles as buffer layer in OSC is perhaps the most used approach [313–320]. Chen et al. [315] reported an improvement in device performance after the addition of various concentrations of Au nanoparticles into the PEDOT:PSS layer, as shown in Fig. 12. The average Au particle size estimated from scanning electron microscopy was ~30–40 nm. The buffer layer was prepared by spin-coating a mixture of Au nanoparticles solution with the PEDOT:PSS solution on top of the ITO substrate. The reference device prepared with pristine PEDOT:PSS exhibited a Voc of 0.59 V, Jsc of 8.95 mA cm^{-2}, a FF of 65.9% and efficiency of 3.48%. After the addition of Au nanoparticles to the buffer layer, the values of Voc remained unchanged, but FF was improved. The maximum PCE of 4.19% was obtained using 20% of Au (Jsc = 10.18 mA cm^{-2}, FF = 69.8%). Further increases in the concentration of the metal nanoparticles to 30% led to a decrease in the value of Jsc, due to enhanced backward scattering and/or increased resistivity of the buffer layer. The trends in IPCE followed the trends of Jsc, as displayed in Fig. 12b. The authors suggested that the surface plasmon resonance increased not only the rate of exciton generation, but also the probability of exciton dissociation [315].

In a similar way, Au nanostructures were fabricated through the layer-by-layer deposition of Au nanorods onto the ITO substrates and transformed into nanodots through a thermally induced shape transition. The incorporation of plasmonic Au nanodots on the ITO surface was found to result in an increase in the PCE of P3HT/PCBM devices from 3.04 to 3.65%. The strong coupling between the organic excitons and plasmons of the Au nanostructures results in more efficient charge transfer in the BHJ system [316].

Ag nanoparticles of ~13 nm were electrodeposited onto the ITO and covered with PEDOT:PSS as buffer layer in a P3HT/PCBM device. The overall PCE

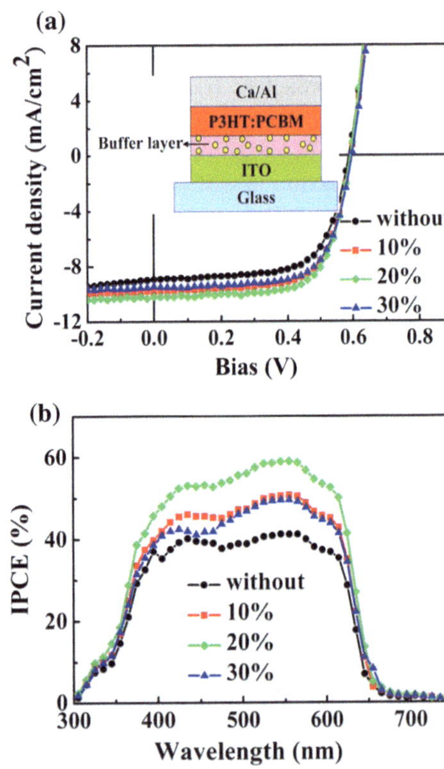

Fig. 12 Color online **a** *J–V* characteristics, recorded under 100 mW cm-^{2} of devices incorporating PEDOT:PSS doped with various concentrations of Au nanoparticles solutions: without doping (*circle*); 10% (*black square*); 20% and (*black diamond*) 30% Au nanoparticles solution doping (*black triangle*). **b** Corresponding IPCE curves of these devices. *Inset* schematic representation of the device assembly. Reprinted from Ref. [315], with the permission of AIP Publishing

increased from 3.05 to 3.69%, mainly because of the improved photocurrent density, as a result of enhanced absorption of the photoactive conjugate polymer due to the high electromagnetic field strength in the vicinity of the excited surface plasmons. Improved IPCE was also observed at wavelength regions longer than 400 nm, related to the surface plasmon resonance band of electrodeposited Ag nanoparticles [317].

Ag nanoparticle layers were fabricated using vapor-phase deposition on ITO electrodes and the influence of the Ag film thickness in P3HT/PCBM solar cells was investigated. The highest increase in Jsc was observed for devices incorporating a layer of 2 nm of Ag, and, as the film thickness was increased, a downward trend in Jsc was observed. The PCE was improved from 1.3 to 2.2%. The Voc decreased slightly with increasing the Ag layer thickness, which was related to a decrease in the work function of the transparent electrode [318].

Chuang et al. [321] synthesized nanocomposites of Au nanoparticles and graphene oxide, which were applied as buffer layers in BHJ OSC based on P3HT:PCBM or P3HT:ICBA mixtures. The incorporation of Au nanoparticles into graphene oxide in the buffer layer increased the PCE from 3.26 to 3.98% for P3HT:PCBM devices, and from 4.02 to 5.05% for P3HT:ICBA devices. The authors attributed the improved device performance to local field enhancement induced by the localized surface plasmon resonance as Au nanoparticles with optimized characteristics (control of agglomeration and concentration) were added to the buffer layer.

The beneficial effects in photocurrent and device efficiency from the plasmon resonance of Au, Ag or Cu nanoparticles as buffer layers have also been shown for OSC assembled with Cu-based phtalocyanines [321–323] and Zn-based phthalocyanines [324]. The exploration of these metals as interfacial layers in tandem OSC devices assembled with small organic molecules [325, 326] was also shown to be successful. Yakimov et al. [326] reported significant enhancement in the photovoltaic efficiency of a tandem device containing Ag nanoparticles in the middle of two photoactive layers. The optical intensity near the Ag nanoparticle was significantly enhanced due to scattering [327].

The incorporation of Au or Ag nanoparticles in the BHJ active layer, on the other hand, has been less explored, and there has been some controversy about the effect of metal nanoparticles in BHJ OSC using this approach [327–333].

In 2005, Carroll and Kim [328] reported the incorporation of Au ($\sim$6 nm) or Ag ($\sim$5 nm) nanoparticles stabilized with dodecylamine in the P3OT/C_{60} active layer. The BHJ was prepared from a chlorobenezene co-solution of all the materials. The authors showed an increase in the electrical conductivity of the films upon the incorporation of the metal nanoparticles, observed from the *J–V* characteristics in the dark (Fig. 13). They proposed that hole transfer occurs from P3OT to Ag or Au, attributed to the "dopant" states introduced by the metal nanoparticles in the band gap of the polymer. These authors also showed that the contribution to light aborsption enhancement was minimal, less than 10%. Despite the different plasmon absorption of the two nanoparticles (443 nm for Ag and 523 nm for Au), they did

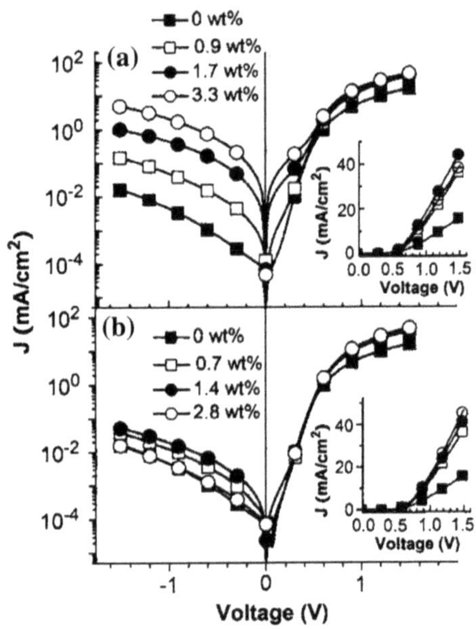

Fig. 13 *J–V* characteristics in the dark of P3OT-C_{60} solar cells containing: **a** Ag and **b** Au nanoparticles. Reprinted from Ref. [328], with the permission of AIP Publishing

not observe significant changes in the absorption spectra. The best efficiency was obtained using Ag nanoparticles.

In 2008, Park et al. [329] added small amounts of Au nanoparticles (3–6 nm) to the BHJ of P3HT/PCBM. The films were prepared by adding different amounts of a toluene solution containing the nanoparticles to a chlorobenzene solution of P3HT/PCBM, and then annealed at 120 °C for 30 min under nitrogen. Contrary to the work of Carroll and Kim [328], the authors did not observe enhanced dark conductivity for this system. Under illumination, increased PCE was observed at small Au contents, which was mainly related to an improvement of the Jsc. Voc and FF remained almost the same. At a weight fraction of 6.25×10^{-8} Au nanoparticles/P3HT, about 50% enhancement was shown in PCE (from 1.43 to 2.17%). They attributed the increase in Jsc to improved light-harvesting, although only a minor change was observed in the absorption spectra.

As illustrated in Fig. 14, there is a small barrier for hole injection at the PEDOT: PSS contact [329]. Thus, it is considered that the Au nanoparticles might assist such injection (and therefore, charge collection) at the electrode in these systems, especially in cases where the concentration of these particles is large.

In 2009, Shen et al. [330] and Duche et al. [331] investigated the influence of Ag nanoparticles on the light-absorption characteristics of OSC by theoretical simulations. Shen et al. [330] found that near-field enhancement is the main reason for absorption enhancement in the active layer, and that the optimum conditions are dependent both on the film thickness and the nanoparticle size. For a thin film (33 nm) of P3HT/PCBM, nanoparticles with diameters of ~24 nm are necessary for optimum absorption enhancement. These conditions would lead to an

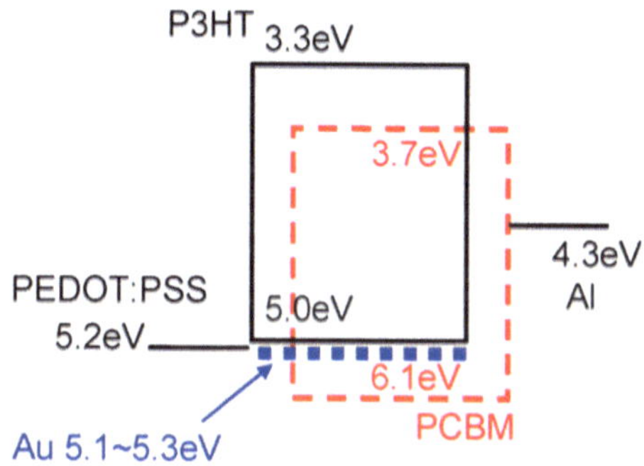

Fig. 14 Energy levels of HOMO and LUMO of P3HT and PCBM, and work function of Au, PEDOT:PSS and Al. Reprinted from Ref. [329]. Copyright 2008, with permission from Elsevier

enhancement factor of 1.56, which should compensate and bring the performance close to that expected for a much thicker device prepared without metal nanoparticles. Duche et al. [331] compared the modeling results with experimental results for MEH-PPV/PCBM BHJ OSC. Enhanced absorption of light up to 50% was experimentally obtained in a 50 nm-thick layer including silver nanospheres with 40 nm of diameter, in agreement with the high values expected from calculations.

In 2010, Topp et al. [331] investigated the incorporation of different amounts of P3HT-capped, dodecyl amine-capped, or pyridine-capped Au nanoparticles into P3HT/PCBM mixtures. The use of Au directly stabilized with P3HT seems to be a promising approach to incorporate these materials in OSC without further addition of organic components. Moreover, P3HT ligands strongly bound to Au nanoparticles might lead to direct electron transfer from the polymer to Au. However, the authors observed a slightly decrease in the device performance (see Table 4) after addition of 3 wt% of P3HT-capped Au nanoparticles, and a strong decrease of performance after addition of 16 wt% of this material. No significant changes were observed in the absorption spectra, suggesting that the absorption was dominated by the polymer (even at 16 wt% of Au), and there was no disruption in the crystalline order of P3HT. The field effect mobility of holes was estimated to be the same ($\sim 1 \times 10^{-4}$ cm^2 V^{-1} s^{-1}) before and after Au incorporation. Photoinduced absorption spectroscopy was used to investigate charge transfer between P3HT and Au nanoparticles. The authors found no indications for Au-enhanced formation of long-lived polarons in P3HT, although the results presented were very noisy and therefore should be interpreted carefully. Since other effects were ruled out, the authors suggested that the loss of performance in these devices could be explained by a quenching of the excited state of the polymer in the presence of Au. Phase segregation was not excluded, although the authors considered that unlikely due to the strong interaction between the S-containing polymer and Au. OSC assembled with 3 wt% or $\sim$50 wt% of dodecyl amine-stabilized Au nanoparticles also showed decreased performance, attributed to the ligand shell which is an insulating barrier for charge transport, as evidenced in *J–V* characteristics in the dark (the current under forward bias was reduced after introduction of Au). Interestingly, the cells containing DDA-capped Au nanoparticles seemed to work better than those containing P3HT-capped Au nanoparticles. This is in agreement with the proposed mechanism, where the quenching of the excited state in the polymer was the main cause for performance loss in the P3HT-capped Au nanoparticle system, while the presence of the DDA shell reduces the quenching probability. In a third approach, the authors used pyridine-capped Au nanoparticles, following strategies typically used for BHJ OSC containing CdSe nanoparticles. This approach did not improve the efficiency of devices after incorporation of Au nanoparticles, probably related to the quenching of the excited state and to Au segregation phenomena.

In 2011, Wang et al. [333] demonstrated positive effects arising from the addition of 5 wt% of larger Au nanoparticles (size ~70 nm) into BHJ OSC composed of mixtures of polymer/$PC_{71}BM$. The improvement in Jsc, FF and IPCE resulted from a combination of enhanced light absorption caused by the light scattering of Au nanoparticles and improved charge transport. The authors also found that the increase in device efficiency depended on the size of the metal nanoparticles and the weight ratio of these materials in the BHJ film.

In 2015, de León et al. [334] investigated the incorporation of Au nanoparticles stabilized with thiols containing a phenyl, biphenyl or terphenyl group, into the BHJ of P3HT:PCBM. The nanoparticles were synthesized by two different routes, with or without tetraoctadecylammonium bromide (TOAB) as co-stabilizer. The aromatic nature of the ligand (thiol) was expected to promote a good interaction with the organic conjugated mixture. Nevertheless, for almost all of the samples, a decrease of the device performance was observed in comparison to the reference cell prepared without Au nanoparticles. The exception was the device containing the nanoparticles capped with the thiol with terphenyl group, synthesized via the one-phase method without TOAB, which delivered a PCE of 1.84%, in comparison to 1.77% obtained for the reference cell. The main effect of nanoparticle incorporation was related to a change of the morphology of the active layer, being the changes in optical and electrical properties a consequence of the morphologic effect.

Table 4 summarizes the results obtained by different authors for BHJ OSC before and after incorporation of Au or Ag nanoparticles into the active layer. From Table 4 it is evident that there is some controversy between the results involving metal nanoparticle incorporation into BHJ OSC reported by different authors. Interestingly, none of these papers investigated systematically a possible morphological effect introduced by the presence of metal nanoparticles. As observed in polymer/PCBM/CdSe BHJ OSC (Sect. 3), morphology can be expected to play a crucial role in BHJ OSC containing metal nanoparticles as well.

In an earlier work, Conturbia showed that the morphology of the P3HT/PCBM system is significantly changed upon incorporation of Au nanoparticles [334]. Figure 15 shows HRTEM images of a P3HT/PCBM/Au system containing 1 wt% of Au nanoparticles. In Fig. 15a can be seen that the Au nanoparticles are distributed in the polymer and PCBM domains, while Fig. 15b shows, in higher magnification, the existence of large PCBM domains/crystallites in the sample. The PCBM domain shown is estimated to have ~40 nm of size. The d-spacing of the nanocrystals at different orientations shown in regions 1, 2, and 3 correspond to 0.26, 0.28, and 0.19 nm. These results are similar to those found by Reyes-Reyes

Table 4 Performance parameters reported by different authors for OSC containing metal nanoparticles in the bulk heterojunction

Active layer	Film thickness /active area	NP size/passivation	NP Concentration /solvent	PV parameters	Reference
P3OT/C_{60}	–/0.19 cm^2	–	–/CB	Jsc = 5.1 mA cm^{-2} Voc = 0.55 V FF = 0.32 PCE = 1.1%	[328]
P3OT/C_{60}/Au	150 nm/0.19 cm^2	5.3 ± 1.1/DDA	1.7 wt%/CB	Jsc = 5.9 mA cm^{-2} Voc = 0.56 FF = 0.41 PCE = 1.7%	[328]
P3OT/C_{60}/Ag	150 nm/0.19 cm^2	6.1 ± 1.3/DDA	1.4 wt%/CB	Jsc = 6.2 mA cm^{-2} Voc = 0.56 V FF = 0.43 PCE = 1.9%	[328]
P3HT/PCBM	110 nm/0.09 cm^2	–	–/CB	Jsc = 6.15 mA cm^{-2} Voc = 0.61 V FF = 0.37 PCE = 1.43%	[329]
P3HT/PCBM/Au	110 nm/0.09 cm^2	3–6 nm/–	3×10^{-7} wt%/ CB + toluene	Jsc = 9.17 mA cm^{-2} Voc = 0.61 V FF = 0.38 PCE = 2.17%	[329]
P3HT/PCBM	90 nm/0.07 cm^2	–	–/CB	Jsc = 8 mA cm^{-2} Voc = 0.6 FF = 0.45 PCE = 2.5%	[331]
P3HT/PCBM/Au	80 nm/0.07 cm^2	3.7 nm/P3HT	3 wt%/CB	Jsc = 6.0 mA cm^{-2} Voc = 0.54 V FF = 0.35	[331]

(continued)

Table 4 (continued)

Active layer	Film thickness /active area	NP size/passivation	NP Concentration /solvent	PV parameters	Reference
				PCE = 1.5%	
P3HT/PCBM/Au	75 nm/0.07 cm^2	3.7 nm/P3HT	16 wt%/CB	Jsc = 2.6 mA cm^{-2} Voc = 0.40 V FF = 0.35 PCE = 0.4%	[331]
P3HT/PCBM	70 nm/0.04 cm^2	–	–/CB	Jsc = 9.7 mA cm^{-2} Voc = 0.65 FF = 0.52 PCE = 3.2%	[331]
P3HT/PCBM/Au	90 nm/0.04 cm^2	7.0 nm/DDA	3 wt%/CB	Jsc = 10.8 mA cm^{-2} Voc = 0.59 FF = 0.46 PCE = 2.9%	[331]
P3HT/PCBM/Au	85 nm/0.06 cm^2	7.0 nm/DDA	52 wt%/CB	Jsc = 5.3 mA cm^{-2} Voc = 0.27 FF = 0.25 PCE = 0.4%	[331]
P3HT/PCBM	65 nm/0.05 cm^2	–	–/CB + py	Jsc = 8.7 mA cm^{-2} Voc = 0.62 FF = 0.45 PCE = 2.4%	[331]
P3HT/PCBM/Au	45 nm/0.08 cm^2	15 nm/py	23 wt%/CB + py	Jsc = 7.1 mA cm^{-2} Voc = 0.42 FF = 0.31 PCE = 1.1%	[331]

(continued)

Table 4 (continued)

Active layer	Film thickness /active area	NP size/passivation	NP Concentration /solvent	PV parameters	Reference
P3HT/ $PC_{71}BM$/Au	220 nm/9.84 mm^2	70 nm/–	5 wt%/DCB	Jsc = 11.18 mA cm^{-2} Voc = 0.63 FF = 0.61 PCE = 4.36%	[333]
PCDTBT/ $PC_{71}BM$/Au	120 nm/9.84 mm^2	70 nm/–	5 wt%/DCB + CB	Jsc = 11.16 mA cm^{-2} Voc = 0.89 FF = 0.65 PCE = 6.45%	[333]
Si–PCPDTBT/ $PC_{71}BM$/Au	150 nm/9.84 mm^2	70 nm/–	5 wt%/DCB	Jsc = 13.13 mA cm^{-2} Voc = 0.57 FF = 0.61 PCE = 4.54%	[333]

CB chlorobenzene; *DCB* dichlorobenzene; *DDA* dodecyl amine; *NP* nanoparticles; *P3HT* poly(3-hexylthiophene); *P3OT* poly(3-octylthiophene); *PCBM* [6,6]-phenyl C_{61}-butyric acid methyl ester; *$PC_{71}MB$* [6,6]-phenyl C_{61}-butyric acid methyl ester; *PCDTBT* poly[*N*-9″-hepta-decanyl-2,7-carbazole-alt-5,5-(4′,7′-di-2-thienyl-2′,1′,3′-benzothiadiazole); *PV* photovoltaic; *py* pyridine; *Si–PCPDTBT* poly{[4, 4′-bis(2-ethylhexyl) dithieno(3,2-b:2′,3′-d)silole]-2,6-diyl-alt-[4,7-bis(2-thienyl)-2,1,3-benzothiadiazole]-5,5′-diyl}

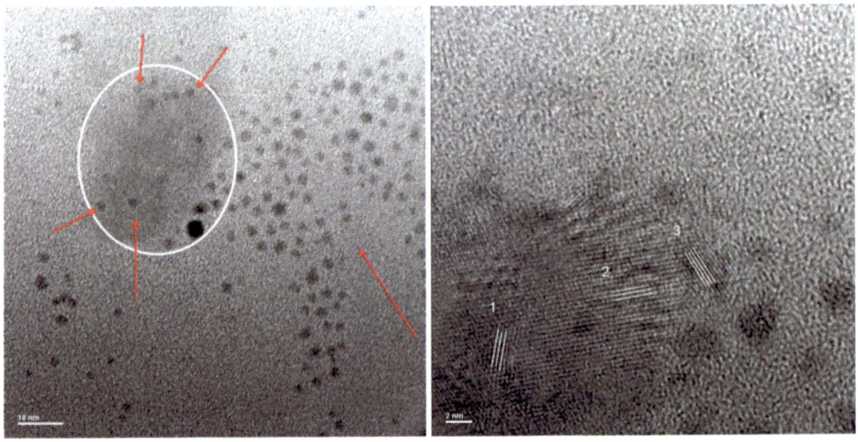

Fig. 15 HRTEM images obtained for the P3HT/PCBM/Au nanoparticles system, containing 1 wt% of Au. (*Left*) *White circle* indicates the PCBM region, while *red arrows* indicate Au nanoparticles dispersed in the P3HT matrix and in the PCBM domain. (*Right*) The high-resolution image shows PCBM crystallites oriented in distinct directions, as indicated by the *white lines* (regions 1, 2, and 3)

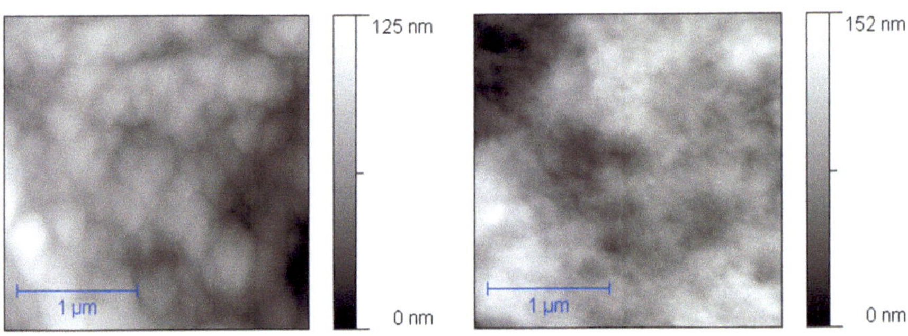

Fig. 16 AFM images in the tapping mode for (*left*) P3HT/PCBM and (*right*) P3HT/PCBM/Au nanoparticle films

et al. [335] for P3HT/PCBM unannealed films containing high concentrations of PCBM, or films containing low concentrations of PCBM and annealed for short times. These authors correlated the aspect of the PCBM crystallites with the device efficiency. The size and shape of PCBM crystallites seem to be affected by annealing time and temperature [336], and by the solvent and solvent evaporation ratio during film deposition [336]. Here, it is suggested that the incorporation of Au nanoparticles might also affect the crystallization of PCBM.

AFM images in the tapping mode are shown in Fig. 16. The RMS surface roughness is estimated to be 21 and 26 nm for P3HT/PCBM and P3HT/PCBM/Au films, respectively, both deposited from 1,2,4-trichlorobenzene. Some authors have associated the increase in roughness observed in AFM images with a higher

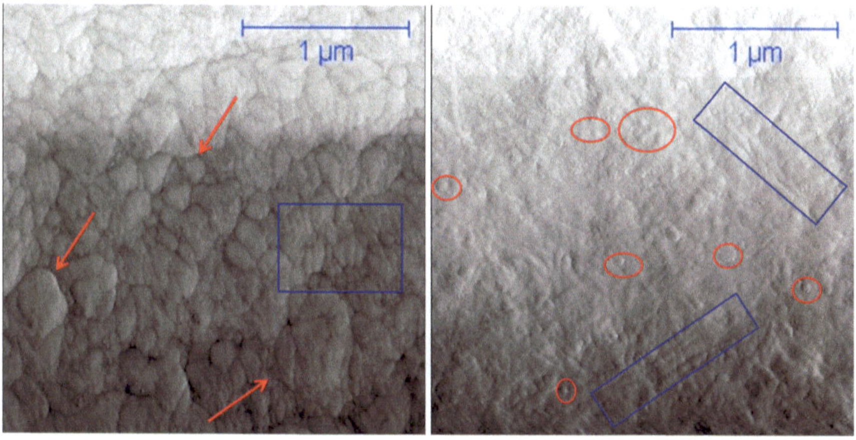

Fig. 17 AFM images in the phase mode for (*left*) P3HT/PCBM and (*right*) P3HT/PCBM/Au nanoparticles films. (*Left*) The *red arrows* indicate large P3HT aggregates (mainly amorphous phase), while the *blue rectangle* indicates an area with fibrillar structures (associated with P3HT crystalline phase). (*Right*) *Red circles* indicate possible PCBM domains, while *blue rectangles* indicate areas with fibrillar structures (P3HT crystallites)

organization of P3HT chains [75, 81, 218]. Figure 17 shows AFM images of the same regions obtained in the phase mode. These figures show the different morphological aspects between the samples with and without Au nanoparticles. For example, in Fig. 17a the presence of large agglomerates can be seen, possibly attributed to the amorphous phase of P3HT. In some points it is possible to see fibrillar structures, attributed to crystalline regions of P3HT. In Fig. 17b (P3HT/PCBM/Au sample) the presence of fibrillar structures is more evident, and there are highly ordered regions, with anisotropy, suggesting that this sample is richer in P3HT crystallites [337]. In this image, well-defined regions with $\sim$50 nm dimensions can be seen, which are associated to PCBM crystallites/domains, in agreement with the PCBM phases observed in HRTEM. These results suggest that the morphologic changes induced by incorporation of Au nanoparticles may act in the way of the organization/crystallization of both PCBM and P3HT, and this feature could be responsible for improving or decreasing the device efficiency.

An interesting aspect of these systems, as observed by Contrubia [335], is that the incorporation of Au nanoparticles does not seem to disrupt the crystalline structure of P3HT (the alpha-axis orientation signal of the polymer in X-ray diffractograms was the same before and after Au incorporation). Also, the absorption characteristics of the film are dominated by the characteristics of P3HT, and remain almost unchanged after addition of Au [335]. These data are in agreement with the observations reported by Park et al. [329] for X-ray patterns and absorption characteristics of P3HT/PCBM containing small amounts of Au nanoparticles.

From all the data discussed, it is suggested that the performance of BHJ OSC containing Au nanoparticles depends strongly on the morphology of the film. Although a careful analysis was not carried out considering the effects of the nanoparticle concentration and size, it is expected that these parameters also affect morphology and, therefore, solar cell performance.

5 Conclusions and Perspectives

In this review, recent results on the incorporation of inorganic nanoparticles into organic solar cells were highlighted. Such materials are not only easily synthesized in a great variety of sizes and shapes, but also possess important optical features that can be explored to improve the photocurrent and efficiency of solar cells. In this sense, semiconductor quantum dots, such as CdSe or PbS, may act as electron acceptors, at the same time contributing to increase light-harvesting, generating more charge carriers in the active layer. Metal nanoparticles, especially Au and Ag, are more suitable to act as light scattering structures in the buffer layer.

One of the main drawbacks associated with the use of inorganic nanoparticles in the active layer of organic solar cells is the presence of bulky organic molecules (surfactants) on the surface of the nanoparticles, which prevent their aggregation but also hinder the charge transfer and charge transport processes. Much work still needs to be done to change the nanoparticles surface in order to improve their dispersion in polymer matrices. It is clear that the morphology of the hybrid films plays a key role in the route to achieve devices with higher efficiencies. Several reports present evidences that the morphological effects influence the charge carrier generation and transport, therefore affecting the photocurrent of these devices. If the formation of a percolation network of nanoparticles at high loadings could be achieved at a more controlled condition, without or with minimum phase segregation, would also be beneficial for these systems, especially because it would contribute to a better control of the device parameters and to the reproducibility of the results. Thus, efforts contributing to increase the solubility/miscibility of organic and inorganic counterparts is encouraged.

There are also other parameters to be considered, as optimization of the morphology alone might not be sufficient to produce hybrid polymer/nanoparticle devices with efficiencies similar to those achieved for polymer/fullerene devices. The presence of traps on the nanoparticle surface, either by incomplete ligand coverage or by oxidation, can also affect charge generation and charge transport, therefore increasing charge recombination. Furthermore, in the case of semiconductor nanoparticles, the alignment of energy levels between donor and acceptor species is also a fundamental aspect affecting the performance of photovoltaic devices, thus the development of new polymers and the synthesis of nanoparticles with different sizes are frequently used approaches in the pursuit of higher efficiencies. An interesting approach that has received growing attention is the incorporation of quantum dots in polymer/fullerene mixtures. In such ternary systems, the inorganic nanoparticles may contribute to light-harvesting and to the generation of free charge carriers, while also play an important role in the extend of phase separation between the polymer and fullerene derivative.

Considering metal nanoparticles, the effects of the incorporation of these materials into polymer/fullerene mixtures is still controversial. Some authors have shown an improvement of device efficiency when adding small amounts of Au or Ag nanoparticles into the bulk heterojunction active layer, attributed to an enhancement of charge carrier transport. On the other hand, other authors found that the incorporation of these materials decreased device efficiency. The morphology effect induced in the solar cell active layer by these nanoparticles is an important parameter, which should be further explored and investigated, since it may clarify the reasons for different performances found in such devices. Meanwhile, the use of metal nanoparticles as buffer layers under the active layer has been presented as a way of exploring the optical and electronic properties of these materials in organic solar cells.

The results discussed here show that hybrid solar cells are complex systems. There is plenty room for optimization of the devices and materials, and the study of fundamental phenomena occurring in these systems continue to be necessary for a bright future. Moreover, advantage can be taken of the versatility of inorganic nanoparticles and their interactions with different materials, such as fullerenes, polymers and metal oxides, in order to improve even further the efficiency of hybrid devices. Efforts in the pursuit of a new generation of devices, with different architectures and/or exploring the combination of several layers and novel materials are welcome to the field.

Acknowledgements The authors acknowledge FAPESP (fellowship 2009/15428-0), CNPq (458413/2014-3) and Capes for financial support, LME/LNNano/CNPEM for the technical support during HR-TEM and AFM work, and Prof. N. Serdar Sariciftci, Prof. Mônica A. Cotta, João H. Clerice and Giovanni Conturbia for scientific discussions.

References

1. Fthenakis V, Alsema E (2006) Photovoltaics energy payback times, greenhouse gas emissions and external costs: 2004—early 2005 status. Prog Photovolt 14:275–280
2. Zhao J, Wang A, Altermatt P, Green MA (1995) 24 percent efficient silicon solar-sells with double-layer antireflection coatings and reduced resistance loss. Appl Phys Lett 66:3636–3638
3. Zhao J, Wang A, Green MA, Ferrazza F (1998) 19.8% efficient "honeycomb" textured multicrystalline and 24.4% monocrystalline silicon solar cells. Appl Phys Lett 73:1991–1993
4. Shockley W, Queisser HQ (1961) Detailed balance limit of efficiency of p–n junction solar cells. J Appl Phys 32:510–519
5. Knapp K, Jester T (2001) Empirical investigation of the energy payback time for photovoltaic modules. Sol Energy 71:165–172
6. Goetzberger A, Luther J, Willeke G (2002) Solar cells: past, present, future. Sol Energy Mater Sol Cells 74:1–11

7. Liu Y, Zhao J, Li Z, Mu C, Ma W, Hu H, Jiang K, Lin H, Ade H, Yan H (2014) Aggregation and morphology control enables multiple cases of high-efficiency polymer solar cells. Nat Commun 5(5293):1–8
8. He Z, Xiao B, Liu F, Wu H, Yang Y, Xiao S, Wang C, Russel TP, Cao Y (2015) Single-junction polymer solar cells with high efficiency and photovoltage. Nat Photon 9:174–179
9. Heeger AJ (2010) Semiconducting polymers: the third generation. Chem Soc Rev 39:2354–2371
10. Liz-Marzán LM (2004) Nanometals: formation and color. Mater Today 7:26–31
11. Scholes GD, Rumbles G (2006) Excitons in nanoscale systems. Nat Mater 5:683–696
12. Xia Y, Yang P, Sun Y, Wu Y, Mayers B, Gates B, Yin Y, Kim F, Yan H (2003) One-dimensional nanostructures: synthesis, characterization, and applications. Adv Mater 15:353–389
13. Manna L, Sher EC, Alivisatos AP (2002) Shape control of colloidal semiconductor nanocrystals. J Clust Sci 13:521–532
14. Cozzoli PD, Pellegrino T, Manna L (2006) Synthesis, properties and perspectives of hybrid nanocrystal structures. Chem Soc Rev 35:1195–1208
15. Biju V, Itoh T, Anas A, Sujith A, Ishikawa M (2008) Semiconductor quantum dots and metal nanoparticles: syntheses, optical properties, and biological applications. Anal Bioanal Chem 391:2469–2495
16. Moriarty P (2001) Nanostructured materials. Rep Prog Phys 64:297–381
17. Sau TK, Rogach AL (2010) Nonspherical noble metal nanoparticles: colloid-chemical synthesis and morphology control. Adv Mater 22:1781–1804
18. Sau TK, Rogach AL, Jäckel F, Klar TA, Feldmann J (2010) Properties and applications of colloidal nonspherical noble metal nanoparticles. Adv Mater 22:1805–1825
19. Henzie J, Lee J, Lee MH, Hasan W, Odom TW (2009) Nanofabrication of plasmonic structures. Annu Rev Phys Chem 60:147–165
20. Eustis S, El-Sayed MA (2006) Why gold nanopartilces are more precious than pretty gold: noble metal surface plasmon resonance and its enhancement of the radiative and nonradiative properties of nanocrystals of different shapes. Chem Soc Rev 35:209–217
21. Noguez C, Garzon IL (2009) Optically active metal nanoparticles. Chem Soc Rev 38:757–771
22. Zhang JZ, Noguez C (2008) Plasmonic optical properties and applications of metal nanostructures. Plasmonics 3:127–150
23. Arici E, Meissner D, Schaffler F, Sariciftci NS (2003) Core/shell nanomaterials in photovoltaics. Int J Photoenergy 5:199–208
24. Saunders BR, Turner ML (2008) Nanoparticle-polymer photovoltaic cells. Adv Colloid Interface Sci 138:1–23
25. Skompska M (2010) Hybrid conjugated polymer/semiconductor photovoltaic cells. Synth Met 160:1–15
26. Tang CW (1986) 2-layer organic photovoltaic cell. Appl Phys Lett 48:183–185
27. Meskers SCJ, Huebner M, Oestreich M, Baessler H (2001) Dispersive relaxation dynamics of photoexcitations in a polyfluorene film involving energy transfer: experiment and Monte Carlo simulations. J Phys Chem B 105:9139–9149
28. Mayer AC, Scully SR, Hardin BE, Rowell MW, McGehee MD (2007) Polymer-based solar cells. Mater Today 10:28–33
29. Pope M, Swenberg CE (1999) Electronic Processes in Organic Crystals and Polymers, 2nd edn. Oxford University Press, New York
30. Miranda PB, Moses D, Heeger AJ (2001) Ultrafast photogeneration of charged polarons in conjugated polymers. Phys Rev B 64:081201-1–081201-4
31. Harrison NT, Hayes GR, Phillips RT, Friend RH (1996) Singlet intrachain exciton generation and decay in poly(*p*-phenylenevinylene). Phys Rev Lett 77:1881–1884
32. Yu G, Zhang C, Heeger AJ (1994) Dual function semiconducting polymer devices—light-emitting and photodetecting diodes. Appl Phys Lett 64:1540–1542

33. Gregg BA (2003) Excitonic solar cells. J Phys Chem B 107:4688–4698
34. Savenije TJ, Warman JM, Goossens A (1998) Visible light sensitisation of titanium dioxide using a phenylene vinylene polymer. Chem Phys Lett 287:148–153
35. Nelson J (2002) Organic photovoltaic films. Curr Opin Solid State Mater Sci 6:87–95
36. Yu G, Gao J, Hummelen JC, Wudl F, Heeger AJ (1995) Polymer photovoltaic cells—enhanced efficiencies via a network of internal donor-acceptor heterojunctions. Science 270:1789–1791
37. Halls JJM, Walsh CA, Greenham NC, Marseglia EA, Friend RH, Moratti SC, Holmes AB (1995) Efficient photodiodes from interpenetrating polymer networks. Nature 376:498–500
38. Sariciftci NS, Smilowitz L, Heeger AJ, Wudl F (1992) Photoinduced electron-transfer from a conducting polymer to Buckminsterfullerene. Science 258:1474–1476
39. Brabec CJ, Zerza G, Cerulo G, De Silvestri S, Luzatti S, Hummelen JC, Sariciftci NS (2001) Tracing photoinduced electron transfer process in conjugated polymer/fullerene bulk heterojunctions in real time. Chem Phys Lett 340:232–236
40. Nogueira AF, Montanari I, Nelson J, Durrant JR, Winder C, Sariciftci NS, Brabec CJ (2003) Charge recombination in conjugated polymer/fullerene blended films studied by transient absorption spectroscopy. J Phys Chem B 107:1567–1573
41. Kim JS, Granström M, Friend RH, Johansson N, Salaneck WR, Daik R, Feast WJ, Cacialli F (1998) Indium-tin oxide treatments for single- and double-layer polymeric light-emitting diodes: the relation between the anode physical, chemical, and morphological properties and the device performance. J Appl Phys 84:6859–6870
42. Koch N, Kahn A, Ghijsen J, Prieaux JJ, Schwartz S, Johnson RL, Elschner A (2003) Conjugated organic molecules on metal versus polymer electrodes: demonstration of a key energy level alignment mechanism. Appl Phys Lett 82:70–72
43. Brabec CJ, Shaheen SE, Winder C, Sariciftci NS, Denk P (2002) Effect of LiF/metal electrodes on the performance of plastic solar cells. Appl Phys Lett 80:1288–1290
44. Malliaras GG, Salem JR, Brock PJ, Scott JC (1998) Photovoltaic measurement of the built-in potential in organic light emitting diodes and photodiodes. J Appl Phys 84:1583–1587
45. Markvart T, Castafier L (2005) Principles of solar cell operation. In: Markvart T, Castafier L (eds) Solar cells: materials, manufacture and operation. Elsevier, Amsterdam
46. Meissner D, Ronstalski J (2001) Photovoltaics of interconnected networks. Synth Met 121:1551–1552
47. Gregg BA, Hanna MC (2003) Comparing organic to inorganic photovoltaic cells: theory, experiment, and simulation. J Appl Phys 93:3605–3614
48. Hoppe H, Glatzel T, Niggemann M, Schwinger W, Schaeffler F, Hinsch A, Lux-Steiner MC, Sariciftci NS (2006) Efficiency limiting morphological factors of MDMO-PPV:PCBM plastic solar cells. Thin Solid Films 511:587–592
49. Hoppe H, Niggemann M, Winder C, Kraut J, Hiesgen R, Hinsch A, Meissner D, Sariciftci NS (2004) Nanoscale morphology of conjugated polymer/fullerene-based bulk-heterojunction solar cells. Adv Funct Mater 14:1005–1011
50. Rispens MT, Meetsma A, Rittberger R, Brabec CJ, Sariciftci NS, Hummelen JC (2003) Influence of the solvent on the crystal structure of PCBM and the efficiency of MDMO-PPV: PCBM "plastic" solar cells. Chem Commun 17:2116–2118
51. Gebeyehu D, Brabec CJ, Padinger F, Fromherz T, Hummelen JC, Badt D, Schindler H, Sariciftci NS (2001) The interplay of efficiency and morphology in photovoltaic devices based on interpenetrating networks of conjugated polymers with fullerenes. Synth Met 118:1–9
52. Yang X, Alexeev A, Michels MAJ, Loos J (2005) Effect of spatial confinement on the morphology evolution of thin poly(p-phenylenevinylene)/methanofullerene composite films. Macromolecules 38:4289–4295
53. Choulis SA, Nelson J, Kim Y, Poplavskyy D, Kreouzis T, Durrant JR, Bradley D (2003) Investigation of transport properties in polymer/fullerene blends using time-of-flight photocurrent measurements. Appl Phys Lett 83:3812–3814

54. Snaith HJ, Arias AC, Morteani AC, Silva C, Friend RH (2002) Charge generation kinetics and transport mechanisms in blended polyfluorene photovoltaic devices. Nano Lett 2:1353–1357
55. Brabec CJ, Cravino A, Meissner D, Saricftci NS, Fromherz T, Rispens MT, Sanchez L, Hummelen JC (2001) Origin of the open circuit voltage of plastic solar cells. Adv Funct Mater 11:374–380
56. Brabec CJ (2004) Organic photovoltaics: technology and market. Sol Energy Mater Sol Cells 83:273–292
57. Brabec CJ, Cravino A, Meissner D, Sariciftci NS, Rispens MT, Sanchez L, Hummelen JC, Fromherz T (2002) The influence of materials work function on the open circuit voltage of plastic solar cells. Thin Solid Films 403:368–372
58. Sharber MC, Mühlbacher D, Koppe M, Denk P, Waldauf C, Heeger AJ, Brabec CJ (2006) Design rules for donors in bulk-heterojunction solar cells—towards 10% energy-conversion efficiency. Adv Mater 18:789–794
59. Gadisa A, Svensson M, Andersson M, Inganäs O (2004) Correlation between oxidation potential and open-circuit voltage of composite solar cells based on blends of polythiophenes/fullerene derivative. Appl Phys Lett 84:1609–1611
60. Yamanari T, Taima T, Sakai J, Saito K (2009) Origin of the open-circuit voltage of organic thin-film solar cells based on conjugated polymers. Sol Energy Mater Sol Cells 93:759–761
61. Liu J, Shi Y, Yang Y (2001) Solvation-induced morphology effects on the performance of polymer-based photovoltaic devices. Adv Funct Mater 11:420–424
62. Mihailetchi VD, Blom PWM, Hummelen JC, Rispens MT (2003) Cathode dependence of the open-circuit voltage of polymer:fullerene bulk heterojunction solar cells. J Appl Phys 94:6849–6854
63. Ramsdale CM, Barker JA, Arias AC, MacKenzie JD, Friend RH (2002) The origin of the open-circuit voltage in polyfluorene-based photovoltaic devices. J Appl Phys 92:4266–4270
64. Eo YS, Rhee HW, Chin BD, Yu J-W (2009) Influence of metal cathode for organic photovoltaic device performance. Synth Met 159:1910–1913
65. Alem S, Gao J, Wantz G (2009) Photovoltaic response of symmetric sandwich polymer cells with identical electrodes. J Appl Phys 106:044505-1–044505-5
66. Frohne H, Shaheen S, Brabec CJ, Müeller D, Sariciftci NS, Meerholz K (2002) Influence of the anodic work function on the performance of organic solar cells. Chem Phys Chem 3:795–799
67. Shaheen SE, Brabec CJ, Saricftci NS, Padinger F, Fromherz T, Hummelen JC (2001) 2.5% efficient organic plastic solar cells. Appl Phys Lett 78:841–843
68. Alem S, de Bettignies R, Nunzi J-M, Cariou M (2004) Efficient polymer-based interpenetrated network photovoltaic cells. Appl Phys Lett 84:2178–2180
69. Schilinsky P, Waldauf C, Brabec CJ (2002) Recombination and loss analysis in polythiophene based bulk heterojunction photodetectors. Appl Phys Lett 81:3885–3887
70. Padinger F, Rittberger R, Sariciftci NS (2003) Effects of postproduction treatment on plastic solar cells. Adv Funct Mater 13:85–88
71. Dennler G, Mozer AJ, Juska G, Pivrikas A, Osterbacka R, Fucnsbauer A, Sariciftci NS (2006) Charge carrier mobility and lifetime versus composition of conjugated polymer/fullerene bulkheterojunction solar cells. Org Electron 7:229–234
72. Kline RJ, Mcgehee MD, Kadnikova EN, Liu J, Fréchet JM (2003) Controlling the field-effect mobility of regioregular polythiophene by changing the molecular weight. Adv Mater 15:1519–1522
73. Hoppe H, Sariciftci NS (2004) Organic solar cells: an overview. J Mater Res 19:1924–1945
74. Li G, Shrotriya V, Huang J, Yao Y, Moriarty T, Emery K, Yang Y (2005) High-efficiency solution processable polymer photovoltaic cells by self-organization of polymer blends. Nat Mater 4:864–868
75. Yang X, Loos J, Veenstra SC, Verhees WJH, Wienk MM, Kroon JM, Michaels MAJ, Janssen RAJ (2005) Nanoscale morphology of high-performance polymer solar cells. Nano Lett 5:579–583

76. Erb T, Zhokkavets U, Gobsch G, Raleva S, Stühn B, Schilinsky P, Waldauf C, Brabec CJ (2005) Correlation between structural and optical properties of composite polymer/fullerene films for organic solar cells. Adv Funct Mater 15:1193–1196
77. Zhokhavets U, Erb T, Hoppe H, Gobsch G, Sariciftci NS (2006) Effect of annealing of poly (3-hexylthiophene)/fullerene bulk heterojunction composites on structural and optical properties. Thin Solid Films 496:679–682
78. Reyes-Reyes M, Kim K, Dewald J, López-Sandoval R, Avadhanula A, Curran S, Carroll DL (2005) Meso-structure formation for enhanced organic photovoltaic cells. Org Lett 7:5749–5752
79. Ma W, Yang C, Gong X, Lee K, Heeger AJ (2005) Thermally stable, efficient polymer solar cells with nanoscale control of interpenetrating network morphology. Adv Funct Mater 15:1617–1622
80. Li G, Shrotriya V, Yao Y, Yang Y (2005) Investigation of annealing effects and film thickness dependence of polymer solar cells based on poly(3-hexylthiophene). J Appl Phys 98:043704-1–043704-5
81. Clarke TM, Ballantyne AM, Nelson J, Bradley DDC, Durrant JR (2008) Free energy control of charge photogeneration in polythiophene/fullerene solar cells: the influence of thermal annealing on P3HT/PCBM blends. Adv Funct Mater 18:4029–4035
82. Moon JS, Takacs CJ, Cho S, Coffin RC, Kim H, Bazan GC, Heeger AJ (2010) Effect of processing additive on the nanomorphology of a bulk heterojunction material. Nano Lett 10:4005–4008
83. Privikas A, Stadler P, Neugebauer H, Sariciftci NS (2008) Substituting the postproduction treatment for bulk-heterojunction solar cells using chemical additives. Org Electron 9:775–782
84. Peet J, Heeger AJ, Bazan GC (2009) "Plastic" solar cells: self-assembly of bulk heterojunction nanomaterials by spontaneous phase separation. Acc Chem Res 42:1700–1708
85. Ameri T, Dennler G, Lungenschmied C, Brabec CJ (2009) Organic tandem solar cells: a review. Energy Environ Sci 2:347–363
86. Helgesen M, Sondergaard R, Krebs FC (2010) Advanced materials and processes for polymer solar cell devices. J Mater Chem 20:36–60
87. Liang Y, Xu Z, Xia J, Tsai S-T, Wu Y, Li G, Ray C, Yu L (2010) For the bright future-bulk heterojunction polymer solar cells with power conversion efficiency of 7.4%. Adv Mater 22: E135–E138
88. Kim JY, Lee K, Coates NE, Moses D, Nguyen TQ, Dante M, Heeger AJ (2007) Efficient tandem polymer solar cells fabricated by all-solution processing. Science 317:222–225
89. Wong WY, Wang XZ, He Z, Djurišić AB, Yip CT, Cheung KY, Wang H, Mak CSK, Chan WK (2007) On the efficiency of polymer solar cells. Nat Mater 6:704–705
90. He Z, Zhong C, Huang X, Wong W-Y, Wu H, Chen L, Su S, Cao Y (2011) Simultaneous enhancement of open-circuit voltage, short-circuit current density, and fill factor in polymer solar cells. Adv Mater 23:4636–4643
91. He Z, Zhong C, Su S, Xu M, Wu H, Cao Y (2012) Enhanced power conversion efficiency in polymer solar cells using an inverted device structure. Nat Photon 6:591–595
92. Zhang S, Ye L, Zhao W, Liu D, Yao H, Hou J (2014) Side chain selection for designing highly efficient photovoltaic polymers with 2D-conjugated structure. Macromolecules 47:4653–4659
93. Nam S, Seo J, Woo S, Kim WH, Bradley DDC, Kim Y (2015) Inverted polymer fullerene solar cells exceeding 10% efficiency with poly(2-ethyl-2-oxazoline) nanodots on electron-collecting buffer layers. Nat Commun 6(8929):1–9
94. Nozik AJ (2010) Nanoscience and nanostructures for photovoltaics and solar fuels. Nano Lett 10:2735–2741
95. Nozik AJ (2008) Multiple exciton generation in semiconductor quantum dots. Chem Phys Lett 457:3–11
96. Nozik AJ (2002) Quantum dot solar cells. Phys E 14:115–120

97. Nozik AJ, Beard MC, Luther JM, Law M, Ellingson RJ, Johnson JC (2010) Semiconductor quantum dots and quantum dot arrays and applications of multiple exciton generation to third-generation photovoltaic solar cells. Chem Rev 110:6873–6890
98. Rabani E, Baer R (2010) Theory of multiexciton generation in semiconductor nanocrystals. Chem Phys Lett 496:227–235
99. Beard MC, Midgett AG, Hanna MC, Luther JM, Hughes BK, Nozik AJ (2010) Comparing multiple exciton generation in quantum dots to impact ionization in bulk semiconductors: implications for enhancement of solar energy conversion. Nano Lett 10:3019–3027
100. Kang MS, Sahu A, Norris DJ, Frisbie D (2010) Size-dependent electrical transport in CdSe nanocrystal thin films. Nano Lett 10:3727–3732
101. Arici E, Sariciftci NS, Meissner D (2003) Hybrid solar cells based on nanoparticles of CuInS2 in organic matrices. Adv Funct Mater 13:165–171
102. Arici E, Hoppe H, Schaffler F, Meissner D, Malik MA, Sariciftci NS (2004) Morphology effects in nanocrystalline CuInSe2-conjugated polymer hybrid systems. Appl Phys A—Mater Sci Process 79:59–64
103. Arici E, Hoppe H, Schaffler F, Meissner D, Malik MA, Sariciftci NS (2004) Hybrid solar cells based on inorganic nanoclusters and conjugated polymers. Thin Solid Films 451:612–618
104. Yue W, Han S, Peng R, Shen W, Geng H, Wu F, Tao S, Wang M (2010) CuInS2 quantum dots synthesized by a solvothermal route and their application as effective electron acceptors for hybrid solar cells. J Mater Chem 20:7570–7578
105. Maier E, Rath T, Haas W, Werzer O, Saf R, Hofer F, Meissner D, Volobujeva O, Bereznev S, Mellikov E, Amenitsch H, Resel R, Trimmel G (2011) CuInS2-poly(3-(ethyl-4-butanoate)thiophene) nanocomposite solar cells: preparation by na in situ formation route, performance and stability issues. Sol Energy Mater Sol Cells 95:1354–1361
106. Rath T, Edler M, Haas W, Fischereder A, Moscher S, Schenk A, Trattnig R, Sezen M, Mauthner G, Pein A, Meischler D, Bartl K, Saf R, Bansal N, Haque SA, Hofer F, List EJW, Trimmel G (2011) A direct route towards polymer/copper indium sulfide nanocomposite solar cells. Adv Energy Mater 1:1046–1050
107. Radychev N, Scheunemann D, Kruszynska M, Frevert K, Miranti R, Kolny-Olesiak J, Borchert H, Parisi J (2012) Investigation of the morphology and electrical characteristics of hybrid blends based on poly(2-hexylthiophene) and colloidal CuInS2 nanocrystals of different shapes. Org Electron 13:3154–3164
108. Arar M, Gruber M, Edler M, Haas W, Hofer F, Bansal N, Reynolds LX, Haque SA, Zojer K, Trimmel G, Rath T (2013) Influence of morphology and polymer:nanoparticle ratio on device performance of hybrid solar cells—an approach in experiment and simulation. Nanotechnology 24:484005
109. Arar M, Pein A, Haas W, Hofer F, Norrman K, Krebs FC, Rath T, Trimmel G (2012) Comprehensive investigation of silver nanoparticle/aluminum electrodes for copper indium sulfide/polymer hybrid solar cells. J Phys Chem C 116:19191–19196
110. Yue W, Zhang G, Wang S, Sun W, Lan M, Nie G (2014) Influence of crystal phase for CuInS2 on device performance of polymer-CuInS2/oxide nanoarrays solar cells. Mater Sci Semicond Process 25:337–343
111. Yang Y, Zhong H, Bai H, Zou B, Li Y, Scholes GD (2012) Transition from photoconductivity to photovoltaic effect in P3HT/CuInSe2 composites. J Phys Chem C 116:7280–7286
112. Lin Y-Y, Wang D-Y, Yen H-C, Chen H-L, Chen C-C, Chen C-M, Tang C-Y, Chen C-W (2009) Extended red light harvesting in a poly(3-hexylthiophene)/iron disulfide nanocrystal hybrid solar cell. Nanotechnology 20:405207
113. Steinhagen C, Harvey TB, Stolle CJ, Harris J, Korgel BA (2012) Pyrite nanocrystal solar cells: promising, or fool's gold? J Phys Chem Lett 3:2352–2356
114. Layek A, Middya S, Ray PP (2013) Increase in open circuit voltage by the incorporation of band gap engineered FeS2 nanoparticle within MEHPPV solar cell. J Mater Sci Mater Electron 24:3749–3755

115. Bansal N, O'Mahony FTF, Lutz T, Haque SA (2013) Solution processed polymer-inorganic semiconductor solar cells employing Sb2S3 as a light harvesting and electron transporting material. Adv Energy Mater 3:986–990
116. O'Mahony FTF, Cappel UB, Tokmoldin N, Lutz T, Lindblad R, Rensmo H (2013) Low-temperature solution processing of mesoporous metal-sulfide semiconductors as light-harvesting photoanodes. Angew Chem Int Ed 52:12047–12051
117. Saha SK, Pal AJ (2015) Schottky diodes between Bi2S3 nanorods and metal nanoparticles in a polymer matrix as hybrid bulk-heterojunction solar cells. J Appl Phys 118:014503
118. Wang Z, Qu S, Zeng X, Liu J, Zhang C, Tan F, Jin L, Wang Z (2009) The application of SnS nanoparticles to bulk heterojunction solar cells. J Alloys Compd 482:203–207
119. Du Pasquier A, Mastrogiovanni DDT, Klein LA, Wang T, Garfunkel E (2007) Photoinduced charge transfer between poly(3-hexylthiophene) and germanium nanowires. Appl Phys Lett 91:183501-1–183501-3
120. Novotny CJ, Yu ET, Yu PKL (2008) InP nanowire/polymer hybrid photodiode. Nano Lett 8:775–779
121. Liu C-Y, Holman ZC, Kortshagen UR (2009) Hybrid solar cells from P3HT and silicon nanocrystals. Nano Lett 9:449–452
122. Huynh WU, Dittmer JJ, Alivisatos AP (2002) Hybrid nanorod-polymer solar cells. Science 295:2425–2427
123. Huynh WU, Dittmer JJ, Libby WC, Whiting GL, Alivisatos AP (2003) Controlling the morphology of nanocrystal-polymer composites for solar cells. Adv Funct Mater 13:73–79
124. Sun BQ, Greenham NC (2006) Improved effciency of photovoltaics based on CdSe nanorods and poly(3-hexylthiophene) nanofibers. Phys Chem Chem Phys 8:3557–3560
125. Hindson JC, Saghi Z, Hernandez-Garrido J-C, Midgley PA, Greenham NC (2011) Morphological study of nanoparticle-polymer solar cells using high-angle annular dark-field electron tomography. Nano Lett 11:904–909
126. Zhou Y, Li YC, Zhong HZ, Hou JH, Ding YQ, Yang CH, Li YF (2006) Hybrid nanocrystal/polymer solar cells based on tetrapod-shaped CdSexTe1-x nanocrystals. Nanotechnology 17:4041–4047
127. Sun BQ, Marx E, Greenham NC (2003) Photovoltaic devices using blends of branched CdSe nanoparticles and conjugated polymers. Nano Lett 3:961–963
128. Gur I, Fromer NA, Chen C-P, Kanaras AG, Alivisatos AP (2007) Hybrid solar cells with prescribed nanoscale morphologies based on hyperbranched semiconductor nanocrystals. Nano Lett 7:409–414
129. Grancini G, Biasiucci M, Mastria R, Scotognella F, Tassone F, Polli D, Gigli G, Lanzani G (2012) Dynamic microscopy study of ultrafast charge transfer in a hybrid P3HT/hyperbranched CdSe nanoparticle blend for photovoltaics. J Phys Chem Lett 3:517–523
130. Lee K-S, Kim I, Gullapalli S, Wong MS, Jabbour GE (2011) Enhanced performance of hybrid solar cells using longer arms of quantum cadmium selenide tetrapods. Appl Phys Lett 99:223515
131. Dayal S, Reese MO, Ferguson AJ, Ginley DS, Rumbles G, Kopidakis N (2010) The effect of nanoparticle shape on the photocarrier dynamics and photovoltaic device performance of poly(3-hexylthiophene):CdSe nanoparticle bulk heterojunction solar cells. Adv Funct Mater 20:2629–2635
132. Sun BQ, Snaith HJ, Dhoot AS, Westenhoff S, Greenham NC (2005) Vertically segregated hybrid blends for photovoltaic devices with improved efficiency. J Appl Phys 97:014914-1–014914-6
133. Dayal S, Kopidakis N, Olson DC, Ginley DS, Rumbles G (2010) Photovoltaic devices with a low band gap polymer and CdSe nanostructures exceeding 3% efficiency. Nano Lett 10:239–242
134. Kuo C-Y, Su M-S, Chen G-Y, Ku C-S, Lee H-Y, Wei K-H (2011) Annealing treatment improves the morphology and performance of photovoltaic devices prepared from thieno

[3,4-c]pyrrole-4,6-dione-based donor/acceptor conjugated polymers and CdSe nanostructures. Energy Environ Sci 4:2316–2322
135. Zhou R, Zheng Y, Qian L, Yang Y, Holloway PH, Xue J (2012) Solution-processed, nanostructured hybrid solar cells with broad spectral sensitivity and stability. Nanoscale 4:3507–3514
136. Qiao F (2013) Improved performance of photovoltaic devices based on poly (3-hexylthiophene) nanofibers and CdSe quantum dots through ligand exchange and annealing treatment. Solid-State Electron 82:25–28
137. Peng Y, Song G, Hu X, He G, Chen Z, Xu X, Hu J (2013) In situ synthesis of P3HT-capped CdSe superstructures and their application in solar cells. Nanoscale Res Lett 8:106
138. Celik D, Krueger M, Veit C, Schleiermacher HF, Zimmermann B, Allard S, Dumsch I, Scherf U, Rauscher F, Niyamakom P (2012) Performance enhancement of CdSe nanorod-polymer based hybrid solar cells utilizing a novel combination of post-synthetic nanoparticle surface treatments. Sol Energy Mater Sol Cells 98:433–440
139. Wu Y, Zhang G (2010) Performance enhancement of hybrid solar cells through chemical vapor annealing. Nano Lett 10:1628–1631
140. Choi S-H, Song HJ, Park IK, Yum J-H, Kim S-S, Lee SH, Sung Y-E (2006) Synthesis of size-controlled CdSe quantum dots and characterization of CdSe-conjugated polymer blends for hybrid solar cells. J Photochem Photobiol A 179:135–141
141. Tang A-W, Teng F, Jui H, Gao Y-H, Hou Y-B, Liang C-J, Wang Y-S (2007) Investigation on photoconductive properties of MEH-PPV/CdSe-nanocrystal nanocomposites. Mater Lett 61:2178–2181
142. Han LL, Qin DH, Jiang X, Liu YS, Wang L, Chen JW, Cao Y (2006) Synthesis of high quality zinc-blende CdSe nanocrystals and their application in hybrid solar cells. Nanotechnology 17:4736–4742
143. Jiang X, Chen F, Qiu W, Yan Q, Nan Y, Xu H, Yang L, Chen H (2010) Effects of molecular interface modification in CdS/polymer hybrid bulk heterojunction solar cells. Sol Energy Mater Sol Cells 94:2223–2229
144. Chen F, Qiu W, Chen X, Yang L, Jiang X, Wang M, Chen H (2011) Large-scale fabrication of CdS nanorod arrays on transparent conductive substrates from aqueous solutions. Sol Energy 85:2122–2129
145. Mohamed NBH, Haouari M, Ebdelli R, Zaaboud Z, Habchi MM, Hassen F, Maaref H, Ouada HB (2015) Role of surface modification of CdS nanoparticles on the performance of hybrid photovoltaic devices based on p-phenylenevinylene derivative. Phys E 69:145–152
146. Zhou YF, Riehle FS, Yuan Y, Schleiermacher H-F, Niggemann M, Urban GA, Krüger M (2010) Improved efficiency of hybrid solar cells based on non-ligand-exchanged CdSe quantum dots and poly(3-hexylthiophene). Appl Phys Lett 96:013304-1–013304-3
147. Zhou Y, Eck M, Veit C, Zimmermann B, Rauscher F, Niyamakom P, Yilmaz S, Dumsch I, Allard S, Scherf U, Kruger M (2011) Efficiency enhancement for bulk-heterojunction hybrid solar cells based on acid treated CdSe quantum dots and low bandgap polymer PCPDTBT. Sol Energy Mater Sol Cells 95:1232–1237
148. Zhou Y, Eck M, Men C, Rauscher F, Niyamakom P, Yilmaz S, Dumsch I, Allard S, Scherf U, Krüger M (2011) Efficient polymer nanocrystal hybrid solar cells by improved nanocrystal composition. Sol Energy Mater Sol Cells 95:3227–3232
149. Fu W-F, Shi Y, Qiu WM, Wang L, Nan YX, Shi M-M, Li H-Y, Chen H-Z (2012) High efficiency hybrid solar cells using post-deposition ligand exchange by monothiols. Phys Chem Chem Phys 14:12094–12098
150. Fu W-F, Shi Y, Wang L, Shi M-M, Li H-Y, Chen H-Z (2013) A geen, low-cost, and highly effective strategy to enhance the performance of hybrid solar cells: post-deposition ligand exchange by acetic acid. Sol Energy Mater Sol Cells 117:329–335
151. Ren S, Chang L-Y, Lim S-K, Zhao J, Smith M, Zhao N, Bulovic V, Bawendi M, Gradecak S (2011) Inorganic-organic hybrid solar cell: bridging quantum dots to conjugated polymer nanowires. Nano Lett 11:3998–4002

152. Dixit SK, Madan S, Madhwal D, Kumar J, Sihgh I, Bhatia CS, Bhatnagar PK, Mathur PC (2012) Bulk heterojunction formation with induced concentration gradient from a bilayer structure of P3HT:CdSe/ZnS quantum dots using inter-diffusion process for developing high efficiency solar cell. Org Electron 13:710–714
153. de Freitas JN, Pivrikas A, Nowacki BF, Akcelrud LC, Sariciftci NS, Nogueira AF (2010) Investigation of new PPV-type polymeric materials containing fluorene and thiophene units and their application in organic solar cells. Synth Met 160:1654–1661
154. de Freitas JN, Grova IR, Akcelrud LC, Arici E, Sariciftci NS, Nogueira AF (2010) The effects of CdSe incorporation into bulk heterojunction solar cells. J Mater Chem 20:4845–4853
155. Huynh WU, Dittmer JJ, Teclemariam N, Milliron DJ, Alivisatos AP, Barnham KWJ (2003) Charge transport in hybrid nanorod-polymer composite photovoltaic cells. Phys Rev B 67:115326-1–115326-12
156. Lin Y-Y, Chen C-W, Chang J, Lin TY, Liu IS, Su W-F (2006) Exciton dissociation and migration in enhanced order conjugated polymer/nanoparticle hybrid materials. Nanotechnology 17:1260–1263
157. Wang P, Abrusci A, Wong HMP, Svensson M, Andersson MR, Greenham NC (2006) Photoinduced charge transfer and efficient solar energy conversion in a blend of a red polyfluorene copolymer with CdSe nanoparticles. Nano Lett 6:1789–1793
158. Ginger DS, Greenham NC (1999) Charge separation in conjugated-polymer/nanocrystal blends. Synth Met 101:425–428
159. Ginger DS, Greenham NC (1999) Photoinduced electron transfer from conjugated polymers to CdSe nanocrystals. Phys Rev B 59:10622–10629
160. Kucur E, Riegler J, Urban G, Nann T (2004) Charge transfer efficiency in hybrid bulk heterojunction composites. J Chem Phys 121:1074–1079
161. Greenham NC, Peng XG, Alivisatos AP (1996) Charge separation and transport in conjugated-polymer/semiconductor-nanocrystal composites studied by photoluminescence quenching and photoconductivity. Phys Rev B 54:17628–17637
162. Ginger DS, Greenham NC (2000) Charge injection and transport in films of CdSe nanocrystals. J Appl Phys 87:1361–1368
163. Baker DR, Kamat PV (2010) Tuning the emission of CdSe quantum dots by controlled trap enhancement. Langmuir 13:11272–11276
164. Talforn E, Moysidou E, Abellon RD, Savenije TJ, Goossens A, Houtepen AJ, Siebbeles LDA (2010) Highly photoconductive CdSe quantum-dot films: influence of capping molecules and film preparation procedure. J Phys Chem C 114:3441–3447
165. Lokteva I, Radychev N, Witt F, Borchert H, Parisi J, Kolny-Olesiak J (2010) Surface treatment of CdSe nanoparticles for application in hybrid solar cells: the effect of multiple ligand exchange with pyridine. J Phys Chem C 114:12784–12791
166. Dasgupta U, Bera A, Pal AJ (2015) pn-Junction nanorods in a polymer matrix: a paradigm shift from conventional hybrid bulk-heterojunction solar cells. Sol Energy Mater Sol Cells 143:319–325
167. Cappel UB, Dowland SA, Reynolds LX, Dimitrov S, Haque SA (2013) Charge generation dynamics in CdS:P3HT blends for hybrid solar cells. J Phys Chem Lett 4:4253–4257
168. Wood K, Garnett O, Tokmoldin N, Tsoi WC, Haque SA, Kim J-S (2014) In situ formation of organic-inorganic hybrid nanostructures for photovoltaic applications. Faraday Discuss 174:267–279
169. Wengeler L, Schmitt M, Peters K, Scharfer P, Schabel W (2013) Comparison of large scale coating techniques for organic and hybrid films in polymer based solar cells. Chem Eng Process 68:38–44
170. Kumar N, Dutta V (2014) Fabrication of polymer/cadmium sulfide hybrid solar cells [P3HT: CdS and PCPDTBT:CdS] by spray deposition. J Colloid Interface Sci 434:181–187
171. Liu JS, Tanaka T, Sivula K, Alivisatos AP, Fréchet JMJ (2004) Employing end-functional polythiophene to control the morphology of nanocrystal-polymer composites in hybrid solar cells. J Am Chem Soc 126:6550–6551

172. Albero J, Martınez-Ferrero E, Ajuria J, Waldauf C, Paciosc R, Palomares E (2009) Photo-induced eléctron recombination dynamics in CdSe/P3HT hybrid heterojunctions. Phys Chem Chem Phys 11:9644–9647
173. Truong NTN, Kim WK, Park C (2011) Effect of CdSe/P3HT composition on electrical and structural properties of bulk heterojunction solar cell active layer. Sol Energy Mater Sol Cells 95:167–170
174. Dowland SA, Reynolds LX, McLachlan A, Cappel UB, Haque SA (2013) Photoinduced electron and hole transfer in CdS:P3HT nanocomposite films: effect of nanomorphology on charge separation yield and solar cell performance. J Mater Chem A 1:13896–13901
175. Querner C, Reiss P, Bleuse J, Pron A (2004) Chelating ligands for nanocrystals' surface functionalization. J Am Chem Soc 126:11574–11582
176. Milliron DJ, Gur L, Alivisatos AP (2005) Hybrid organic: nanocrystal solar cells. MRS Bull 30:41–44
177. Advincula RC (2006) Hybrid organic-inorganic nanomaterials based on polythiophene dendronized nanoparticles. Dalton Trans 23:2778–2784
178. Sih BC, Wolf M (2007) CdSe nanorods functionalized with thiol-anchored oligothiophenes. J Phys Chem C 111:17184–17192
179. Aldakov D, Querner C, Kervella Y, Jousselme B, Demadrille R, Rossitto E, Reiss P, Pron A (2008) Oligothiophene-functionalized CdSe nanocrystals: preparation and electrochemical properties. Microchim Acta 160:335–344
180. Skaff H, Sill K, Emrick T (2004) Quantum dots tailored with poly(para-phenylene vinylene). J Am Chem Soc 126:11322–11352
181. Odoi MY, Hammer NI, Sill K, Emrick T, Barnes MD (2006) Observation of enhanced energy transfer in individual quantum dot-oligophenylene vinylene nanostructures. J Am Chem Soc 128:3506–3507
182. Pokrop R, Pamula K, Deja-Drogomirecka S, Zagorska M, Reiss P, Louarn G, Chandezon F, Pron A (2010) Molecular hybrids of CdSe semiconductor nanocrystals with terthiophene carboxylic acid or its polymeric analogue. Mater Chem Phys 123:756–760
183. Shallcross RC, D'Ambruoso GD, Pyun J, Armstrong NR (2010) Photoelectrochemical processes in polymer-tethered CdSe nanocrystals. J Am Chem Soc 132:2622–2632
184. Zhang QL, Russel TP, Emrick T (2007) Synthesis and characterization of CdSe nanorods functionalized with regioregular poly(3-hexylthiophene). Chem Mater 19:3712–3716
185. Xu J, Wang J, Mitchell M, Mukherjee P, Jeffries-EL M, Petrich JW, Lin Z (2007) Organic-inorganic nanocomposites via directly grafting conjugated polymers onto quantum dots. J Am Chem Soc 129:12828–12833
186. Wang T-L, Yang C-H, Shieh Y-T, Yeh A-C, Juan L-W, Zeng HC (2010) Synthesis of new nanocrystal-polymer nanocomposite as the electron acceptor in polymer bulk jeterojunction solar cells. Eur Polym J 46:634–642
187. Robel J, Kuno M, Kamat PV (2007) Size-dependent electron injection from excited CdSe quantum dots into TiO_2 nanoparticles. J Am Chem Soc 129:4136–4137
188. Kongkanand A, Tvrdy K, Takechi K, Kuno M, Kamat PV (2008) Quantum dot solar cells. Tuning photoresponse through size and shape control of CdSe-TiO_2 architecture. J Am Chem Soc 130:4007–4015
189. Kamat PV (2008) Quantum dot solar cells. Semiconductor nanocrystals as light harvesters. J Phys Chem C 112:18737–18753
190. Baker DR, Kamat PV (2009) Photosensitization of TiO_2 nanostructures with CdS quantum dots: particulate versus tubular support architectures. Adv Funct Mater 19:805–811
191. Bang JH, Kamat PV (2010) Solar cells by design: photoelectrochemistry of TiO_2 nanorod arrays decorated with CdSe. Adv Funct Mater 20:1970–1976
192. Sambur JB, Riha SC, Choi D, Parkinson BA (2010) Influence of surface chemistry on the binding and electronic coupling of CdSe quantum dots to single crystal TiO_2 surfaces. Langmuir 26:4839–4847

193. Shin K, Seok SI, Im SH, Park JH (2010) CdS or CdSe decorated TiO_2 nanotube arrays from spray pyrolysis deposition: use in photoelectrochemical cells. Chem Commun 46:2385–2387
194. Mora-Seró I, Likodimos V, Gimenez S, Martínez-Ferrero E, Albero J, Palomares E, Kontos AG, Falaras P, Bisquert J (2010) Fast regeneration of CdSe quantum dots by Ru dye in sensitized TiO_2 electrodes. J Phys Chem C 114:6755–6761
195. Shalom M, Albero J, Tachan Z, Martinez-Ferrero E, Zaban A, Palomares E (2010) Quantum dot-bilayer-sensitized solar cells: breakng the limits imposed by the low absorbance of dye monolayers. J Phys Chem Lett 1:1134–1138
196. Gao X-F, Sun W-T, Ai G, Peng L-M (2004) Photoelectric performance of TiO_2 nanotube array photoelectrodes cosensitized with CdS/CdSe quantum dots. Appl Phys Lett 96:153104-1–153104-3
197. Huang S, Zhang Q, Huang X, Guo X, Deng M, Li D, Luo Y, Shen Q, Toyoda T, Meng Q (2010) Fibrous CdS/CdSe quantum dot co-sensitized solar cells based on ordered TiO_2 nanotube arrays. Nanotechnology 21:375201-1–375201-7
198. Talgorn E, Abellon RD, Kooyman PJ, Piris J, Savenije TJ, Goossens A, Houtepen AJ, Siebbeles LDA (2010) Supercrystals of CdSe quantum dots with high charge mobility and efficient electron transfer to TiO_2. ACS Nano 4:1723–1731
199. Kniprath R, Rabe JP, McLeskey JT Jr, Wang D, Kirstein S (2009) Hybrid photovoltaic cells with II–VI quantum dot sensitizers fabricated by layer-by-layer deposition of water-soluble components. Thin Solid Films 518:295–298
200. Hamada M, Nakanishi S, Itoh T, Ishikawa M, Biju V (2010) Blinking suppression in CdSeCdSe/ZnS single quantum dots by TiO_2 nanoparticles. ACS Nano 4:4445–4454
201. Liu Z, Miyauchi M, Uemura Y, Cui Y, Hara K, Zhao Z, Sunahara K, Furube A (2010) Enhancing the performance of quantum dots sensitized solar cell by SiO2 surface coating. Appl Phys Lett 96:233107-1–233107-3
202. Luo L, Lv G, Li B, Hu X, Jin L, Wang J, Tang Y (2010) Formation of aligned ZnO nanotube arrays by chemical etching and coupling with CdSe for photovoltaic application. Thin Solid Films 518:5146–5152
203. Timp BA, Zhu X-Y (2010) Electronic energy alignment at the PbSe quantum dots/ZnO (1010) interface. Surf Sci 604:1335–1341
204. Huang J, Huang Z, Yang Y, Zhu H, Lian T (2010) Multiple exciton dissociation in CdSe quantum dots by ultrafast electron transfer to adsorbed methylene blue. J Am Chem Soc 132:4858–4864
205. Guchhait A, Rath AK, Pal AJ (2009) Hybrid core-shell nanoparticles: photoinduced electron-transfer for charge separation and solar cell applications. Chem Mater 21:5292–5299
206. Narayanan SS, Sinhá SS, Verma PK, Pal SK (2008) Ultrafast energy transfer from 3-mercaptopropionic acid capped CdSe/ZnS QDs to dye-labelled DNA. Chem Phys Lett 463:160–165
207. Deepa M, Gakhar R, Joshi AG, Singh BP, Srivastava AK (2010) Enhanced photoelectrochemistry and interactions in cadmium selenide-functionalized multiwalled carbon nanotube composite films. Electrochim Acta 55:6731–6742
208. Zhang L, Jia Y, Wang S, Li Z, Ji C, Wei J, Zhu H, Wang K, Wu D, Shi W, Fang Y, Cao A (2010) Carbon nanotube and CdSe nanobelt Schottky junction solar cells. Nano Lett 10:3583–3589
209. Schulz-Drost C, Sgobba V, Gerhardsm C, Leubner S, Calderon RMK, Ruland A, Guldi DM (2010) Innovative inorganic-organic nanohybrid materials: coupling quantum dots to carbon nanotubes. Angew Chem Int Ed 49:6425–6429
210. Chen Z, Berciaud S, Nukolls C, Heinz TF, Brus LE (2010) Energy transfer from individual semiconductor nanocrystals to graphene. ACS Nano 4:2964–2968
211. Biebersdorf A, Dietmuller R, Susha AS, Rogach AL, Poznyak SK, Talapin DV, Weller H, Klar TA, Feldmann J (2006) Semiconductor nanocrystals photosensitize C-60 crystals. Nano Lett 6:1559–1563

212. Chen H-Y, Lo MKF, Yang G, Monbouquette HG, Yang Y (2008) Nanoparticle-assisted high photoconductive gain in composites of polymer and fullerene. Nat Nanotechnol 3:543–547
213. de Freitas JN, Nogueira AF (2010) Hybrid nanostructured solar cells based on the incorporation of inorganic nanoparticles in polymer-fullerene mixtures. Proc SPIE Int Soc Opt Eng 7772:77721K. doi:10.1117/12.862510
214. Xue B, Vaughan B, Poh C-H, Burke KB, Thomsen L, Stapleton A, Zhou X, Bryant GW, Belcher W, Dastoor PC (2010) Vertical stratification and interfacial structure in P3HT: PCBM organic solar cells. J Phys Chem C 114:15797–15805
215. Huang Y-C, Liao Y-C, Li S-S, Wu M-C, Chen C-W, Su W-F (2009) Study of the effect of annealing process on the performance of P3HT/PCBM photovoltaic devices using scanning-probe microscopy. Sol Energy Mater Sol Cells 93:888–892
216. Dante M, Peet J, Nguyen T-Q (2008) Nanoscale charge transport and internal structure of bulk heterojunction conjugated polymer/fullerene solar cells by scanning probe microscopy. J Phys Chem C 112:7241–7249
217. Zhao Y, Xie Z, Qu Y, Geng Y, Wang L (2007) Solvent-vapor treatment induced performance enhancement of poly(3-hexylthiophene):methanofullerene bulk heterojunction photovoltaic cells. Appl Phys Lett 90:043504-1–043504-3
218. Watts B, Belcher WJ, Thomsen L, Ade H, Dastoor PC (2009) A quantitative study of PCBM diffusion during annealing of P3HT:PCBM blend films. Macromolecules 42:8392–8397
219. Alves JPD, de Freitas JN, Atvars TDZ, Nogueira AF (2013) Photophysical and photovoltaic properties of a polymer-fullerene system containing CdSe nanoparticles. Synth Met 164:69–77
220. Cao F, Wang H, Xia Z, Dai X, Cong S, Dong C, Sun B, Lou Y, Sun Y, Zhao J, Zou G (2015) An alternative route towards monodisperse CdS quantum dots for hybrid solar cells. Mater Chem Phys 149–150:124–128
221. Sharma |R, Bhalerao S, Gupta D (2016) Effect of incorporation of CdS NPs on performance of PTB7: PCBM organic solar cells. Org Electron 33:274–280
222. Leventis HC, King SP, Sudlow A, Hill MS, Molloy KC, Haque SA (2010) Nanostructured hybrid polymer-inorganic solar cell active layers formed by controllable in situ growth of semiconducting sulfide networks. Nano Lett 10:1253–1258
223. Xi DJ, Zhang H, Furst S, Chen B, Pei Q (2008) Electrochemical synthesis and photovoltaic property of cadmium sulfide-polybithiophene interdigitated nanohybrid thin films. J Phys Chem C 112:19765–19769
224. Aldakov D, Jiu T, Zagorska M, de Bettignies R, Jouneau P-H, Pron A, Chandezon F (2010) Hybrid nanocomposites of CdSe nanocrystals distributed in complexing thiophene-based copolymers. Phys Chem Chem Phys 12:7497–7505
225. Wise F (2000) Lead salt quantum dots: the limit of strong quantum confinement. Acc Chem Res 33:773–780
226. Ma W, Swisher SL, Ewers T, Engel J, Ferry VE, Atwater HA, Alivisatos AP (2011) Photovoltaic performance of ultrasmall PbSe quantum dots. ACS Nano 5:8140–8147
227. Scholes GD, Rumbles G (2006) Exciton in nanoscales systems. ACS Nano 5:683–693
228. Morrels I, Lambert K, Smeets D, De Muynck D, Nollet T, Martins JC, Vanhaecke F, Vantomme A, Delerue C, Allan G, Hens Z (2009) Size-dependent optical properties of colloidal PbS quantum dots. ACS Nano 3:3023–3030
229. Hines MA, Scholes D (2003) Colloidal PbS nanocrystal with size-tunable near-infrared emission: observation of post-synthesis self-narrowing of the particle size distribution. Adv Mater 15:1844–1849
230. Watt AAR, Blake D, Warner JH, Thomsen EA, Tavenner AL, Rubinsztein-Dunlop H (2005) Lead sulphide nanocrystal: conducting polymer solar cells. J Phys D Appl Phys 38:2006–2012
231. Cui D, Xu J, Zhu T, Paradee G, Ashok S, Gerhold M (2006) Harvest of near infrared light in PbSe nanocrystal-polymer hybrid photovoltaic solar cells. Appl Phys Lett 88:183111-1–183111-3

232. Jiang X, Schaller RD, Lee SB, Pietryga JM, Klimov VI, Zakhidov AA (2007) PbSe nanocrystal/conducting polymer solar cells with an infrared response to 2 micron. J Mater Res 22:2204–2210
233. Thapa R, Choudhury KR, Kim WJ, Sahoo Y, Cartwright AN, Prasad PN (2007) Polymeric nanocomposite infrared photovoltaics enhanced by pentacene. Appl Phys Lett 90:252112-1–252112-3
234. McDonald SA, Konstantatos G, Zhang S, Cyr PW, Klem EJD, Levina L, Sargent EH (2005) Solution-processed PbS quantum dot infrared photodetectors and photovoltaics. Nat Mater 4:138–142
235. Maria A, Cyr PW, Klem EJD, Levina L, Sargent EH (2005) Solution-processed infrared photovoltaic devices with >10% momochomatic internal quantum efficiency. Appl Phys Lett 87:213112–213113
236. Seo J, Kim SJ, Kim WJ, Singh R, Samoc M, Cartweight AN, Prasad PN (2009) Enchancement of the photovoltaic performance in PbS nanocrystal:P3HT hybrid composite devices by post-treatment-driven ligand exchange. Nanotechnology 20:095202-1–095202-6
237. de Freitas JN, Gonçalves AS, Nogueira AF (2014) A comprehensive review of the application of chalcogenide nanoparticles in polymer solar cells. Nanoscale 6:6371–6397
238. Noone KM, Strein E, Anderson NC, Wu P-T, Jenekhe SA, Ginger DS (2010) Broadband absorbing bulk heterojunction photovoltaics using low-bandgap solution-processed quantum dots. Nano Lett 10:2635–2639
239. Klem EJD, MacNeil DD, Cyr PW, Levina L, Sargent EH (2007) Efficient solution-processed infrared photovoltaic cells: planarized all-inorganic bulk heterojunction devices via inter-quantum-dot bridging during growth from solution. Appl Phys Lett 90:183113-1–183113-3
240. Luther JM, Law M, Beard MC, Song Q, Reese MO, Ellingson RJ, Nozik AJ (2008) Schottky solar cells based on colloidal nanocrystal films. Nano Lett 8:3488–3492
241. Greaney MJ, Brutchey RL (2015) Ligand engineering in hybrid polymer:nanocrystal solar cells. Mater Today 18:31–38
242. Seo J, Cho MJ, Lee D, Cartwright Prasad PN (2011) Efficient heterojunction photovoltaic cell utilizing nanocomposites of lead sulphide nanocrystals and a low-bandgap polymer. Adv Mater 8:3984–3988
243. Zhang Y, Li Z, Ouyang J, Tsang S-W, Lu J, Yu K, Ding J, Tao Y (2012) Hole transfer from PbS nanocrystal quantum dots to polymers and efficient hybrid solar cells utilizing infrared photons. Org Electron 13:2773–2780
244. Pilliego C, Manca M, Kroon R, Yarema M, Szendrei K, Andersson MR, Heiss W, Loi MA (2012) Charge separation dynamics in a narrow band gap polymer-PbS nanocrystal blend for efficient hybrid solar cells. J Mater Chem 22:24411–24416
245. Liu Z, Sun Y, Yuan J, Wei H, Huang X, Han L, Wang W, Wang H, Ma W (2013) High-efficiency hybrid solar cells based on polymer/PbS_xSe_{1-x} nanocrystals benefiting from vertical phase segregation. Adv Mater 25:5772–5778
246. Nam M, Kim S, Kim S, Kim S-W, Lee K (2013) Efficient hybrid solar cells using PbS_xSe_{1-x} quantum dots and nanorods for broad-range photon absorption and well-assembled charge transfer networks. Nanoscale 5:8202–8209
247. Colbert AE, Janke EM, Hsieh ST, Subramaniyan S, Schlenker CW, Jenekhe SA, Ginger DS (2013) Hole transfer from low band gap quantum dots to conjugated polymers in organic/inorganic hybrid photovoltaics. J Phys Chem Lett 4:280–284
248. Nam M, Park J, Kim S-W, Lee K (2014) Broadband-absorbing hybrid solar cells with efficiency greater than 3% based on a bulk heterojunction of PbS quantum dots and a low-bandgap polymer. J Mater Chem A 2:3978–3985
249. Yuan J, Gallagher A, Liu Z, Sun Y, Ma W (2015) High-efficiency polymer-PbS hybrid solar cells via molecular engineering. J Mater Chem A 3:2572–2579
250. Firdaus Y, Vandenplas E, Justo Y, Gehlhaar R, Cheyns D, Hens Z, Van der Aueraer M (2014) Enhancement of the photovoltaic performance in P3HT:PbS hybrid solar cells using small size PbS quantum dots. J Appl Phys 118:094305-1–094305-7

251. Firdaus Y, Miranti R, Fron E, Khetubol A, Vandenplas E, Cheyns D, Borchert H, Parisi J, Van der Aueraer M (2015) Charge separation dynamics at bulk heterojunctions between poly(3-hexylthiophene) and PbS quantum dots. J Appl Phys 118:055502-1–055502-16
252. Borrielo C, Bruno A, Diana R, Di Luccio T, Morvillo P, Ricciardi R, Villani F, Minarini C (2015) PbS nanocrystals in hybrid systems for solar cell applications. Phys Status Solidi A 212:245–251
253. Giansante C, Mastria R, Lerario G, Moretti L, Kriegel I, Scotognella F, Lanzini G, Carallo S, Esposito M, Biasiucci M, Rizzo A, Gigli G (2015) Molecular-level switching of polymer/nanocrystal non-covalent interactions and application in hybrid solar cells. Adv Funct Mater 25:111–119
254. Alves JPC (2017) Troca de ligantes em nanopartículas de PbS: influência sobre as propriedades fotofísicas, morfológicas e fotovoltaicas de filmes híbridos. PhD thesis. Universidade Estadual de Campinas
255. Kahmann S, Mura A, Protesescu L, Kovalenko MV, Brabec CJ, Loi MA (2015) Opto-electronics of PbS quantum dot and a narrow bandgap polymer blens. J Mater Chem C 5:5499–5505
256. Choi JJ, Luria J, Hyun B-R, Bartnik AC, Sun L, Lim Y-F, Marohn JA, Wise FW, Hanrath T (2010) Photogenerated exciton dissociation in highly coupled lead salt nanocrystal assemblies. Nano Lett 10:1805–1811
257. Mastria R, Rizzo A, Giansante C, Ballarini D, Dominici L, Igañas O, Gigli G (2015) Role of polymer in hybrid polymer/PbS quantum dot solar cells. J Phys Chem C 119:14972–14979
258. Colbert AE, Wu W, Janke EM, Ma F, Ginger DS (2015) Effects of ligands on charge generation and recombination in hybrid polymer/quantum dots solar cells. J Phys Chem C 119:24733–24739
259. Lu H, Joy J, Gaspar RL, Bradforth SE, Brutchey RL (2016) Iodide-passivated colloidal PbS nanocrystals leading to highly efficient polymer:nanocrystal hybrid solar cells. Chem Mater 28:1897–1906
260. Kisslinger R, Hua W, Shankar K (2017) Bulk heterojunction solar cells based on blends of conjugated polymers with II–IV and IV–VI inorganic semiconductor quantum dots. Polymers 9:35-1–35-29
261. Lan X, Voznyy O, de Arquer FPG, Liu M, Xu J, Proppe AH, Walters G, Fan F, Tan H, Liu M, Yang Z, Hoogland S, Sargent EH (2016) 10.6% certified colloidal quantum dot solar cells via solvent-polarity-engineered halide passivation. Nano Lett 16:4630–4634
262. http://www.britishmuseum.org/explore/highlights/highlight_objects/pe_mla/t/the_lycurgus_cup.aspx. Accessed 10 Jan 2011
263. Henglein A (1993) Physical properties of small metal particles in solution—microelectrode reactions, chemisorption, composite metal particles, and the atom-to-metal transition. J Phys Chem 97:5457–5471
264. Kreibig U, Vollmer M (1996) Optical Properties of Metal Clusters. Springer, Berlin
265. Kelly KL, Coronado E, Zhao LL, Schatz GC (2003) The optical properties of metal nanoparticles: the influence of size, shape, and dielectric environment. J Phys Chem B 107:668–677
266. Mulvaney P (1996) Surface plasmon spectroscopy of nanosized metal particles. Langmuir 12:788–800
267. Underwood S, Mulvaney P (1994) Effect of the solution refractive-index on the color of gold colloids. Langmuir 10:3427–3430
268. Ung T, Liz-Marzan LM, Mulvaney P (2001) Optical properties of thin films of Au@SiO2 particles. J Phys Chem B 105:3441–3452
269. Link S, El-Sayed MA (1999) Spectral properties and relaxation dynamics of surface plasmon electronic oscillations in gold and silver nanodots and nanorods. J Phys Chem B 103:8410–8426
270. Alvarez MM, Khoury JT, Schaaff TG, Shafigullin MN, Vezmar I, Whetten RL (1997) Optical absorption spectra of nanocrystal gold molecules. J Phys Chem B 101:3706–3712

271. Kreibig U, Genzel L (1985) Optical-absorption of small metallic particles. Surf Sci 156:678–700
272. Schaadt DM, Feng B, Yu ET (2005) Enhanced semiconductor optical absorption via surface plasmon excitation in metal nanoparticles. Appl Phys Lett 86:063106-1–063106-3
273. Catchpole KR, Polman A (2008) Plasmonic solar cells. Opt Express 16:21739–21800
274. Akimov YA, Koh WS (2010) Resonant and nonresonant plasmonic nanoparticle enhancement for thin-film silicon solar cells. Nanotechnology 21:235201-1–235201-6
275. Pillai S, Catchpole KR, Trupke T, Green MA (2007) Surface plasmon enhanced silicon solar cells. J Appl Phys 101:093105–1–093105-8
276. Hägglund C, Zäch M, Petersson G, Kasemo B (2008) Electromagnetic coupling of light into a silicon solar cell by nanodisk plasmons. Appl Phys Lett 92:053110–1–053110-3
277. Akimov YA, Koh WS, Ostrikov K (2009) Enhancement of optical absorption in thin-film solar cells through the excitation of higher-order nanoparticle plasmon modes. Opt Express 17:10195–10205
278. Temple TL, Mehanama GDK, Reehal HS, Bagnall DM (2009) Influence of localized surface plasmon excitation in silver nanoparticles on the performance of silicon solar cells. Sol Energy Mater Sol Cells 93:1978–1985
279. Akimov YA, Ostrikov K, Li EP (2009) Surface plasmon enhancement of optical absorption in thin-film silicon solar cells. Plasmonics 4:107–113
280. Ferry VE, Verschuuren MA, Li HBT, Verhagen E, Walters RJ, Schropp REI, Atwater HA, Polman A (2010) Light trapping in ultrathin plasmonic solar cells. Opt Express 18:A237–A245
281. Pala RA, White J, Barnard E, Liu J, Brongersma ML (2009) Design of plasmonic thin-film solar cells with broadband absorption enhancements. Adv Mater 21:3504–3509
282. Catchpole KR, Polman A (2008) Design principles for particle plasmon enhanced solar cells. Appl Phys Lett 93:191113-1–191113-3
283. Nakayama K, Tanabe K, Atwater HA (2008) Plasmonic nanoparticle enhanced light absorption in GaAs solar cells. Appl Phys Lett 93:121904-1–121904-3
284. Pryce IM, Koleske DD, Fischer AJ, Atwater HA (2010) Plasmonic nanoparticle enhanced photocurrent in GaN/InGaN/GaN quantum well solar cells. Appl Phys Lett 96:153501-1–153501-3
285. Hägglund C, Zach M, Kasemo B (2008) Enhanced charge carrier generation in dye sensitized solar cells by nanoparticle plasmons. Appl Phys Lett 92:013113-1–013113-3
286. Standridge SD, Schatz GC, Hupp JT (2009) Distance dependence of plasmon-enhanced photocurrent in dye-sensitized solar cells. J Am Chem Soc 131:8407–8409
287. Du L, Furube A, Yamamoto K, Hara K, Katoh R, Tachiya M (2009) Plasmon-induced charge separation and recombination dynamics in gold-TiO_2 nanoparticle systems: dependence on TiO_2 particle size. J Phys Chem C 113:6454–6462
288. Sudeep PK, Takechi K, Kamat PV (2007) Harvesting photons in the infrared. Electron injection from excited tricarbocyanine dye (IR-125) into TiO_2 and Ag@TiO_2 core-shell nanoparticles. J Phys Chem C 111:488–494
289. Kathiravan A, Kumar PS, Renganathan R, Anandan S (2009) Photoinduced electron transfer reactions between meso-tetrakis(4-sulfonatophenyl)porphyrin and colloidal metal-semiconductor nanoparticles. Colloids Surf A 333:175–181
290. Grätzel M (2003) Solar cells to dye for. Nature 421:586–587
291. McFarland EW, Tang J (2003) A photovoltaic device structure based on internal electron emission. Nature 421:616–618
292. Hussain AM, Neppolian B, Kim SH, Kim JY, Choi H-C, Lee K, Park S-J, Heeger AJ (2009) Improved performance of polymer light-emitting diodes with nanocomposites. Appl Phys Lett 94:073306-1–073306-3
293. Dhas V, Muduli S, Lee W, Han S-H, Ogale S (2008) Enhanced conversion efficiency in dye-sensitized solar cells based on ZnO bifunctional nanoflowers loaded with gold nanoparticles. Appl Phys Lett 93:243108-1–243108-3

294. Chen ZH, Tang YB, Liu CP, Leung YH, Yun GD, Chen LM, Wang YQ, Bello I, Zapien JA, Zhang WJ, Lee CS, Lee ST (2009) Vertically aligned ZnO nanorod arrays sensitized with gold nanoparticles for Schottky barrier photovoltaic cells. J Phys Chem C 113:13433–13437
295. Peh CKN, Ke L, Ho GW (2010) Modification of ZnO nanorods through Au nanoparticles surface coating for dye-sensitized solar cells applications. Mater Lett 64:1372–1375
296. Jakob M, Levanon H, Kamat PV (2003) Charge distribution between UV-irradiated TiO_2 and gold nanoparticles: determination of shift in the Fermi level. Nano Lett 3:353–358
297. Furube A, Du L, Hara K, Katoh R, Tachiya M (2007) Ultrafast plasmon-induced electron transfer from gold nanodots into TiO_2 nanoparticles. J Am Chem Soc 129:14852–14853
298. Guduru S, Singh VP, Rajaputra S, Mishra S, Mangu R, St. Omer I (2010) Characteristics of gold/cadmium sulfide nanowire Schottky diodes. Thin Solid Films 518:1809–1814
299. Haberer ED, Joo JH, Hodelin JF, Hu EL (2009) Enhanced photogenerated carrier collection in hybrid films of bio-templated gold nanowires and nanocrystalline CdSe. Nanotechnology 29:415206-1–415206-7
300. Yang T-T, Chen W-T, Hsu Y-J, Wei KH, Lin TY, Lin TW (2010) Interfacial charge carrier dynamics in core-shell Au-CdS nanocrystals. J Phys Chem C 114:11414–11420
301. Arakawa T, Munaoka T, Akiyama T, Yamada S (2009) Effects of silver nanoparticles on photoelectrochemical responses of organic dyes. J Phys Chem C 113:11830–11835
302. Nicholson PG, Ruiz V, Macpherson JV, Unwin PR (2005) Enhanced visible photoluminescence in ultrathin poly(3-hexylthiophene) films by incorporation of Au nanoparticles. Chem Commun 12:1052–1054
303. Park JH, Lim YT, Park OO, Kim JK, Yu J-W, Kim YC (2004) Polymer/gold nanoparticle nanocomposite light-emitting diodes: enhancement of electroluminescence stability and quantum efficiency of blue-light-emitting polymers. Chem Mater 16:688–692
304. Parfenov A, Gryczynski I, Malicka J, Geddes CD, Lakowicz JR (2003) Enhanced fluorescence from fluorophores on fractal silver surfaces. J Phys Chem B 107:8829–8833
305. Saranthy KV, Narayan KS, Kim J, White JO (2000) Novel fluorescence and morphological structures in gold nanoparticle-polyoctylthiophene based thin films. Chem Phys Lett 318:543–548
306. Chen XC, Green PF (2010) Control of morphology and its effects on the optical properties of polymer nanocomposites. Langmuir 26:3659–3665
307. Li F, Zhou Y, Zhang F, Liu X, Zhan Y, Fahlman M (2009) Tuning work function of noble metals as promising cathodes in organic electronic devices. Chem Mater 21:2798–2802
308. Nakamura M, Yang C, Tajima K, Hashimoto K (2009) High-performance polymer photovoltaic devices with inverted structure prepared by thermal lamination. Sol Energy Mater Sol Cells 93:1681–1684
309. Chen X, Zhao C, Rothberg L, Ng MK (2008) Plasmon enhancement of bulk heterojunction organic photovoltaic devices by electrode modification. Appl Phys Lett 93:123302-1–123302-3
310. Tjeng LH, Hesper R, Heessels ACL, Heers A, Jonkman HT, Sawatzky GA (1997) Development of the electronic structure in a K-doped C-60 monolayer on a Ag(111) surface. Solid State Commun 103:31–35
311. Hunt MRC, Modesti S, Rudolf P, Palmer RE (1995) Charge-transfer and structure in C60 adsorption on metal-sufaces. Phys Rev B 51:10039–10047
312. Chase SJ, Bacsa WS, Mitch MG, Pilione LJ, Lannin JS (1992) Surface-enhanced Raman-scattering and photoemission of C60 on noble-metal surfaces. Phys Rev B 46:7873–7877
313. Morioka R, Yasui K, Ozawa M, Odoi K, Ichikawa H, Fujita K (2010) Anode buffer layer containing Au nanoparticles for high stability organic solar cells. J Photopolym Sci Technol 23:313–316
314. Chen F-C, Wu J-L, Lee C-L, Hong Y, Kuo C-H, Huang MH (2009) Plasmonic-enhanced polymer photovoltaic devices incorporating solution-processable metal nanoparticles. Appl Phys Lett 95:013305-1–013305-3

315. Lee JH, Park JH, Kim JS, Lee DY, Cho K (2009) High efficiency polymer solar cells with wet deposited plasmonic gold nanodots. Org Electron 10:413–420
316. Kim S-S, Na S-I, Jo J, Kim D-Y, Nah Y-C (2008) Plasmon enhanced performance of organic solar cells using electrodeposited Ag nanoparticles. Appl Phys Lett 93:073307-1–073307-3
317. Morfa AJ, Rowlen KL, Reilly TH III, Romero MJ, van de Lagemaat J (2008) Plasmon-enhanced solar energy conversion in organic bulk heterojunction photovoltaics. Appl Phys Lett 92:013504-1–013504-3
318. Tvingstedt K, Persson N-K, Inganas O, Rahachou A, Zozoulenko IV (2007) Surface plasmon increase absorption in polymer photovoltaic cells. Appl Phys Lett 91:113514
319. Nemes CT, Vikapurapu DK, Petoukhoff CE, Cheung GZ, O'Carroll DM (2013) Absorption and scattering effects by silver nanoparticles near the interface of organic/inorganic semiconductor tandem films. J Nanopart Res 15:1801
320. Chuang M-K, Lin SW, Chen FC, Chu CW, Hsu CS (2014) Gold nanoparticle-decoratd graphene oxides for plasmonic-enhanced polymer photovoltaic devices. Nanoscale 6:1573–1579
321. Stenzel O, Stendal A, Voigtsberger K, von Borczykowski C (1995) Enhancement of the photovoltaic conversion efficiency of copper phthalocyanine thin-film devices by incorporation of metal-clusters. Sol Energy Mater Sol Cells 37:337–348
322. Mapel JK, Singh M, Baldo MA, Celebi K (2007) Plasmonic excitation of organic double heterostructure solar cells. Appl Phys Lett 90:121102-1–121102-3
323. Lindquist NC, Luhman WA, Oh S-W, Holmes RJ (2008) Plasmonic nanocavity arrays for enhanced efficiency in organic photovoltaic cells. Appl Phys Lett 93:123308-1–123308-3
324. Westphalen M, Kreibig U, Rostalski J, Luth H, Meissner D (2000) Metal cluster enhanced organic solar cells. Sol Energy Mater Sol Cells 61:97–105
325. Yakimov A, Forrest SR (2002) High photovoltage multiple-heterojunction organic solar cells incorporating interfacial metallic nanoculsters. Appl Phys Lett 80:1667–1669
326. Rand BP, Peumans P, Forrest SR (2004) Long-range absorption enhancement in organic tandem thin-film solar cells containing silver nanoclusters. J Appl Phys 96:7519–7526
327. Kim K, Carroll DL (2005) Roles of Au and Ag nanoparticles in efficiency enhancement of poly(3-octylthiophene)/C60 bulk heterojunction photovoltaic devices. Appl Phys Lett 87:203113-1–203113-3
328. Park M, Chin BD, Yu J-W, Chun M-S, Han S-H (2008) Enhanced photocurrent and efficiency of poly(3-hexylthiophene)/fullerene photovoltaic devices by the incorporation of gold nanoparticles. J Ind Eng Chem 14:382–386
329. Shen H, Bienstman P, Maes B (2009) Plasmonic absorption enhancement in organic solar cells with thin active layers. J Appl Phys 106:073109-1–073109-5
330. Duche D, Torchio P, Escoubas L, Monestier F, Simon J-J, Flory F, Mathian G (2009) Improving light absorption in organic solar cells by plasmonic contribution. Sol Energy Mater Sol Cells 93:1377–1382
331. Topp K, Borchert H, Johnen F, Tunc AV, Knipper M, von Hauff E, Parisi J, Al-Shamery K (2010) Impact of the incorporation of Au nanoparticles into polymer/fullerene solar cells. J Phys Chem A 114:3981–3989
332. Wang DH, Kim DY, Choi KW, Seo JH, Im SH, Park JH, Park OO, Heeger AJ (2011) Enhancement of donor-acceptor polymer bulk heterojunction solar cell power conversion efficiencies by addition of Au nanoparticles. Angew Chem Int Ed 50:1–6
333. de León A, Arias E, Moggio I, Gallardo-Vega C, Ziolo R, Rodríguez O, Trigari S, Giorgetti E, Leibig C, Evans D (2015) Synthesis of mercaptopropyl-(phenylene)s-benzoates passivated gold nanoparticles: implications for plasmonic photovoltaic cells. J Colloid Interface Sci 456:182–189
334. Conturbia GLC (2009) Células solares baseadas em nanotubos de carbono modificado e nanopartículas de ouro. Dissertation. Universidade Estadual de Campinas

335. Reyes-Reyes M, López-Sandoval R, Arenas-Alatorre J, Garibay-Alonso R, Carroll DL, Lastras-Martinez A (2007) Methanofullerene elongated nanostructure formation for enhanced organic solar cells. Thin Solid Films 516:52–57
336. Yang X, van Duren JKK, Rispens MT, Hummelen JC, Hanssen RAJ, Michels MAJ, Loos J (2004) Crystalline organization of a methanofullerene as used for plastic solar-cell applications. Adv Mater 16:802–806
337. Hugger S, Thomann R, Heinzel T, Thurn-Albrecht T (2004) Semicrystalline morphology in thin films of poly(3-hexylthiophene). Colloid Polym Sci 282:932–938

Nanomaterials for Solar Energy Conversion: Dye-Sensitized Solar Cells Based on Ruthenium(II) *tris*-Heteroleptic Compounds or Natural Dyes

Juliana dos Santos de Souza, Leilane Oliveira Martins de Andrade, Andressa Vidal Müller and André Sarto Polo

1 Aims and Scope

Dye-sensitized solar cells, DSSCs, gained much attention since it is a simple and cheap device capable of converting the sunlight into electricity through a regenerative photoelectrochemical process. DSSCs overall efficiency attained up to 14% and it is estimated to last around 20 years. Besides the economic advantages, these devices can be transparent, which allows their use for distinct architectonic purposes such as facades of buildings. DSSCs are based on a nanocrystalline mesoporous semiconductor films sensitized by dyes, which are responsible for light harvesting and electron transfer, these processes start the energy conversion and are directly responsible for its overall efficiency.

This paper aims to review a specific class of synthetic dye, the *tris*-heteroleptic ruthenium sensitizers, which have been attracting much attention on the last years due to the possibility of tune their spectroscopic and electrochemical properties as well as to improve the stability of the device. The recent advances on the use of natural dyes as semiconductor sensitizers, from 2003 to 2016, are also reviewed.

2 Introduction

The use of fossil fuel based technologies is the major responsible for the continuous increase in the pollution and in the concentration of greenhouse gases. Renewable sources must have higher contribution on the energetic matrix in providing more energy available for the humanity in a short period, having low environmental

J.d.S. de Souza · L.O.M. de Andrade · A.V. Müller · A.S. Polo (✉)
Centro de Ciências Naturais e Humanas, Universidade Federal do ABC,
Av. dos Estados - 5001, Santo André - SP 09210-170, Brazil
e-mail: andre.polo@ufabc.edu.br

F.L. de Souza and E. Leite (eds.), *Nanoenergy*,
DOI 10.1007/978-3-319-62800-4_2

impact [1, 2]. The interest on the conversion of environmentally friendly energy sources led to the development of several devices that took the advantage of the continuous evolution on several fields of research, which can result in new materials for already developed devices. For instance, the performance of direct methanol fuel cells, a well-known technology [3, 4] was improved due to the development of nanomaterials especially designed for the energy conversion process [5, 6] and their evolution allows the use of light to boost the process through a synergic arrangement [7–10].

The use of sunlight has been gaining much attention due to its abundance. For instance, it is possible to supply human energy needs up to year 2050 covering only 0.16% of the earth surface with 10% efficiency solar devices [1, 11]. There are several investigations on the conversion of sunlight in substances with more chemical energy than the reactants in a process that mimics the photosynthesis, this approach is known as artificial photosynthesis [12]. Most recently, the investigation on this research field is being called solar fuels and several papers were published describing photochemical approaches to produce high energy content substances, or fuels, from simple reactants such as water or CO_2 [13–19].

Great interest is dedicated to an especially attractive device, the dye-sensitized solar cells, DSSC, since they are capable of converting the sunlight into electricity based on photoelectrochemical principles. The materials employed for the construction of these new solar cells are common and cheap and the procedures do not require controlled environment, thus clean rooms or any other sophisticated control can be avoided, consequently a very low production cost is estimated (less than 1 € per W_{peak}) [20]. The use of new nanomaterials allows interesting features of these devices, such as transparency, possibility to have distinct colors, among others. These characteristics are very interesting for new applications of solar cells, since it can substitute glass windows and promote the co-generation of energy, or for any other architecture design.

Albeit the possible use of sensitization effect for solar energy conversion is known for a long time [21], the breakthrough of these solar cells was in 1991 when B. O'Reagan and M. Grätzel published the use of nanocrystalline and mesoporous TiO_2 film [22]. This film enhanced the light absorption due to its sponge-like characteristic which increases the surface area. The nanocrystallinity plays an important role on the electron injection and transport in these devices [23].

Since the paper of 1991, this field has been growing very fast and all the aspects of these solar cells are investigated [24–27]. In this review, the focus is on the development of *tris*-heteroleptic ruthenium (II) dyes as well as the use of natural extracts as a source of sensitizers. The absorption spectra and photoelectrochemical parameters published for these compounds since 2003 will be reviewed and discussed.

2.1 Dye-Sensitized Solar Cells—Principles and Operation

Dye-sensitized solar cells are prepared in a sandwich arrangement and are comprised by two electrodes, the photoanode and the counter electrode, Fig. 1. The photoanode is a conducting glass covered by a mesoporous and nanocrystalline TiO_2 film, sensitized by the dye-sensitizers. The counter electrode is a conducting glass covered by a thin film of catalyst, such as platinum or graphite. Between these electrodes is placed a mediator layer, usually a solution of I_3^-/I^- in nitriles.

In order to promote the energy conversion, the sunlight is harvested by the dye-sensitizers leading to an excited state capable of inject an electron into the semiconductor conducting band. The oxidized dye is immediately regenerated by the mediator and the injected electron percolates through the semiconductor film, reaches the conducting glass, and flows by the external circuit to the counter electrode. The counter electrode is responsible for regenerating the oxidized specie of the mediator, reducing it by a catalyzed reaction using electrons from the external circuit. Since there is not a permanent chemical change for dye-sensitized solar cells, the estimated lifetime of these devices is 20 years [23].

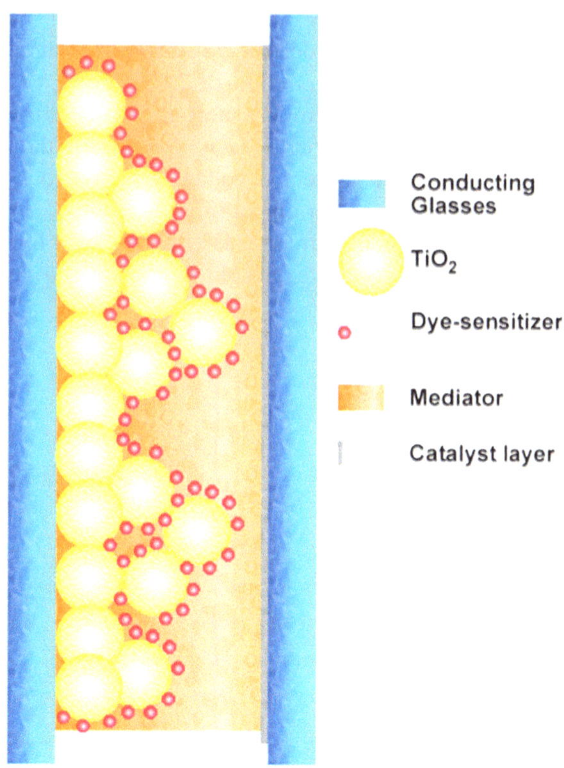

Fig. 1 Schematic arrangement of a dye-sensitized solar cell

2.2 Performance Experiments

Dye-sensitized solar cells are evaluated by several experimental approaches. For instance, the recombination processes or electron injection dynamics are investigated by time-resolved experiments [27–35], information about electron transport and electrical characteristics of TiO_2 film can be obtained by electrochemical impedance spectroscopy [36]. Among several experiments used in evaluation of DSSCs, two experiments play an important role for investigation of dye performance, the current–voltage curves and photocurrent action spectra. Due to their importance, they are detailed in the next sections.

3 Current–Voltage ($I \times V$) Curves

Current–voltage curves allow the access to one of the most important information about the prepared solar cells, the overall efficiency, η. Other important parameters such as the short-circuit current density, J_{SC}, open-circuit potential, V_{OC}, and fill factor, *ff*, are also determined by this experiment. In most cases, $I \times V$ curves determined experimentally for dye-sensitized solar cells are similar to the schematic one, Fig. 2.

Short-circuit current density, J_{SC}, and open-circuit potential, V_{OC}, are the values determined by the intersection of $I \times V$ curve to the current density axis. The voltage at this axis is zero, the short-circuit condition, thus the current is named for this condition. Analogous idea is applied for the determination of open-circuit potential, since the current at voltage axis is zero, open-circuit condition.

The maximum power output of a DSSC, P_{max}, is the highest value obtained for the multiplication of current density and voltage for each point of the $I \times V$ curve and can be graphically expressed as the area covered by the orange rectangle in

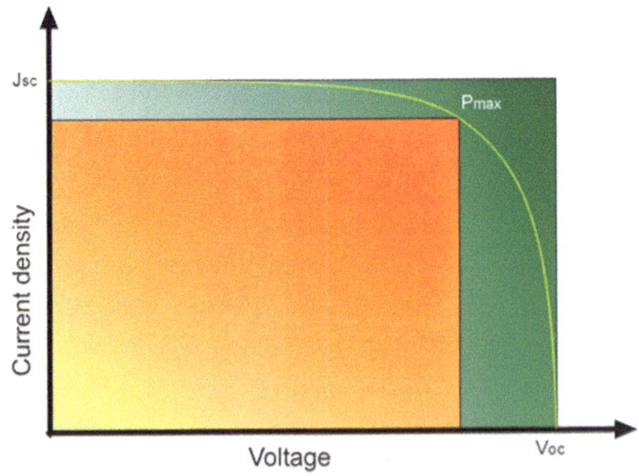

Fig. 2 Schematic current–voltage curve

Fig. 2. On the other hand, the multiplication of V_{OC} by J_{SC} results in the maximum power output possible to be achieved for this DSSC and it can also be represented by the green rectangle of Fig. 2. The fill factor,*ff*, is named for the amount of the green rectangle which is filled by the orange one. Thus, *ff* express the electrical losses of DSSCs. Mathematically, *ff* can be determined by the ratio of P_{max} and the multiplication of J_{SC} by V_{OC}, Eq. 1.

$$ff = \frac{P_{max}(\mathrm{mW\,cm^{-2}})}{J_{SC}(\mathrm{mA\,cm^{-2}}) \cdot V_{OC}(V)} \tag{1}$$

Under simulated solar irradiation condition, (1 sun = P_{irr} = 100 mW cm^{-2}), the overall efficiency, η_{Cell}, can be determined by dividing P_{max} by the total incident light power, P_{irr}, Eq. 2, resulting in the percentage amount of solar light converted in electrical output.

$$\eta_{Cell}\% = \frac{P_{max}}{P_{irr}} \cdot 100\% \tag{2}$$

4 Photocurrent Action Spectra

Photocurrent action spectra exhibit the photoelectrochemical behavior of solar cells as a function of wavelength. For each wavelength, the incident photon-to-current conversion efficiency, IPCE, can be determined and the spectra are valuable to analyze the performance of new dyes prepared. IPCE values can be determined by a relationship that considers the energy and intensity of the incident light, the J_{SC} and Planck's constant, Eq. 3.

$$\mathrm{IPCE}(\lambda) = \frac{J_{SC}}{P_{irr} \cdot e} \cdot \frac{hc}{\lambda} \tag{3}$$

J_{sc} Short-circuit photocurrent density (A m^{-2});
h Planck's constant (J s);
c Speed of light (m s^{-1});
λ Irradiation wavelength (nm);
P_{irr} Power of the incident light (W m^{-2});
e Elementary charge (C).

For practical purposes, this equation can be simplified to Eq. 4.

$$\mathrm{IPCE}\%(\lambda) = \left(1239.8 \cdot \frac{J_{SC}(\mathrm{mA\,cm^{-2}})}{P_{irr}(\mathrm{mW\,cm^{-2}}) \cdot \lambda(\mathrm{nm})}\right) \cdot 100\% \tag{4}$$

IPCE values are also related to some important parameters for DSSCs, such as light harvesting efficiency, LHE, electron injection quantum efficiency, Φ_{EI}, and the efficiency of collecting electrons in the external circuit, η_{EC}, Eq. 5 [37]. The simple measurements, such as J_{SC} and P_{irr} allow the access to important information such as the electron injection quantum yield.

$$\mathrm{IPCE}(\lambda) = \mathrm{LHE}\Phi_{EI}\eta_{EC} \quad (5)$$

Photocurrent action spectra are valuable experiments to evaluate new dye-sensitizers since it is possible to directly associate the absorption response of the dye with the conversion efficiency. This is important information for designing new sensitizers.

4.1 Molecular Engineering

The design of new dye-sensitizers is based on joining in just one specie components capable of performing specific tasks. Using different ligands, it is possible to have excellent light harvesting, electron injection on semiconductor conducting band and fast regeneration by the mediator. A new molecule to be employed in DSSCs should fulfill some basic requirements such as having an intense absorption in the visible region, which corresponds to 44% of the incident sunlight on the earth's surface, having an anchoring group capable of promoting the chemical adsorption onto TiO_2 surface, improving the electronic coupling between dye and semiconductor interface.

The first DSSC that exhibited $\eta > 10\%$ employed *cis*-di(isothiocyanato)*bis*-(2,2′-bipyridyl-4,4′-dicarboxylic acid)ruthenium(II), N3, as dye-sensitizer [38]. After this dye, the complex *mer-tri*(isothiocyanato)(2,2′,2″-terpyridyl-4,4′,4″-tricarboxylic acid)ruthenium(II), black dye, was prepared and also successfully used as sensitizer [39], Fig. 3.

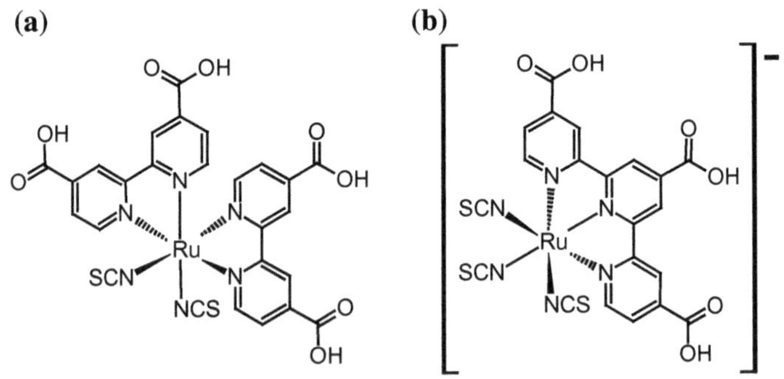

Fig. 3 Structures of the N3 (**a**) and black dye (**b**) sensitizers

Due to the outstanding performance of N3 and black dye as dye-sensitizers, they can be used as models for molecular engineering of new dyes. Their chemical attachment onto TiO_2 surface occur through the carboxylic acid groups of the 2,2′-bipyridine or of the 2,2′,2″-terpyridine ligands. Particularly, the 4,4′-dicarboxylic acid-2,2′-bipyridine anchoring ligand is been widely employed among several other possible groups investigated and it has been considered the best one for ruthenium(II) sensitizers [40]. This ligand allows intimate electronic coupling between the dye excited state wave function and the semiconductor conducting band. Its lowest unoccupied orbital, LUMO, is the lowest one of the coordination compound and facilitates an efficient electronic transfer from excited dye molecules to titania nanocrystals [41].

Great influence on the absorption spectra and molar absorptivities of compounds, emission maxima and quantum yields, as well as excited state lifetimes, in addition to the redox properties was observed as a function of the degree of protonation of the carboxylic acids of the ligand. These changes are directly responsible for the increase on photovoltaic performance of solar cells sensitized by N719, Fig. 4, which is the di-deprotonated N3 specie [42]. As a natural consequence, the use of compounds having one or more deprotonated carboxylic groups in the $dcbH_2$ has been increasing [32, 41, 43–48].

In the case of N3, consequently of N719, the presence of two $dcbH_2$ ligands results in absorption spectra which overlaps the visible region of the incident sunlight. The absorption bands have high molar absorptivity ($\varepsilon \sim 10^4$ L mol^{-1} cm^{-1}), typical of metal-to-ligand charge transfer transitions, $MLCT_{d\pi Ru-\pi^* dcbH_2}$. The high molar absorptivity improves the light harvesting efficiency, allowing the absorption of almost all incident light in a few micrometers of optical length of the TiO_2 film. Besides the bipyridine, the two isothiocyanate ligands in these complexes are valuable to promote the stabilization of the t_{2g} orbitals and result in a fine tuning of the energy levels of the complex.

5 Ruthenium *tris*-Heteroleptic Complexes

The knowledge acquired understanding the structure of the N3 dye can be used for the development of several other complexes by using the molecular engineering [49]. Among several classes of compounds developed, ruthenium *tris*-heteroleptic

Fig. 4 Structure of N719

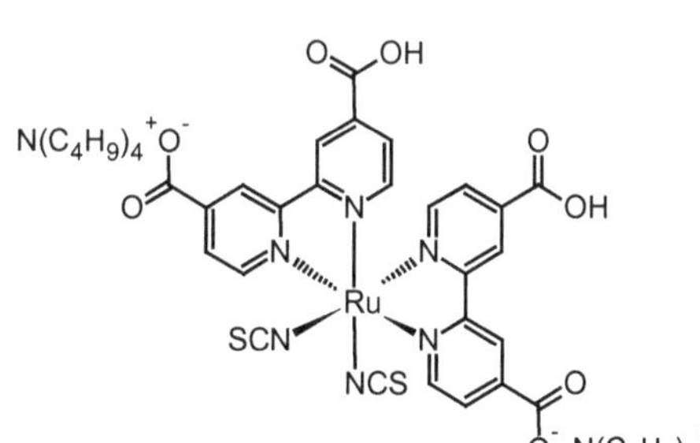

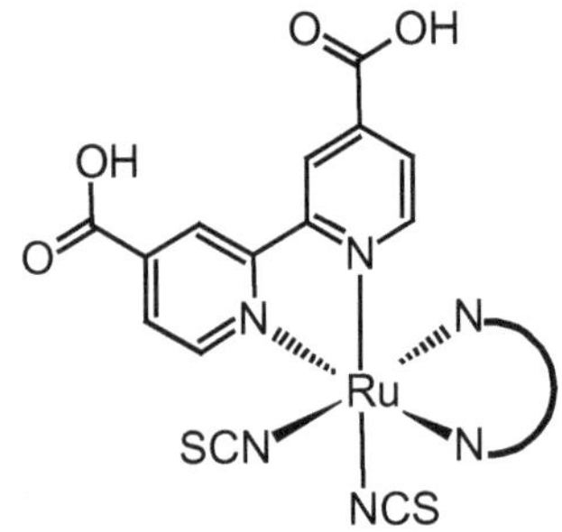

Fig. 5 General structure of *cis*-[Ru(dcbH$_2$)(L)(NCS)$_2$] dyes

complexes have been gaining attention on the last years due to the possibility of modulate their properties, just changing one of the polypyridinic ligands. This approach is very interesting for the development of new sensitizers. A comprehensive review about these compounds was also published in 2016 [50].

There are several classes of ruthenium *tris*-heteroleptic compounds described by the general formula *cis*-[Ru(dcbH$_2$)(L)(NCS)$_2$], Fig. 5, since each new ligand L and its derivatives can be a new class. In this work, our focus will be on 2,2′-bipyridine derivatives and 1,10-phenanthroline derivatives, even that several other compounds of this general formula are known [34, 51–53].

5.1 2,2′-Bipyridine Derivative Ligands

The search for high-efficiency ruthenium(II) dyes is focused on the development of complexes having high molar absorptivity, mainly in visible and near infrared region [54, 55]. A good light harvesting yield and a reduction on the film thickness, which imply reduction of transport losses in the nanoporous environment, result in higher open-circuit potentials and more efficient devices [56, 57]. Another approach is the development of dye-sensitizers capable of improving the lifetime performance of a dye-sensitized solar cell.

The first *tris*-heteroleptic ruthenium compounds investigated as dye-sensitizers are based on 2,2′-bipyridine derivatives and it is possible to observe three different approaches, following the bipyridine substituent. These subclasses are the amphiphilic, donor antenna, and thiophene compounds.

6 Amphiphilic Compounds

In 2003, a thermally stable DSSC was disclosed employing the amphiphilic Z907 sensitizer. Using this dye it was possible to prepare stable devices under prolonged thermal stress at 80°. However, the molar extinction coefficient of this sensitizer is somewhat lower than that of the standard N719 dye. Meanwhile, a compromise between efficiency and high temperature stability has been noted for the Z907 sensitizer [58]. Subsequently, the concept of developing a high molar extinction coefficient, amphiphilic ruthenium sensitizer, was followed by other groups, with a motivation to enhance device efficiency [34, 59–62]. The absorption properties as well as the performance parameters determined for ruthenium *tris*-heteroleptic complexes having amphiphilic derivatives of 2,2′-bipyridine are listed in Table 1.

The absorption spectra of amphiphilic compounds usually exhibit two MLCT bands in the visible region, typical of ruthenium *bis*-bipyridyl compounds. Molar absorptivity values listed on Table 1 are similar to those determined for the complexes N3 or N719. This behavior is expected since the aliphatic substituents do not have significant influence in the chromophoric properties of the complexes.

Amphiphilic ruthenium *tris*-heteroleptic dye-sensitizers exhibit lower photoelectrochemical performance than determined for N3. The highest efficiency achieved by this class of dyes is 8.6% [60]. The advantage of these compounds is their long-term stability. These amphiphilic heteroleptic sensitizers have the ground-state pK_a of 4,4′-dicarboxy-2,2′-bipyridine higher than determined for N3, enhancing the chemical adsorption of the complex onto the TiO_2 surface [61, 70, 71]. The structure of amphiphilic ligands decreases the charge density on the sensitizer, resulting in less electrostatic repulsion and higher amount of dye adsorbed. The hydrophobic substituent of 2,2′-bipyridine does not allow the presence of water molecules close to TiO_2 surface, improving the stabilization of solar cells toward water-induced desorption of the dye. The redox potentials of these complexes are shifted toward a more positive electrochemical potential in comparison to the N3 sensitizer, increasing the reversibility of the ruthenium III/II couple, leading to higher electrochemical stability [61, 70, 71].

7 Donor Antenna Compounds

Complexes prepared with donor antenna substituents of 2,2′-bipyridine are an approach to improve the light absorption at the same time that the hydrophobic character is enhanced. The use of aromatic substituents can have this function since the aromaticity increases the light absorption and the existence of the hydrophobic chain allows the protection to dye desorption caused by water. The spectral and photoelectrochemical parameters of this class of dyes are listed in Table 2.

In most cases, it is observed higher molar absorptivities values in comparison to amphiphilic compounds or N3 or N719 dyes which can be ascribed to an extended π-cloud delocalized in the substituent. The higher light harvesting efficiency results directly in higher IPCE values as well as overall efficiency of the solar cell, Table 2.

There are a few investigations on the use of π-excessive heteroaromatic rings as end groups in substituted bpy ligands [43, 44, 72]. The use of conjugated π-excessive heteroaromatic rings as end substituents donors directs the electron

Table 1 Absorption properties and photoelectrochemical performance of ruthenium(II) *tris*-heteroleptic compounds having amphiphilic derivatives of 2,2′-bipyridine

Abbreviation	Structure	Absorption properties		Photoelectrochemical performance					References
		Solvent	λ_{MAX}/nm (ε_{MAX}/ 10^4 L mol^{-1} cm^{-1})	V_{OC}/ V	J_{SC}/ mA cm^{-2}	$IPCE_{MAX}$ % (λ/nm)	*ff*	η_{cell} %	
Z907	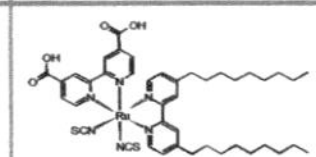	Ethanol	295 (4.24); 312 (3.01); 385 (1.09); 526 (1.16)	0.73	12.5	80 (540)	0.67	6.2	[58]
CS9	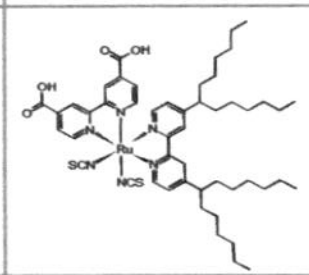	DMF	296 (4.17); 366 (1.03); 518 (0.7)	0.63	14.59	60 (540)	0.62	5.68	[59]
NMK-2[a]	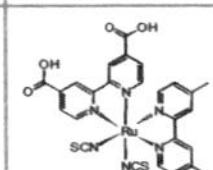	Ethanol	295 (4.54); 312 (3.35); 383 (1.13); 524 (1.16)	0.7	14.7	—	—	6.8	[60]
NMK-3[a]	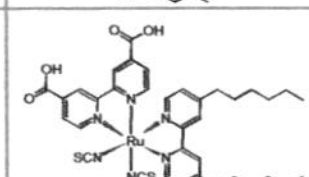	Ethanol	296 (4.26); 312 (3.2); 384 (1.01); 525 (1.11)	0.7	15.5	—	—	7.4	[60]
NMK-5[a]	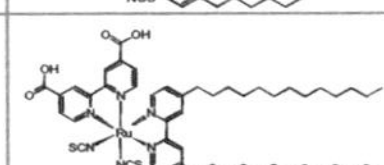	Ethanol	296 (4.21); 312 (3.02); 384 (1.08); 525 (1.15)	0.75	16.2	90	—	8.6	[60]
KC-8[a]	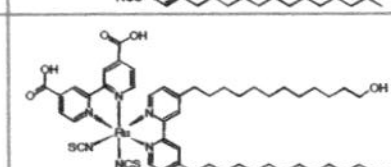	DMF	297 (4.54); 309 (2.74); 370 (1.25); 522 (1.26)	0.673	17.13	86	0.72	8.3	[61]
A597	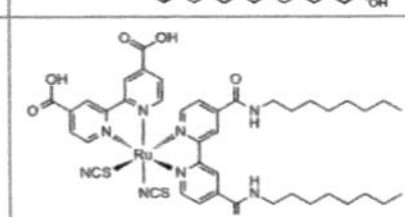	DMF	313 (5.2); 397 (1.6); 539 (1.8)	0.778	11.83	57 (540)	0.78	7.25	[63]

(continued)

Table 1 (continued)

Abbreviation	Structure	Absorption properties		Photoelectrochemical performance					References
		Solvent	λ_{MAX}/nm (ε_{MAX}/ 10^4 L mol^{-1} cm^{-1})	V_{OC}/ V	J_{SC}/ mA cm^{-2}	$IPCE_{MAX}$ % (λ/nm)	*ff*	η_{cell} %	
C4	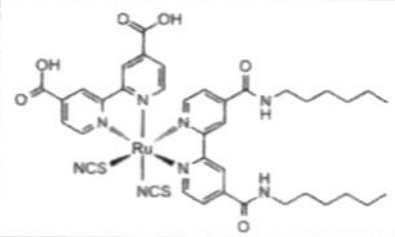	Ethanol	312 (2.6); 391 (0.9); 535 (0.9)	0.71	15.5	75 (500)	0.66	7.3	[64]
C5	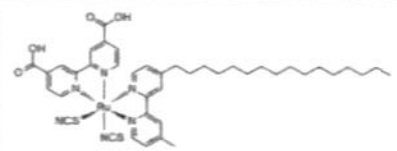	Ethanol	296 (3.65); 314 (2.53); 384 (1.0); 528 (1.01)	0.72	16.7	—	0.64	7.7	[64]
CS27	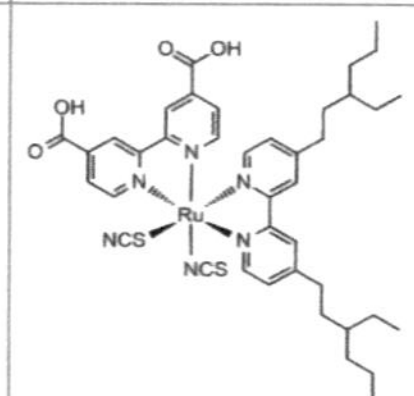	DMF	299 (4.78); 369 (1.25); 517 (1.13)	0.63	3.0	45	0.77	1.46	[65]
CS28	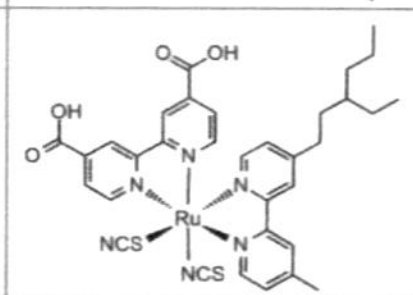	DMF	299 (4.48); 370 (1.20); 518 (1.08)	0.66	7.0	80	0.72	3.28	[65]
CS32	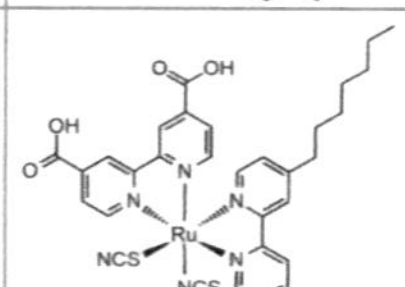	DMF	298 (4.50); 370 (1.17); 518 (1.05)	0.66	6.7	80	0.67	2.95	[65]

(continued)

Table 1 (continued)

Abbreviation	Structure	Absorption properties		Photoelectrochemical performance					References
		Solvent	λ_{MAX}/nm (ε_{MAX}/ 10^4 L mol^{-1} cm^{-1})	V_{OC}/ V	J_{SC}/ mA cm^{-2}	$IPCE_{MAX}$ % (λ/nm)	*ff*	η_{cell} %	
CS43	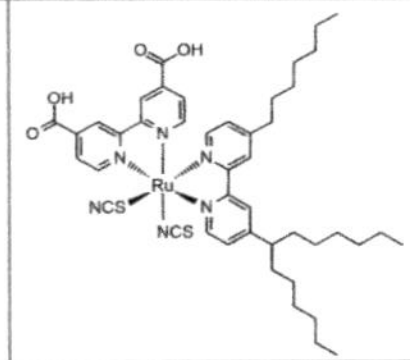	DMF	299 (5.43); 369 (1.34); 518 (1.20)	0.69	6.1	70	0.69	2.91	[65]
K005	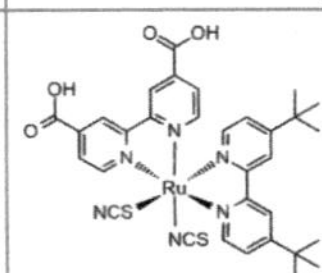	Ethanol	316 (2.68); 418 (0.95); 537 (0.94)	0.553	9.35	—	0.720	3.72	[66]
Ru-C	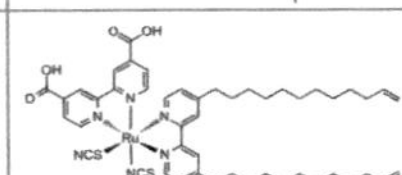	—	—	0.68	11.1	—	0.67	5.1	[67]
RuC9	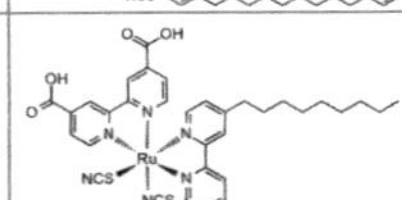	Acetonitrile/ *t*-butanol (1:1 v/v)	215; 294; 379 (0.99); 521 (1.14)	0.673	14.10	—	0.711	6.92	[68]
S8	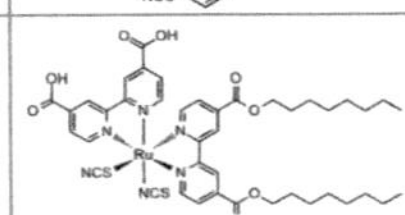	Ethanol	315; 395; 535	0.60	13.02	—	0.69	5.36	[69]

[a]These compounds were named after the initials of the first author of the reference cited

Table 2 Absorption properties and photoelectrochemical performance of ruthenium(II) *tris*-heteroleptic compounds having donor antenna derivatives of 2,2′-bipyridine

Abbreviation	Structure	Absorption properties		Photoelectrochemical performance					References
		Solvent	λ_{MAX}/nm (ε_{MAX}/10^4 L mol^{-1} cm^{-1})	V_{OC}/ V	J_{SC}/ mA cm^{-2}	$IPCE_{MAX}$ % (λ/nm)	*ff*	η_{cell} %	
LXJ1	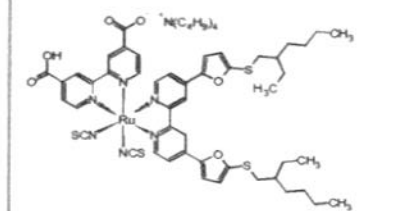	DMF	309 (4.7), 353 (3.3), 549 (1.84)	0.715	16.50	83.7 (550)	0.745	8.80	[43]
IJ-1	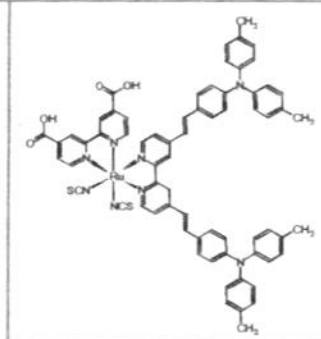	Ethanol	218; 308 (5.0); 432 (4.3), 536 (1.9)	0.748	19.2	87	0.72	10.3	[56]
KW-2[a]	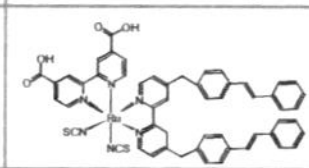	DMF	310 (4.86); 373 (7.95); 550 (2.22)	0.685	3.42	—	0.42	0.99	[57]
KW-3[a]	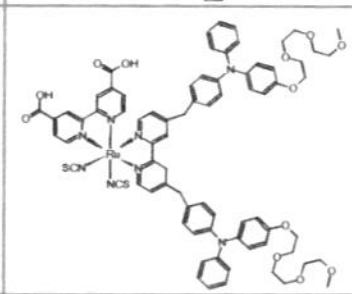	MetOH + 1 wt % KOH	307 (8.13); 429 (5.34); 524 (3.09)	0.735	4.03	—	0.46	1.37	[57]
KW-4[a]	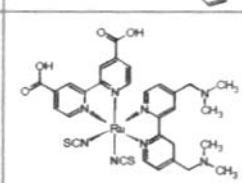	1:1 H_2O: DMF + 1 wt% KOH	307 (3.88); 381 (1.28); 526 (1.13)	0.635	2.15	—	0.42	0.58	[57]

(continued)

Table 2 (continued)

Abbreviation	Structure	Absorption properties		Photoelectrochemical performance					References
		Solvent	λ_{MAX}/nm (ε_{MAX}/ 10^4 L mol^{-1} cm^{-1})	V_{OC}/ V	J_{SC}/ mA cm^{-2}	$IPCE_{MAX}$ % (λ/nm)	*ff*	η_{cell} %	
KW-5[a]	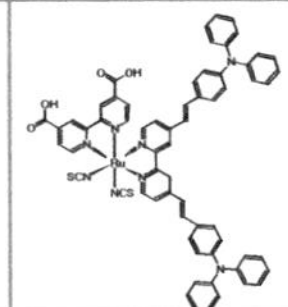	1:1 H_2O: DMF + 1 wt% KOH	304 (6.25); 423 (5.47); 544(2.27)	0.715	4.30	—	0.43	1.31	[57]
C105	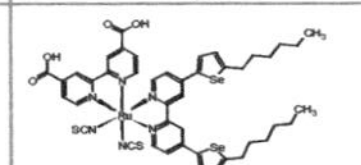	DMF	309 (3.90); 353 (3.2); 420(1.84); 550 (1.84)	0.747	18.9	95 (520)	0.744	10.06	[72]
HRD-1	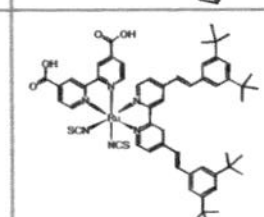	DMF	543 (1.93)	0.59	10.9	60	0.78	4.93	[70]
HRD-2	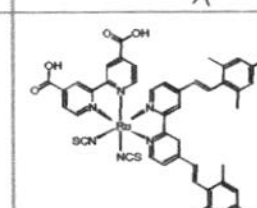	DMF	532 (1.61)	0.60	10.5	64	0.78	4.91	[70]
HRS-2	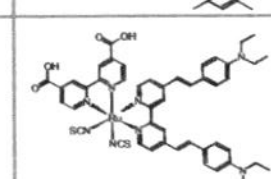	Ethanol	431 (5.93); 542 (2.81)	0.697	17.47	85 (552)	0.711	8.65	[71]
C102	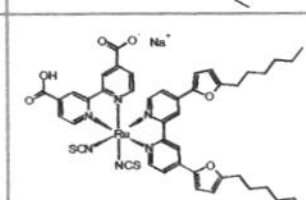	DMF	305; 341; 407; 547 (1.68)	0.740	17.80	82 (550)	—	9.5	[44]

(continued)

Table 2 (continued)

Abbreviation	Structure	Absorption properties		Photoelectrochemical performance					References
		Solvent	λ_{MAX}/nm (ε_{MAX}/ 10^4 L mol^{-1} cm^{-1})	V_{OC}/ V	J_{SC}/ mA cm^{-2}	$IPCE_{MAX}$ % (λ/nm)	ff	η_{cell} %	
AB-1[a]	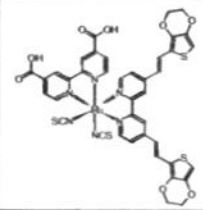	Ethanol	314; 388; 538 (1.6)	0.663	19.1	87	0.72	9.1	[55]
N945	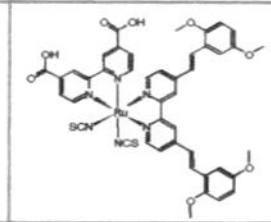	1:1 acetonitrile: *tert*-butanol	400 (3.4); 550 (1.9)	0.728	17.96	61	0.71	9.29	[73]
KCS-1[a]	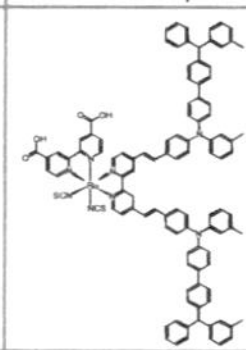	DMF	310 (8.49); 440 (5.34); 540 (2.67)	0.757	9.6	—	0.35	3.4	[74]
K-19	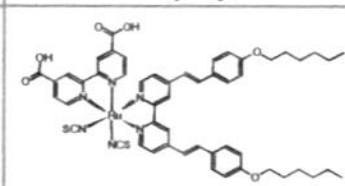	DMF	310 (5.0); 350 (4.8); 410 (1.8); 545 (1.8)	0.718	13.2	—	0.745	7.1	[32]
K-73	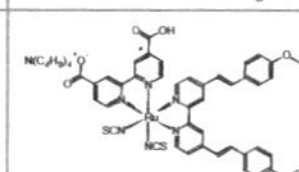	DMF	310; 350; 410; 545	0.748	17.22	85 (540)	0.649	9.5	[32]
K77	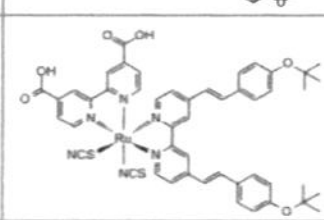	DMF	310; 346; 546 (1.94)	0.737	17.5	90 (550)	0.696	9.0	[75]

(continued)

Table 2 (continued)

Abbreviation	Structure	Absorption properties		Photoelectrochemical performance					References
		Solvent	λ_{MAX}/nm (ε_{MAX}/ 10^4 L mol^{-1} cm^{-1})	V_{OC}/ V	J_{SC}/ mA cm^{-2}	$IPCE_{MAX}$ % (λ/nm)	*ff*	η_{cell} %	
Z-910	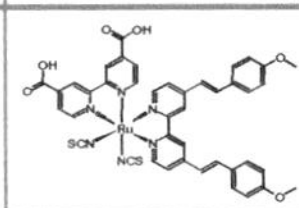	Acetonitrile	410 (1.7); 543 (1.69)	0.777	17.2	87 (520)	0.764	10.2	[33]
KC-5[a]	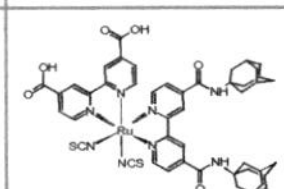	DMF	313 (3.88); 392 (1.17); 537 (1.19)	0.695	15.8	75	0.66	7.01	[61]
KC-6[a]	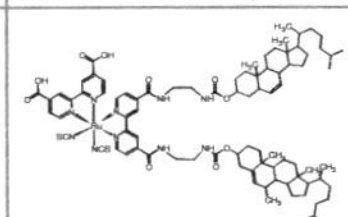	DMF	314 (3.36); 390 (1.11); 531 (1.12)	0.676	15.47	63	0.71	7.42	[61]
KC-7[a]	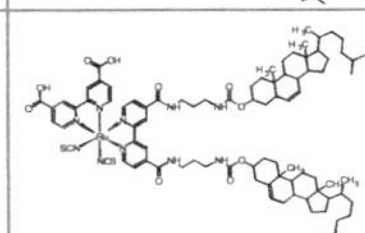	DMF	312 (3.39); 393 (1.12); 533 (1.21)	0.676	16.11	79	0.7	7.62	[61]
MH06	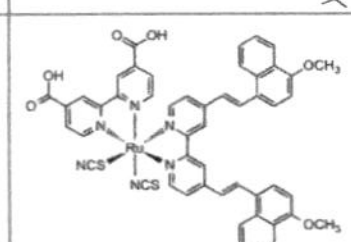	DMF	390 (4.07); 541 (2.15)	0.642	15.07	70	0.692	6.70	[76]
MH11	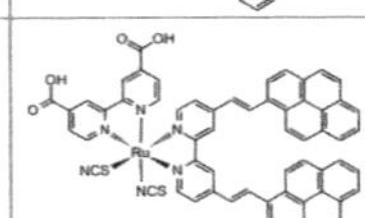	DMF	414 (5.16); 547 (2.7)	0.671	19.70	90	0.664	8.78	[76]

[a]These compounds were named after the initials of the first author of the reference cited

injection in the excited state and enhances the oscillator strength resulting in significant increases in the short-circuit photocurrent [55].

The higher molar absorptivity in the visible region can be understood by the influence of the different delocalized π-systems integrated in the bipyridyl donor antenna ligands. The reason for the lower absorption of the standard N719 dye in this region is the absence of any these groups [57].

8 Thiophene Compounds

Ruthenium(II) sensitizers having 2,2′-bipyridine with thiophene substituents have higher molar absorptivity than observed for the previous classes of compounds. For instance, the compound KW-1 has $\varepsilon_{515} = 3.56\ \text{L mol}^{-1}\ \text{cm}^{-1}$ [57], much higher than the ones determined for N3 or N719 dyes. As it is observed for the donor antenna class of compounds, the higher light harvesting efficiency results in higher IPCE values and consequently improves the overall performance of the solar cell, Table 3.

Ruthenium(II) thiophene compounds gained special attention after C101 dye has set a new DSSC efficiency record of 11.3–11.5% and became the first sensitizer to triumph over the well-known N3 dye [44]. In comparison to its analogues C102 or C105, in which the thiophene is replaced by furan, or selenophene, respectively, the molar absorptivity increases in the order of Se > S > O. This sequence it is consistent with the electropositivity trend and the size of the heteroatoms of five-member conjugated units. The LUMO energy sequence of the spectator ligand is O > S > Se, which explains this behavior [72].

Another important dye employing thiophene derivatives is CYC-B1, which exhibits a remarkably high light-harvesting capacity of up to $2.12 \times 10^4\ \text{L mol}^{-1}\ \text{cm}^{-1}$ [40]. After the development of the CYC-B1 dye, several ruthenium dyes were synthesized by incorporating thiophene derivatives into the ancillary ligand and DSSC cells based on these dyes exhibited excellent photovoltaic performances [45, 46, 77, 78].

The extensive use of polythiophene is due to its similarity to a *cis*-polyacetylene chain bridged with sulfur atoms. The "bridging sulfur atoms" could effectively provide aromatic stability to the polyacetylene chain while preserving the desirable physical properties, such as high charge transport. The facile functionalization of thiophene groups also offers relatively efficient synthetic solutions to solubility, polarity, and energetic tuning. Furthermore, sulfur has greater radial extension in its bonding than the second-row elements, such as carbon. Therefore, thiophene is a more electron-rich moiety and incorporation of thiophene onto bipyridine ligands raises the energy levels of the metal center and the LUMO of the ligands [80]. As a consequence, the band resulting from charge transfer from the metal center to the anchoring ligand is redshifted, and upon illumination of the sample, the electrons on the metal center are transferred to the anchoring $dcbH_2$ ligand, where electrons can move to the outer circuit through the TiO_2 particles more efficiently [40].

Table 3 Absorption properties and photoelectrochemical performance of ruthenium(II) *tris*-heteroleptic compounds having thiophene derivatives of 2,2′-bipyridine

Abbreviation	Structure	Absorption properties		Photoelectrochemical performance					References
		Solvent	λ_{MAX}/nm (ε_{MAX}/ 10^4 L mol^{-1} cm^{-1})	V_{OC}/ V	J_{SC}/ mA cm^{-2}	IPCE$_{MAX}$ % (λ/nm)	*ff*	η_{cell} %	
CYC-B1	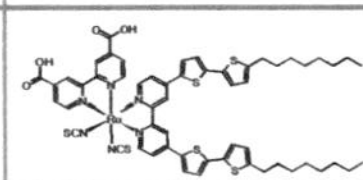	DMF	312 (3.58); 400 (4.64); 553 (2.12)	0.65	23.92	77.5	0.55	8.54	[40]
CYC-B3	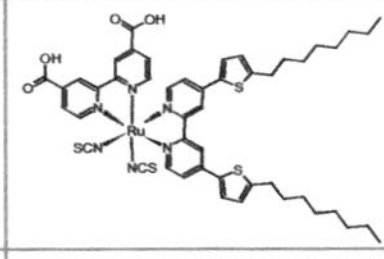	DMF	350; 400; 544 (1.57)	0.669	15.7	64.1 (520)	0.705	7.39	[77]
CYC- B7	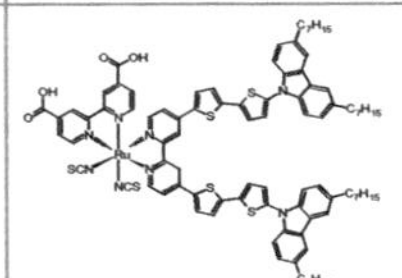	DMF	345 (3.55), 412 (4.35), 551 (2.19)	0.788	17.4	76 (530)	0.654	8.96	[78]
CYC-B11	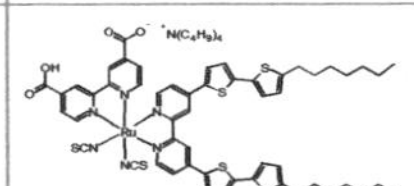	DMF	305(4.5), 388(5.40), 554 (2.42)	0.714	16.1	95 (580)	0.69	7.9	[45]
CYC-B13	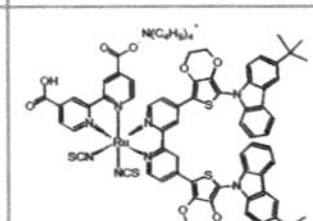	DMF	295 (8.6); 397 (3.4); 547 (1.93)	0.728	10.26	90 (550)	0.68	5.1	[46]
C101	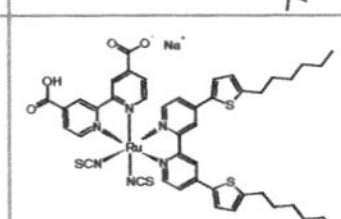	DMF	305; 341; 407; 547 (1.75)	0.746	5.42	89 (580)	0.833	11.3	[44]

(continued)

Table 3 (continued)

Abbreviation	Structure	Absorption properties		Photoelectrochemical performance					References
		Solvent	λ_{MAX}/nm (ε_{MAX}/ 10^4 L mol^{-1} cm^{-1})	V_{OC}/ V	J_{SC}/ mA cm^{-2}	$IPCE_{MAX}$ % (λ/nm)	*ff*	η_{cell} %	
C104	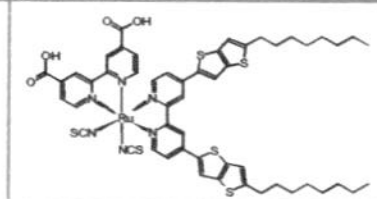	DMF	312 (5.4); 368 (4.5); 553 (2.05)	0.76	17.87	85 (580)	0.776	10.53	[62]
C106	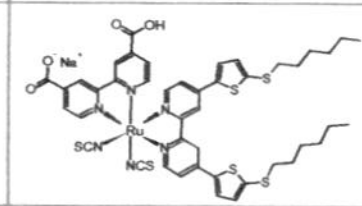	DMF	310 (3.95); 348 (3.1); 550 (1.87)	0.749	18.28	90 (520–640)	0.772	10.57	[41]
KW-1[a]	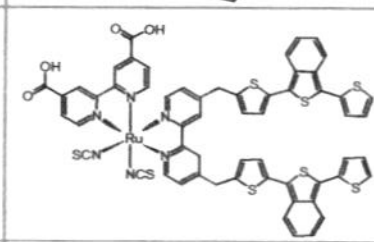	DMF	301 (4.58); 369 (3.51); 515 (3.56)	0.625	1.06	—	0.46	0,31	[57]
RT-1[a]	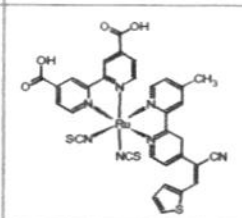	DMF	301 (3.04); 368 (2.28); 529 (1.41)	0.58	6.0	—	0.70	2.45	[79]
RT-2[a]	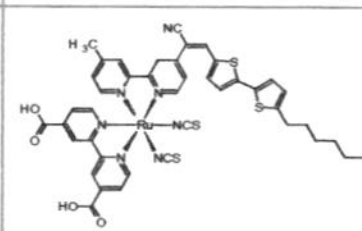	DMF	300 (4.34); 445 (2.88); 551 (1.68)	0.57	5.3	—	0.75	2.84	[79]
RT-3[a]	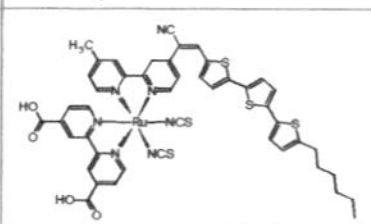	DMF	305 (4.56); 388 (3.15); 563 (2.76)	0.54	4.5	—	0.76	1.85	[79]

(continued)

Table 3 (continued)

Abbreviation	Structure	Absorption properties		Photoelectrochemical performance					References
		Solvent	λ_{MAX}/nm (ε_{MAX}/ 10^4 L mol^{-1} cm^{-1})	V_{OC}/ V	J_{SC}/ mA cm^{-2}	$IPCE_{MAX}$ % (λ/nm)	*ff*	η_{cell} %	
C107	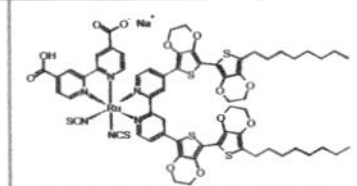	DMF	310 (3.9); 453 (5.43); 552 (2.8)	0.739	19.8	92 (550)	0.751	10.7	[47]
SJW-E1	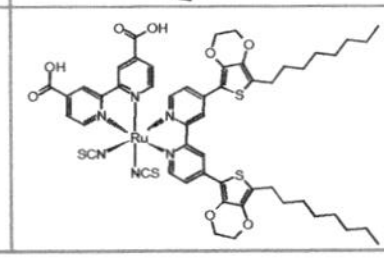	DMF	310; 360; 546 (1.87)	0.669	21.6	72.6 (550)	0.626	9.02	[77]

[a]These compounds were named after the initials of the first author of the reference cited

8.1 1,10-Phenanthroline Derivative Ligands

Besides 2,2′-bipyridine derivatives, 1,10-phenanthroline and its derivatives are gaining attention to be used in *cis*-[Ru(dcbH_2)(L)$(NCS)_2$] sensitizers. Their similarity to 2,2′-bipyridine and the advantage of having an extended π-conjugated structure led to a great potential to be employed as ancillary ligands [81]. This class of compounds still having few complexes reported in DSSCs, and their spectral as well as photoelectrochemical parameters are listed in Table 4.

The use of phenanthroline derivatives in ruthenium(II) sensitizers leads to properties favorable to the energy conversion processes and can increase the DSSCs performances, which have shown promising results [48, 82, 83].

The comparison on the properties of the complex NOK-1 [83] with N3 indicates that the substitution of the 2,2′-bipyridine derivative by 1,10-phenanthroline does not exhibit better performance or absorption properties. On the other hand, the complexes YS5 and AVM-2 exhibit higher absorbance and also had better performance than the complexes N719 and N3 under the same conditions [48, 86], indicating that this is a promising class of compounds to be investigated. Their higher efficiency was ascribed to an enhancement of electron injection. This effect is due to the reduction of dihedral angle between phenantroline and its substituents (phenyl or carbazole) on the excited state, allowing the electron injection through the thermalized triplet excited state.

9 Natural Dyes

Faster, cheaper, low-energy way alternative for ruthenium sensitizers are natural dyes and these compounds have been gaining much attention. Natural dyes can be obtained from fruits, flowers, or leaves and are suitable for educational purposes [91–93] or are an environmentally friendly alternative for dye production since a long-term stability of DSSC using these sensitizers has been demonstrated [94].

On the last years, these natural dyes solar cells experienced a great transformation, focusing on alternatives to improve the performance of such DSSCs [95–102], and special attention is given for dye cocktails [103–106] or co-sensitization approach [107]. There are also several papers describing the use of natural dyes in solid, or quasi-solid, solar cells [108, 109] or using semiconductors such as ZnO [110–112] in such DSSCs, as well as their use on investigations of electron injection/recombination processes. In this chapter, it is presented only the results reported on natural dye-sensitizers used in DSSCs having TiO_2 as semiconductor and liquid mediator.

The absorption properties and photoelectrochemical performance of natural dye-sensitized solar cells reported since 2003 are listed in Table 5.

Table 4 Absorption properties and photoelectrochemical performance of ruthenium(II) *tris*-heteroleptic ruthenium compounds having 1,10-phenanthroline derivatives

Abbreviation	Structure	Absorption properties		Photoelectrochemical performance					References
		Solvent	λ_{MAX}/nm (ε_{MAX}/ 10^4 L mol^{-1} cm^{-1})	V_{OC}/ V	J_{SC}/ mA cm^{-2}	IPCE$_{MAX}$ % (λ/nm)	*ff*	η_{cell} %	
YS5	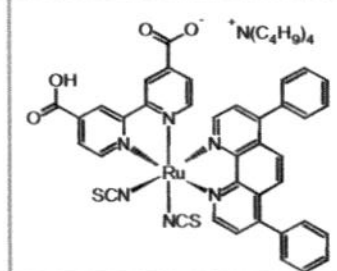	DMF	283(5.68), 308 (3.95), 362 (0,81), 522 (1.71)	0.749	14.52	64.6 (540)	0.557	6.05	[48]
AR25	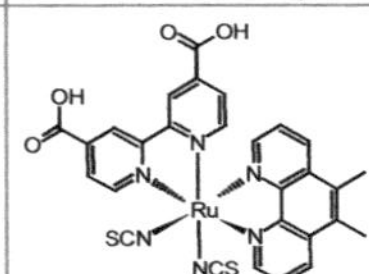	DMF	420; 518(6.58)	0.69	9.6	61 (520)	0.39	2.6	[82]
NOK-1[a]	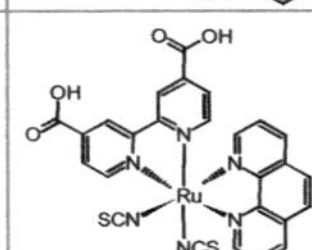	0.01 M NaOH aqueous solution	267 (5.7); 309 (2.9); 400 (1.0); 492(1.2)	0.65	15.3	74	0.67	6.7	[83]
NOK-2[a]	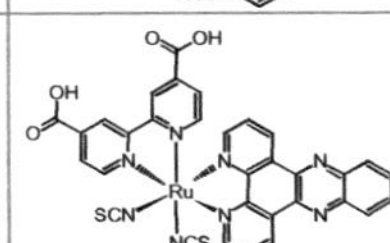	0.01 M NaOH aqueous solution	275 (5.7); 310 (3.5); 374 (1.8); 492 (1.1)	0.62	11.7	54	0.73	5.3	[83]
FC-1[a]	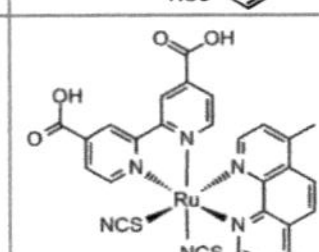	Acetonitrile	430 (1.3); 535 (1.0)	0.683	12.72	60 (515)	0.55	4.60	[84]

(continued)

Table 4 (continued)

Abbreviation	Structure	Absorption properties		Photoelectrochemical performance					References
		Solvent	λ_{MAX}/nm (ε_{MAX}/ 10^4 L mol^{-1} cm^{-1})	V_{OC}/ V	J_{SC}/ mA cm^{-2}	$IPCE_{MAX}$ % (λ/nm)	*ff*	η_{cell} %	
AVM-1[a]	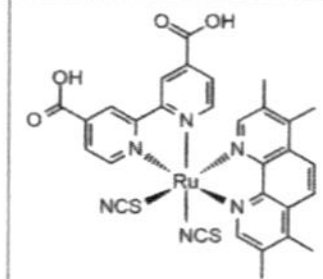	Acetonitrile	420 (1.3); 535 (0.8)	0.627	11.9	50 (505)	0.67	5.0	[85]
AVM-2[a]	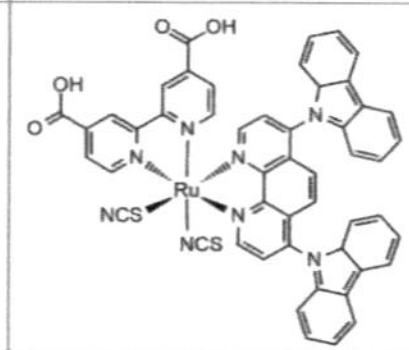	Ethanol	330 (2.0); 375 (1.4); 525 (1.8)	0.76	15.6	79 (500)	0.716	8.6	[86]
AR24a	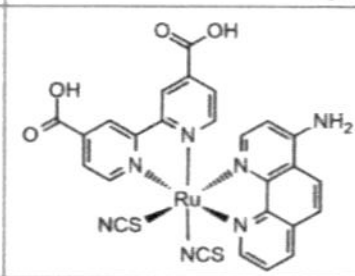	DMF	420 (1.2); 546 (0.9)	0.48	1.45	—	0.31	0.22	[87]
AR24b	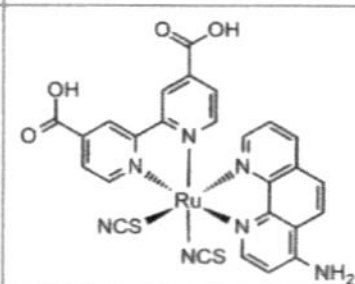	DMF	415 (1.0); 545 (0.9)	0.46	0.98	—	0.31	0.14	[87]
AR27a	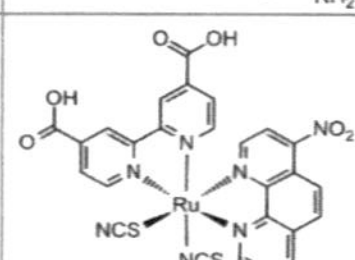	DMF	403 (0.5); 538 (0.7)	0.44	5.26	—	0.34	0.8	[87]

(continued)

Table 4 (continued)

Abbreviation	Structure	Absorption properties		Photoelectrochemical performance					References
		Solvent	λ_{MAX}/nm (ε_{MAX}/ 10^4 L mol^{-1} cm^{-1})	V_{OC}/ V	J_{SC}/ mA cm^{-2}	IPCE$_{MAX}$ % (λ/nm)	*ff*	η_{cell} %	
AR27b	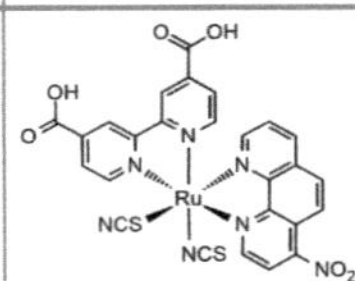	DMF	400 (0.5); 535 (0.7)	0.44	0.79	—	0.35	0.13	[87]
HJ1	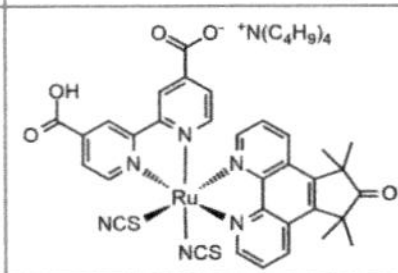	Methanol	304 (1.33); 428 (0.52), 516 (0.57)	0.67	9.65	—	0.44	2.83	[88]
CYC-P1	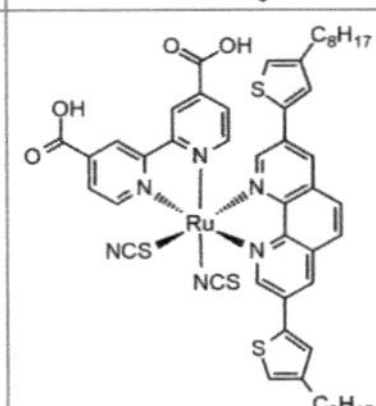	DMF	302 (3.9); 380 (3.8); 518 (0.8)	0.64	14.92	—	0.63	6.01	[89]
CYC-P2	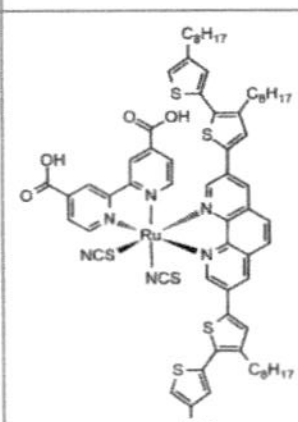	DMF	304 (2.3); 423 (2.2); 530 (0.4)	0.55	9.8	—	0.64	3.42	[89]

(continued)

Table 4 (continued)

Abbreviation	Structure	Absorption properties		Photoelectrochemical performance					References
		Solvent	λ_{MAX}/nm (ε_{MAX}/ 10^4 L mol^{-1} cm^{-1})	V_{OC}/ V	J_{SC}/ mA cm^{-2}	$IPCE_{MAX}$ % (λ/nm)	*ff*	η_{cell} %	
K27	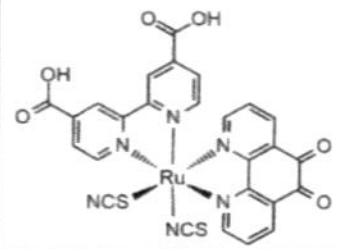	DMF	292 (4.8); 380 (1.3); 524 (1.1)	0.55	7.19	—	0.63	2.51	[90]
K28	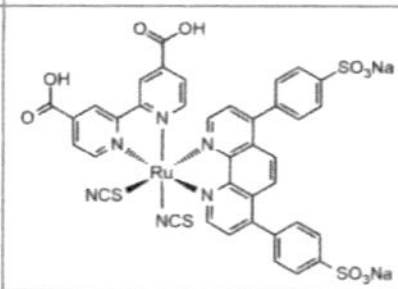	DMF	308 (5.4); 380 (1.0); 476 (1.6); 522 (1.8)	0.50	5.59	—	0.60	1.69	[90]

[a]These compounds were named after the initials of the first author of the reference cited

Table 5 Absorption properties and photoelectrochemical performance of natural dye-sensitized solar cells reported since 2003

Extract of	Dye	Medium	λ_{max} (nm)	V_{OC}/V	J_{SC} (mA cm^{-2})	*ff*	η_{cell} (%)	References
Red cabbage	Anthocyanin	Water	550	0.52	0.68	0.70	0.50	[113]
Rosella flower (*Hibiscus sabdariffa L.*)	Anthocyanin	Water	520	0.404	1.63	0.57	0.37	[114]
Blue pea flower (*Clitoria ternatea*)	Anthocyanin	Water	580, 620	0.372	0.37	0.33	0.05	[114]
Canna indica flower	Anthocyanin	Water and Ethanol	513	0.54	0.82	0.59	0.29	[115]
		TiO_2	530	—	—	—	—	
Salvia splendens flower	Anthocyanin	Water and ethanol	507	0.558	0.7	0.61	0.26	[115]
		TiO_2	516	—	—	—	—	
Cowberry	Anthocyanin	Water and ethanol	522	0.556	0.4	0.54	0.13	[115]
		TiO_2	531	—	—	—	—	
Solanum nigrum	Anthocyanin	Water and ethanol	539	0.54	1.01	0.51	0.31	[115]
		TiO_2	564	—	—	—	—	
Rhododendron arboretum zeylanicum	Anthocyanin	Ethanol	538	0.4020	1.15	0.637	0.29	[116]
Sesbania grandiflora Scarlet	Anthocyanin	Ethanol	544	0.4067	4.40	0.569	1.02	[116]
Hibiscus rosasinensis	Anthocyanin	Ethanol	534	0.4003	4.04	0.633	1.02	[116]
Hibiscus surattensis	Anthocyanin	Ethanol	545	0.3921	5.45	0.535	1.14	[116]
Nerium oleander	Anthocyanin	Ethanol	539	0.4088	2.46	0.591	0.59	[116]
Ixora macrothyrsa	Anthocyanin	Ethanol	537	0.4031	1.31	0.568	0.30	[116]
Black rice (*Oryza sativa L. indica*)	Anthocyanin	Ethanol	560	0.551	1.142	0.52	—	[117]
Rosa xanthina	Anthocyanin	Ethanol	560	0.492	0.637	0.52	—	[117]
Maple leaves	Anthocyanin	Ethanol	536	0.65	1.0	0.60	0.4	[118]
Red Sicilian orange	Anthocyanin	Water	515	0.340	3.84	0.5	—	[119]
Skin of eggplant	Anthocyanin	Water	522	0.350	3.40	0.4	—	[119]
Skin of jaboticaba	Anthocyanin	Ethanol	535	0.66	2.6	0.62	—	[120]
Begonia	Anthocyanin	Ethanol	540	0.537	0.63	0.722	0.24	[121]

(continued)

Table 5 (continued)

Extract of	Dye	Medium	λ_{max} (nm)	V_{OC}/V	J_{SC} (mA cm^{-2})	*ff*	η_{cell} (%)	References
Violet	Anthocyanin	0.1 M HCl—ethanol	546	0.498	1.02	0.645	0.33	[121]
Blackberry	Anthocyanin	—	—	0.320	5.85	0.57	1.07	[122]
Mulberry	Anthocyanin	TiO_2	553	0.42	0.86	0.43	—	[123]
Red mulberry	Anthocyanin	—	—	0.340	4.45	0.64	0.99	[122]
Radicchio	Anthocyanin	—	—	0.322	5.05	0.55	0.90	[122]
Giacchè grapes	Anthocyanin	—	—	0.333	3.06	0.56	0.57	[122]
Erythrina variegata	Anthocyanin	Ethanol	451; 492	0.48	0.78	0.55	—	[124]
Chaste tree fruit	Anthocyanin	TiO_2	555	0.39	1.06	0.48	—	[123]
Cabbage-palm fruit	Anthocyanin	TiO_2	552	0.44	0.37	0.61	—	[123]
Dragon fruit (*Hylocereus polyrhizus*)	Anthocyanin	Water	535	0.22	0.20	0.30	0.22	[125]
Bitter leaf (*Vernonia amygdalin*)	Anthocyanin	Ethanol	400	0.38	0.50	0.89	0.96	[126]
Red onion (*Allium cepa*)	Anthocyanin	Water	520	0.44	0.51	0.48	0.14	[127]
Henna (*Lawsonia inermis*)	Anthocyanin	Ethanol	518	0.61	1.87	0.58	0.66	[128]
Red frangipani flowers	Anthocyanin	Ethanol	530	0.50	0.94	0.65	0.301	[129]
R. chinensis	Anthocyanin	—	—	0.54	0.80	0.66	0.29	[130]
Delonix regia	Anthocyanin	TiO_2	538	0.30	1.33	0.39	0.317	[131]
Eugenia jambolana	Anthocyanin	TiO_2	543	0.35	1.49	0.48	0.505	[131]
Panica granatum pleniflora	Anthocyanin	—	—	0.62	0.40	0.61	0.15	[132]
Consolida orientalis	Anthocyanin	—	—	0.60	0.56	0.53	0.18	[132]
Clemetis orientalis	Anthocyanin	—	—	0.42	0.22	0.49	0.05	[132]
Strawberry	Anthocyanin	—	—	0.47	1.33	0.37	0.229	[133]
Luffa cylyndrica L.	Anthocyanin	Ethanol	420; 450	0.52	0.44	0.60	0.13	[134]
Purple corn (husk)	Anthocyanin	—	—	0.46	3.57	0.64	1.06	[135]

(continued)

Table 5 (continued)

Extract of	Dye	Medium	λ_{max} (nm)	V_{OC}/V	J_{SC} (mA cm^{-2})	*ff*	η_{cell} (%)	References
Purple corn (cob)	Anthocyanin	—	—	0.48	3.42	0.62	1.01	[135]
Purple corn (silk)	Anthocyanin	—	—	0.48	3.25	0.62	0.96	[135]
Red tamarind (*T. indica*)	Anthocyanin	Methanol	525	0.532	0.35	0.67	0.14	[136]
Calafate fruit	Anthocyanin	Water	525	0.47	6.2	0.36	—	[120]
		TiO_2	545	—	—	—	—	
Raw beet	Betalain	TiO_2	470	0.22	2.00	0.51	0.19	[137]
		Water	535	—	—	—	—	
Red turnip	Betalain	Water	484, 536	0.425	9.5	0.37	1.7	[138]
		TiO_2	450	—	—	—	—	
Wild silician prickly pear	Betalain	TiO_2	460	0.375	8.20	0.38	1.19	[138]
Red *Bougainvillea glabra*	Betalain	Water	482; 535	0.26	2.34	0.74	0.45	[139]
Violet *Bougainvillea glabra*	Betalain	Water	547	0.23	1.86	0.71	0.31	[139]
Red *Bougainvillea spectabilis*	Betalain	Water	480	0.28	2.29	0.76	0.48	[139]
Violet *Bougainvillea spectabilis*	Betalain	Water	535	0.25	1.88	0.73	0.35	[139]
Cactus fruits (*O. dillenii*)	Betalain	Methanol	517	0.521	1.09	0.69	0.47	[136]
Kelp	Chlorophyll	Ethanol	460	0.441	0.433	0.62	—	[117]
Wormwood	Chlorophyll	Ethanol	432	0.67	2.3	0.56	0.9	[118]
Bamboo leaves	Chlorophyll	Ethanol	432	0.67	1.9	0.56	0.7	[118]
Pomegranate leaves	Chlorophyll	Ethanol	412, 665	0.560	2.05	0.52	0.597	[140]
Perilla	Chlorophyll	Ethanol	665	0.522	1.36	0.696	0.50	[121]
Petunia	Chlorophyll	Ethanol	665	0.616	0.85	0.605	0.32	[121]
China loropetal	Chlorophyll	Ethanol	665	0.518	0.84	0.626	0.27	[121]
Chinese rose	Chlorophyll	0.1 M HCl—ethanol	516	0.483	0.90	0.619	0.27	[121]

(continued)

Table 5 (continued)

Extract of	Dye	Medium	λ_{max} (nm)	V_{OC}/V	J_{SC} (mA cm^{-2})	*ff*	η_{cell} (%)	References
Spinach	Chlorophyll	Ethanol	437	0.55	0.47	0.51	0.131	[141]
Ipomoea	Chlorophyll	Ethanol	410	0.54	0.91	0.56	0.278	[141]
Festuca ovina	Chlorophyll	Methanol	420; 660	0.54	1.18	0.69	0.46	[142]
Shiso	Chlorophyll	Water and Ethanol	440; 600	0.55	3.56	0.51	1.01	[143]
Ficus reusa	Chlorophyll	Ethanol	670	0.520	7.85	0.29	1.18	[144]
Rhoeo spathacea	Chlorophyll	Ethanol	670	0.496	10.9	0.27	1.49	[144]
Garcinia subelliptica	Chlorophyll	Ethanol	670	0.322	6.48	0.33	0.691	[144]
Anethum graveolens	Chlorophyll	Ethanol	666	0.579	0.965	0.40	0.22	[145]
Parsley (*Petroselinum crispum*)	Chlorophyll	Ethanol	666	0.445	0.535	0.34	0.07	[145]
Arugula	Chlorophyll	Ethanol	666	0.599	0.788	0.42	0.20	[145]
Green algae	Chlorophyll	Ethanol	666	0.559	0.397	0.44	0.10	[145]
U. pinnatifida	Chlorophyll	—	—	0.36	0.8	0.69	0.178	[146]
Curcumin	Polyphenol	Ethanol	430	0.53	0.53	0.72	0.41	[113]
Tangerine peel	Polyphenol	Ethanol	446	0.592	0.74	0.631	0.28	[121]
Flowery knotweed	Polyphenol	Ethanol	435	0.554	0.60	0.627	0.21	[121]
Reseda luteola	Polyphenol	—	—	0.64	0.54	0.64	0.22	[132]
Berberis integerrima	Polyphenol	—	—	0.56	0.004	0.53	0.01	[132]
Reseda gredensis	Polyphenol	—	—	0.53	0.14	0.71	0.07	[132]
Consolida ajacis	Polyphenol	—	—	0.55	1.68	0.65	0.60	[132]
Salvia sclarea	Polyphenol	—	—	0.37	0.10	0.54	0.02	[132]
Mallow	Polyphenol	Ethanol	660	0.60	0.69	0.55	0.215	[133]
Erythrina	Carotenoid	Ethanol	451	0.484	0.776	0.55	—	[117]
Capsicum	Carotenoid	Ethanol	455	0.412	0.225	0.63	—	[117]
Achiode seed	Carotenoid	Chloroform	440, 475, 500	0.56	0.53	0.66	0.19	[147]

(continued)

Table 5 (continued)

Extract of	Dye	Medium	λ_{max} (nm)	V_{OC}/V	J_{SC} (mA cm^{-2})	*ff*	η_{cell} (%)	References
Fructus lycii	Carotenoid	Ethanol	447; 425	0.689	0.53	0.466	0.17	[121]
Marigold	Carotenoid	Ethanol	487	0.542	0.51	0.831	0.23	[121]
Yellow rose	Carotenoid	Ethanol	487	0.609	0.74	0.571	0.26	[121]
Ivy gourd fruits	Carotenoid	Ethanol	458; 480	0.64	0.24	0.49	0.076	[129]
K. japonica	Carotenoid	—	—	0.59	0.56	0.68	0.22	[130]
Adonis flammea	Carotenoid	—	—	0.59	0.40	0.66	0.16	[132]
Red perilla	—	—	—	0.51	0.39	0.67	0.27	[113]
Herba artemisiae scopariae	—	Ethanol	669	0.484	1.03	0.682	0.34	[121]
Bauhinia tree	—	Ethanol	665	0.572	0.96	0.660	0.36	[121]
Lithospermum	—	Ethanol	520	0.337	0.14	0.585	0.03	[121]
Mangosteen pericarp	—	Ethanol	—	0.686	2.69	0.633	1.17	[121]
Rose	—	Water	—	0.595	0.97	0.659	0.38	[121]
Lily	—	Water	—	0.498	0.51	0.667	0.17	[121]
Coffee	—	Water	—	0.559	0.85	0.687	0.33	[121]
Broadleaf holly leaf	—	Water	—	0.607	1.19	0.654	0.47	[121]

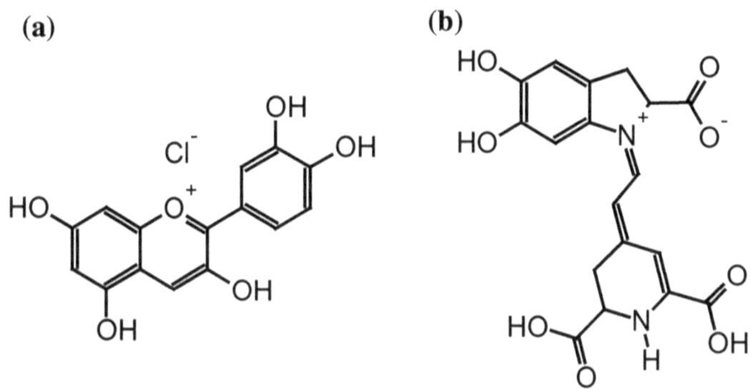

Fig. 6 Structure of anthocyanidin (**a**), a flavinic ion of anthocyanin, and betanidin (**b**), a betalain compound

The most investigated classes of natural dyes are the anthocyanins, commonly found in red-purplish fruits or flowers; or betalains, Fig. 6. Besides these compounds, other sensitizers have also been investigated [148].

Betalain from raw beet, red turnip, and wild sicilian prickly pear have also been used as natural sensitizers and they have presented a good photoelectrochemical response, however these cells have low V_{OC}, with overall efficiency up to 1.7% and reasonable stability [138]. It was also observed the improvement on the photoelectrochemical performance due to changes on pH as well as in the presence of additives in the mediator layer [122].

Other classes of natural dyes, such as chlorophyll, polyphenol, etc., were also investigated, but the photoelectrochemical parameters were not as good as those observed for anthocyanins or betalains. Several reviews were published and describe the performance of these other classes of dyes along to those presented here [149–153].

10 Conclusion

The energy needs will be supplied by alternative sources and dye-sensitized solar cells are one of the most promising devices for this application since they are cheap and environmentally friendly. The investigation of dye-sensitizers is fundamental issue on the development of these devices and one of the most promising alternatives is the use of ruthenium *tris*-heteroleptic dyes sensitizers to modulate or enhance their photoelectrochemical performance. The investigation on natural extracts to be employed as dye-sensitizers has also been attracting much attention in the last years. They can be an alternative to further reduction of the production costs of these revolutionary devices.

Acknowledgements The authors would like to acknowledge to CNPq, FAPESP, and CAPES for financial support.

References

1. Cook TR et al (2010) Solar energy supply and storage for the legacy and nonlegacy worlds. Chem Rev 110(11):6474–6502
2. Nocera DG (2009) Chemistry of personalized solar energy. Inorg Chem 48(21):10001–10017
3. Carrette L, Friedrich KA, Stimming U (2000) Fuel cells: principles, types, fuels, and applications. Chem Phys Chem 1(4):162–193
4. Cameron D, Holliday R, Thompson D (2003) Gold's future role in fuel cell systems. J Power Sources 118(1–2):298–303
5. Lemos SG et al (2007) Electrocatalysis of methanol, ethanol and formic acid using a Ru/Pt metallic bilayer. J Power Sour 163(2):695–701
6. Freitas RG et al (2007) Methanol oxidation reaction on Ti/$RuO_{2(x)}Pt_{(1-x)}$ electrodes prepared by the polymeric precursor method. J Power Sour 171(2):373–380
7. Polo AS et al (2011) Pt-Ru-TiO_2 photoelectrocatalysts for methanol oxidation. J Power Sour 196(2):872–876
8. Gu C, Shannon C (2007) Investigation of the photocatalytic activity of TiO_2-polyoxometalate systems for the oxidation of methanol. J Mol Catal A: Chem 262(1–2):185–189
9. Drew K et al (2005) Boosting fuel cell performance with a semiconductor photocatalyst: TiO_2/Pt-Ru hybrid catalyst for methanol oxidation. J Phys Chem B 109(24):11851–11857
10. Kamat PV (2007) Meeting the clean energy demand: nanostructure architectures for solar energy conversion. J Phys Chem C 111(7):2834–2860
11. Armaroli N, Balzani V (2007) The future of energy supply: challenges and opportunities. Angew Chem Int Ed Engl 46(1–2):52–66
12. Meyer TJ (1989) Chemical approaches to artificial photosynthesis. Acc Chem Res 22 (5):163–170
13. Dubois MR, Dubois DL (2009) Development of molecular electrocatalysts for CO_2 reduction and H_2 production/oxidation. Acc Chem Res 42(12):1974–1982
14. Morris AJ, Meyer GJ, Fujita E (2009) Molecular approaches to the photocatalytic reduction of carbon dioxide for solar fuels. Acc Chem Res 42(12):1983–1994
15. Concepcion JJ et al (2009) Making oxygen with ruthenium complexes. Acc Chem Res 42 (12):1954–1965
16. Walter MG et al (2010) Solar water splitting cells. Chem Rev 110(11):6446–6473
17. Caramori S et al (2010) Photoelectrochemical behavior of sensitized TiO_2 photoanodes in an aqueous environment: application to hydrogen production. Inorg Chem 49(7):3320–3328
18. Koike K et al (2009) Architecture of supramolecular metal complexes for photocatalytic CO_2 reduction: III: effects of length of alkyl chain connecting photosensitizer to catalyst. J Photochem Photobiol, A 207(1):109–114
19. Takeda H et al (2008) Development of an efficient photocatalytic system for CO_2 reduction using rhenium(I) complexes based on mechanistic studies. J Am Chem Soc 130(6):2023–2031
20. Kroon JM et al (2007) Nanocrystalline dye-sensitized solar cells having maximum performance. Prog Photovoltaics Res Appl 15(1):1–18
21. Tributsch H (1972) Reaction of excited chlorophyll molecules at electrodes and in photosynthesis. Photochem Photobiol 16(4):261–269
22. O'Regan B, Gratzel M (1991) A low-cost, high-efficiency solar cell based on dye-sensitized colloidal TiO_2 films. Nature 353(6346):737–740

23. Grätzel M (2001) Photoelectrochemical cells. Nature 414(6861):338–344
24. Katoh R et al (2004) Kinetics and mechanism of electron injection and charge recombination in dye-sensitized nanocrystalline semiconductors. Coord Chem Rev 248(13–14):1195–1213
25. Gregg BA (2004) Interfacial processes in the dye-sensitized solar cell. Coord Chem Rev 248 (13–14):1215–1224
26. Galoppini E (2004) Linkers for anchoring sensitizers to semiconductor nanoparticles. Coord Chem Rev 248(13–14):1283–1297
27. Anderson NA, Lian T (2004) Ultrafast electron injection from metal polypyridyl complexes to metal-oxide nanocrystalline thin films. Coord Chem Rev 248(13–14):1231–1246
28. Asbury JB et al (2003) Parameters affecting electron injection dynamics from ruthenium dyes to titanium dioxide nanocrystalline thin film. J Phys Chem B 107(30):7376–7386
29. Anderson NA, Ai X, Lian T (2003) Electron injection dynamics from ru polypyridyl complexes to ZnO nanocrystalline thin films. J Phys Chem B 107(51):14414–14421
30. Garcia CG et al (2002) Electron injection versus charge recombination in photoelectrochemical solar cells using cis-$[(dcbH_2)_2Ru(CNpy)(H_2O)]Cl_2$ as a nanocrystalline TiO_2 sensitizer. J Photochem Photobiol, A 151(1–3):165–170
31. Garcia CG et al (2002) Time-resolved experiments in dye-sensitized solar cells using $[(dcbH_2)_2Ru(ppy)_2](ClO_4)_2$ as a nanocrystalline TiO_2 sensitizer. J Photochem Photobiol, A 147(2):143–148
32. Kuang DB et al (2006) High molar extinction coefficient heteroleptic ruthenium complexes for thin film dye-sensitized solar cells. J Am Chem Soc 128(12):4146–4154
33. Wang P et al (2004) Stable new sensitizer with improved light harvesting for nanocrystalline dye-sensitized solar cells. Adv Mater 16(20):1806
34. Wang P et al (2004) Amphiphilic polypyridyl ruthenium complexes with substituted 2,2′-dipyridylamine ligands for nanocrystalline dye-sensitized solar cells. Chem Mater 16 (17):3246–3251
35. Pelet S, Moser J-E, Gratzel M (2000) Cooperative effect of adsorbed cations and iodide on the interception of back electron transfer in the dye sensitization of nanocrystalline TiO_2. J Phys Chem B 104(8):1791–1795
36. Patrocinio AOT, Paterno LG, Iha NYM (2010) Role of polyelectrolyte for layer-by-layer compact TiO_2 films in efficiency enhanced dye-sensitized solar cells. J Phys Chem C 114 (41):17954–17959
37. Murakami Iha NY, Arcia CG, Bignozzi CA (2003) Dye-sensitized photoelectrochemical solar cells. In: Nalwa HS (ed) Handbook of photochemistry and photobiology. American Scientific Publishers, Stevenson Ranch, California, USA, pp 49–82
38. Nazeeruddin MK et al (1993) Conversion of light to electricity by cis-X_2bis(2,2′-bipyridyl-4,4′-dicarboxylate)ruthenium(II) charge-transfer sensitizers ($X = Cl^-$, Br^-, I^-, CN^-, and SCN^-) on nanocrystalline TiO_2 electrodes. J Am Chem Soc 115(14):6382–6390
39. Nazeeruddin MK, Gratzel M (2001) Separation of linkage isomers of trithiocyanato (4,4′,4″-tricarboxy-2,2′,6,2″-terpyridine)ruthenium(II) by pH-titration method and their application in nanocrystalline TiO_2-based solar cells. J Photochem Photobiol A-Chem 145(1–2):79–86
40. Chen CY et al (2006) A ruthenium complex with superhigh light-harvesting capacity for dye-sensitized solar cells. Angewandte Chem Int Ed 45(35):5822–5825
41. Cao YM et al (2009) Dye-sensitized solar cells with a high absorptivity ruthenium sensitizer featuring a 2-(Hexylthio)thiophene conjugated bipyridine. J Phys Chem C 113(15):6290–6297
42. Nazeeruddin MK et al (2003) Investigation of sensitizer adsorption and the influence of protons on current and voltage of a dye-sensitized nanocrystalline TiO_2 solar cell. J Phys Chem B 107(34):8981–8987
43. Lv XJ, Wang FF, Li YH (2010) Studies of an extremely high molar extinction coefficient ruthenium sensitizer in dye-sensitized solar cells. ACS Appl Mater Interfaces 2(7):1980–1986

44. Gao F et al (2008) Enhance the optical absorptivity of nanocrystalline TiO_2 film with high molar extinction coefficient ruthenium sensitizers for high performance dye-sensitized solar cells. J Am Chem Soc 130(32):10720–10728
45. Chen CY et al (2009) Highly efficient light-harvesting ruthenium sensitizer for thin-film dye-sensitized solar cells. ACS Nano 3(10):3103–3109
46. Chen CY et al (2009) New ruthenium sensitizer with carbazole antennas for efficient and stable thin-film dye-sensitized solar cells. J Phys Chem C 113(48):20752–20757
47. Yu QJ et al (2009) An extremely high molar extinction coefficient ruthenium sensitizer in dye-sensitized solar cells: the effects of pi-conjugation extension. J Phys Chem C 113 (32):14559–14566
48. Sun YL et al (2010) Viable alternative to N719 for dye-sensitized solar cells. ACS Appl Mater Interfaces 2(7):2039–2045
49. Polo AS, Itokazu MK, Murakami Iha NY (2004) Metal complex sensitizers in dye-sensitized solar cells. Coord Chem Rev 248(13–14):1343–1361
50. Pashaei B et al (2016) Influence of ancillary ligands in dye-sensitized solar cells. Chem Rev 116(16):9485–9564
51. Jin Zhengzhe et al (2008) Triarylamine-functionalized ruthenium dyes for efficient dye-sensitized solar cells. Chem Sus Chem 1(11):901–904
52. Mitsopoulou CA et al (2007) Synthesis, characterization and sensitization properties of two novel mono and bis carboxyl-dipyrido-phenazine ruthenium(II) charge transfer complexes. J Photochem Photobiol A-Chem 191:6–12
53. Huang WK et al (2010) Synthesis and electron-transfer properties of benzimidazole-functionalized ruthenium complexes for highly efficient dye-sensitized solar cells. Chem Commun 46(47):8992–8994
54. Wu SJ et al (2010) An efficient light-harvesting ruthenium dye for solar cell application. Dyes Pigm 84(1):95–101
55. Abbotto A et al (2008) Electron-rich heteroaromatic conjugated bipyridine based ruthenium sensitizer for efficient dye-sensitized solar cells. Chem Commun 42:5318–5320
56. Yum JH et al (2009) High efficient donor-acceptor ruthenium complex for dye-sensitized solar cell applications. Energy Environ Sci 2(1):100–102
57. Willinger K et al (2009) Synthesis, spectral, electrochemical and photovoltaic properties of novel heteroleptic polypyridyl ruthenium(II) donor-antenna dyes. J Mater Chem 19 (30):5364–5376
58. Wang P et al (2003) A stable quasi-solid-state dye-sensitized solar cell with an amphiphilic ruthenium sensitizer and polymer gel electrolyte (vol 2, p 402, 2003). Nat Mater 2(7):498
59. Sahin C et al (2008) Synthesis of an amphiphilic ruthenium complex with swallow-tail bipyridyl ligand and its application in nc-DSC. Inorg Chim Acta 361(3):671–676
60. Nazeeruddin MK et al (2004) Stepwise assembly of amphiphilic ruthenium sensitizers and their applications in dye-sensitized solar cell. Coord Chem Rev 248(13–14):1317–1328
61. Klein C et al (2004) Amphiphilic ruthenium sensitizers and their applications in dye-sensitized solar cells. Inorg Chem 43(14):4216–4226
62. Gao FF et al (2008) A new heteroleptic ruthenium sensitizer enhances the absorptivity of mesoporous titania film for a high efficiency dye-sensitized solar cell. Chem Commun 23:2635–2637
63. Hallett AJ, Jones JE (2011) Purification-free synthesis of a highly efficient ruthenium dye complex for dye-sensitised solar cells (DSSCs). Dalton Trans 40(15):3871–3876
64. Lagref JJ, Nazeeruddin MK, Grätzel M (2003) Molecular engineering on semiconductor surfaces: design, synthesis and application of new efficient amphiphilic ruthenium photosensitizers for nanocrystalline TiO_2 solar cells. Synth Met 138(1–2):333–339
65. Sygkridou D et al (2015) Comparative studies of pyridine and bipyridine ruthenium dye complexes with different side groups as sensitizers in sol-gel quasi-solid-state dye sensitized solar cells. Electrochim Acta 160:227–234
66. Kong F-T, Dai S-Y, Wang K-J (2007) New amphiphilic polypyridyl ruthenium(II) sensitizer and its application in dye-sensitized solar cells. Chin J Chem 25(2):168–171

67. Liu K-Y et al (2010) Synthesis and characterization of cross-linkable ruthenium complex dye and its application on dye-sensitized solar cells. J Polym Sci, Part A: Polym Chem 48 (2):366–372
68. Ni J-S et al (2012) Effects of tethering alkyl chains for amphiphilic ruthenium complex dyes on their adsorption to titanium oxide and photovoltaic properties. J Colloid Interface Sci 386 (1):359–365
69. Song Y-Y et al (2009) Amphiphilic TiO_2 nanotube arrays: an actively controllable drug delivery system. J Am Chem Soc 131(12):4230–4232
70. Giribabu L et al (2009) High molar extinction coefficient amphiphilic ruthenium sensitizers for efficient and stable mesoscopic dye-sensitized solar cells. Energy Environ Sci 2(7):770–773
71. Jiang KJ et al (2008) Efficient sensitization of nanocrystalline TiO_2 films with highmolar extinction coefficient ruthenium complex. Inorg Chim Acta 361(3):783–785
72. Gao FF et al (2009) Conjugation of selenophene with bipyridine for a high molar extinction coefficient sensitizer in dye-sensitized solar cells. Inorg Chem 48(6):2664–2669
73. Nazeeruddin MK et al (2007) A high molar extinction coefficient charge transfer sensitizer and its application in dye-sensitized solar cell. J Photochem Photobiol A Chem 185(2–3):331–337
74. Karthikeyan CS et al (2007) Highly efficient solid-state dye-sensitized TiO_2 solar cells via control of retardation of recombination using novel donor-antenna dyes. Sol Energy Mater Sol Cells 91(5):432–439
75. Kuang D et al (2007) High-efficiency and stable mesoscopic dye-sensitized solar cells based on a high molar extinction coefficient ruthenium sensitizer and nonvolatile electrolyte. Adv Mater 19(8):1133–1137
76. Hussain M et al (2013) Structure-property relationship of extended [small pi]-conjugation of ancillary ligands with and without an electron donor of heteroleptic Ru(ii) bipyridyl complexes for high efficiency dye-sensitized solar cells. Phys Chem Chem Phys 15 (21):8401–8408
77. Chen CY et al (2007) A new route to enhance the light-harvesting capability of ruthenium complexes for dye-sensitized solar cells. Adv Mater 19(22):3888
78. Li J-Y et al (2010) Heteroleptic ruthenium antenna-dye for high-voltage dye-sensitized solar cells. J Mater Chem 20(34):7158–7164
79. Ryu TI et al (2009) Synthesis and photovoltaic properties of novel ruthenium(II) sensitizers for dye-sensitized solar cell applications. Bull Korean Chem Soc 30(10):2329–2337
80. Zhu SS, Kingsborough RP, Swager TM (1999) Conducting redox polymers: investigations of polythiophene-Ru(bpy)(3)(n+) hybrid materials. J Mater Chem 9(9):2123–2131
81. Hara K et al (2001) Dye-sensitized nanocrystalline TiO_2 solar cells based on ruthenium(II) phenanthroline complex photosensitizers. Langmuir 17(19):5992–5999
82. Reynal A et al (2008) A phenanthroline heteroleptic ruthenium complex and its application to dye-sensitised solar cells. Eur J Inorg Chem 12:1955–1958
83. Onozawa-Komatsuzaki N et al (2006) Molecular and electronic ground and excited structures of heteroleptic ruthenium polypyridyl dyes for nanocrystalline TiO_2 solar cells. New J Chem 30(5):689–697
84. Carvalho F et al (2014) Synthesis, characterization and photoelectrochemical performance of a Tris-heteroleptic ruthenium(II) complex having 4,7-dimethyl-1, 10-phenanthroline. Inorg Chim Acta 414:145–152
85. Müller AV et al (2015) Effects of methyl-substituted phenanthrolines on the performance of ruthenium(ii) dye-sensitizers. J Braz Chem Soc 26(11):2224–2232
86. Muller AV et al (2016) A high efficiency ruthenium(ii) Tris-heteroleptic dye containing 4,7-dicarbazole-1,10-phenanthroline for nanocrystalline solar cells. Rsc Adv 6(52):46487–46494
87. Reynal A et al (2008) Interfacial charge recombination between e–TiO_2 and the I^-/I_3^-; electrolyte in ruthenium heteroleptic complexes: dye molecular structure-open circuit voltage relationship. J Am Chem Soc 130(41):13558–13567

88. Oh H et al (2014) Synthesis of heteroleptic Ru(II) complexes ligated with 1,3-dihydro-1,1,3,3-tetramethyl-7,8-diazacyclopenta 1 phenanthren-2-one and application in dye-sensitized solar cells. Synth Met 198:260–266
89. Chen CY et al (2007) New ruthenium complexes containing oligoalkylthiophene-substituted 1,10-phenanthroline for nanocrystalline dye-sensitized solar cells. Adv Func Mater 17 (1):29–36
90. Ocakoglu K et al (2012) The photovoltaic performance of new ruthenium complexes in DSSCs based on nanorod ZnO electrode. Synth Met 162(23):2125–2133
91. Smestad GP, Grätzel M (1998) Demonstrating electron transfer and nanotechnology: a natural dye-sensitised nanocrystalline energy converter. J Chem Edu 75(6):752–756
92. Smestad GP (1998) Education and solar conversion: demonstrating electron transfer. Sol Energy Mater Sol Cells 55(1–2):157–178
93. Sonai GG et al (2015) Solar cells sensitized with natural dyes: an introductory experiment about solar energy for undergraduate students. Quim Nova 38(10):1357–1365
94. Patrocinio AOT, Iha NYM (2010) Toward sustainability: solar cells sensitized by natural extracts. Quim Nova 33(3):574–578
95. Chien CY, Hsu BD (2013) Optimization of the dye-sensitized solar cell with anthocyanin as photosensitizer. Solar Energy Part C 98:203–211
96. Suyitno S et al (2015) Stability and efficiency of dye-sensitized solar cells based on papaya-leaf dye. Spectrochim Acta Part A Mol Biomol Spectrosc 148:99–104
97. Chien CY, Hsu BD (2014) Performance enhancement of dye-sensitized solar cells based on anthocyanin by carbohydrates. Sol Energy 108:403–411
98. Treat NA, Knorr FJ, McHale JL (2016) Templated assembly of betanin chromophore on TiO_2: aggregation-enhanced light-harvesting and efficient electron injection in a natural dye-sensitized solar cell. J Phys Chem C 120(17):9122–9131
99. Teoli F et al (2016) Role of pH and pigment concentration for natural dye-sensitized solar cells treated with anthocyanin extracts of common fruits. J Photochem Photobiol A Chem 316:24–30
100. Akila Y et al (2016) Enhanced performance of natural dye sensitised solar cells fabricated using rutile TiO_2 nanorods. Opt Mater 58:76–83
101. Ananth S et al (2016) Enhanced photovoltaic behavior of dye sensitized solar cells fabricated using pre dye treated titanium dioxide nanoparticles. J Mater Sci Mater Electron 27(1):146–153
102. Kumara N et al (2015) Efficiency enhancement of Ixora floral dye sensitized solar cell by diminishing the pigments interactions. Sol Energy 117:36–45
103. Chang H et al (2013) Characterization of natural dye extracted from wormwood and purple cabbage for dye-sensitized solar cells. Int J Photoenergy
104. Koli P (2014) Photogalvanic cells: comparative study of various synthetic dyes and natural photo sensitizers present in spinach extract. Rsc Adv 4(86):46194–46202
105. Lim A et al (2016) Co-dominant effect of selected natural dye sensitizers in DSSC performance. Spectrochim Acta Part A Mol Biomol Spectroscopy 167:26–31
106. Chang H, Lai XR (2016) Fabrication of natural sensitizer extracted from mixture of purple cabbage, roselle, wormwood and seaweed with high conversion efficiency for DSSC. J Nanosci Nanotechnol 16(2):2072–2075
107. Kumara NTRN et al (2013) Layered co-sensitization for enhancement of conversion efficiency of natural dye sensitized solar cells. J Alloy Compd 581:186–191
108. Bidikoudi M et al (2015) Solidification of ionic liquid redox electrolytes using agarose biopolymer for highly performing dye-sensitized solar cells. Electrochim Acta 179:228–236
109. Adel R et al (2015) Effect of polymer electrolyte on the performance of natural dye sensitized solar cells. Superlattices Microstruct 86:62–67
110. Thambidurai M et al (2011) Dye-sensitized ZnO nanorod based photoelectrochemical solar cells with natural dyes extracted from Ixora coccinea, Mulberry and Beetroot. J Mater Sci Mater Electron 22(11):1662–1666

111. Thambidurai M et al (2014) Rosa centifolia sensitized ZnO nanorods for photoelectrochemical solar cell applications. Sol Energy 106:143–150
112. Thankappan A et al (2015) Highly efficient betanin dye based ZnO and ZnO/Au Schottky barrier solar cell. Thin Solid Films 583:102–107
113. Furukawa S et al (2009) Characteristics of dye-sensitized solar cells using natural dye. Thin Solid Films 518(2):526–529
114. Wongcharee K, Meeyoo V, Chavadej S (2007) Dye-sensitized solar cell using natural dyes extracted from *rosella and blue pea* flowers. Sol Energy Mater Sol Cells 91(7):566–571
115. Luo PH et al (2009) From salmon pink to blue natural sensitizers for solar cells: *Canna indica L.*, *Salvia splendens*, *cowberry* and *Solanum nigrum L.* Spectrochim Acta Part A Mol Biomol Spectro 74(4):936–942
116. Fernando J, Senadeera GKR (2008) Natural anthocyanins as photosensitizers for dye-sensitized solar devices. Curr Sci 95(5):663–666
117. Hao SC et al (2006) Natural dyes as photosensitizers for dye-sensitized solar cell. Sol Energy 80(2):209–214
118. Jin EM et al (2010) Photosensitization of nanoporous TiO_2 films with natural dye. Phys Scripta T139
119. Calogero G, Di Marco G (2008) Red Sicilian orange and purple eggplant fruits as natural sensitizers for dye-sensitized solar cells. Sol Energy Mater Sol Cells 92(11):1341–1346
120. Polo AS, Murakami Iha NY (2006) Blue sensitizers for solar cells: natural dyes from Calafate and Jaboticaba. Sol Energy Mater Sol Cells 90(13):1936–1944
121. Zhou H et al (2011) Dye-sensitized solar cells using 20 natural dyes as sensitizers. J Photochem Photobiol, A 219(2–3):188–194
122. Calogero G et al (2012) Anthocyanins and betalains as light-harvesting pigments for dye-sensitized solar cells. Sol Energy 86(5):1563–1575
123. Garcia CG, Polo AS, Murakami Iha NY (2003) Fruit extracts and ruthenium polypyridinic dyes for sensitization of TiO_2 in photoelectrochemical solar cells. J Photochem Photobiol, A 160(1–2):87–91
124. Hao S et al (2006) Natural dyes as photosensitizers for dye-sensitized solar cell. Sol Energy 80(2):209–214
125. Ali RAM, Nayan N (2010) Fabrication and analysis of dye-sensitized solar cell using natural dye extracted from dragon fruit. Int J Integr Eng 2:55–62
126. Boyo AO et al (2012) Bitter leaf (Vernonia amygdalin) for dye sensitized solar cell. Trends Appl Sci Res 7(7):558–564
127. Dumbravă A et al (2008) Dye-sensitized solar cells based on nanocrystalline TiO_2 and natural pigments. J Optoelectron Adv Mater 10(11):2996–3002
128. Asuloju KA, Shitta MB, Justu S (2011) Effect of extracting solvents on the stability and performances of dye- sensitized solar cell prepared using extract from Lawsonia inermis. Fundam J Mod Phys 1(2):261–268
129. Shanmugam V et al (2013) Performance of dye-sensitized solar cells fabricated with extracts from fruits of ivy gourd and flowers of red frangipani as sensitizers. Spectrochim Acta Part A Mol Biomol Spectrosc 104:35–40
130. Hemalatha KV et al (2012) Performance of *Kerria japonica* and *Rosa chinensis* flower dyes as sensitizers for dye-sensitized solar cells. Spectrochim Acta Part A Mol Biomol Spectrosc 96:305–309
131. Senthil TS et al (2011) Natural dye (cyanidin 3-O-glucoside) sensitized nanocrystalline TiO_2 solar cell fabricated using liquid electrolyte/quasi-solid-state polymer electrolyte. Renew Energy 36(9):2484–2488
132. Hamadanian M et al (2014) Uses of new natural dye photosensitizers in fabrication of high potential dye-sensitized solar cells (DSSCs). Mater Sci Semicond Process 27:733–739
133. Torchani A et al (2015) Sensitized solar cells based on natural dyes. Curr Appl Phys 15 (3):307–312
134. Maurya IC, Srivastava P, Bahadur L (2016) Dye-sensitized solar cell using extract from petals of male flowers *Luffa cylindrica L.* as a natural sensitizer. Opt Mater 52:150–156

135. Phinjaturus K et al (2016) Dye-sensitized solar cells based on purple corn sensitizers. Appl Surf Sci 380:101–107
136. Ramamoorthy R et al (2016) Betalain and anthocyanin dye-sensitized solar cells. J Appl Electrochem 46(9):929–941
137. Zhang D et al (2008) Betalain pigments for dye-sensitized solar cells. J Photochem Photobiol A Chem 195(1):72–80
138. Calogero G et al (2010) Efficient dye-sensitized solar cells using red turnip and purple wild sicilian prickly pear fruits. Int J Mol Sci 11(1):254–267
139. Hernandez-Martinez AR et al (2011) New dye-sensitized solar cells obtained from extracted bracts of bougainvillea glabra and spectabilis betalain pigments by different purification processes. Int J Mol Sci 12(9):5565–5576
140. Chang H, Lo YJ (2010) Pomegranate leaves and mulberry fruit as natural sensitizers for dye-sensitized solar cells. Sol Energy 84(10):1833–1837
141. Chang H et al (2010) Dye-sensitized solar cell using natural dyes extracted from spinach and ipomoea. J Alloy Compd 495(2):606–610
142. Hernández-Martínez AR et al (2012) Natural pigment-based dye-sensitized solar cells. J Appl Res Technol 10(1):38–47
143. Kumara GRA et al (2006) Shiso leaf pigments for dye-sensitized solid-state solar cell. Sol Energy Mater Sol Cells 90(9):1220–1226
144. Lai WH et al (2008) Commercial and natural dyes as photosensitizers for a water-based dye-sensitized solar cell loaded with gold nanoparticles. J Photochem Photobiol, A 195(2–3):307–313
145. Taya SA et al (2013) Dye-sensitized solar cells using fresh and dried natural dyes. Int J Mater Sci Applications 2(2):37–42
146. Calogero G et al (2014) Brown seaweed pigment as a dye source for photoelectrochemical solar cells. Spectrochim Acta Part A Mol Biomol Spectrosc 117:702–706
147. Gomez-Ortiz NM et al (2010) Dye-sensitized solar cells with natural dyes extracted from achiote seeds. Sol Energy Mater Sol Cells 94(1):40–44
148. Zhang D et al (2008) Betalain pigments for dye-sensitized solar cells. J Photochem Photobiol, A 195(1):72–80
149. Narayan MR (2012) Review: dye sensitized solar cells based on natural photosensitizers. Renew Sustain Energy Rev 16(1):208–215
150. Ludin NA et al (2014) Review on the development of natural dye photosensitizer for dye-sensitized solar cells. Renew Sustain Energy Rev 31:386–396
151. Hug H et al (2014) Biophotovoltaics: natural pigments in dye-sensitized solar cells. Appl Energy 115:216–225
152. Al-Alwani MAM et al (2016) Dye-sensitised solar cells: Development, structure, operation principles, electron kinetics, characterisation, synthesis materials and natural photosensitisers. Renew Sustain Energy Rev 65:183–213
153. Escobar MAM, Jaramillo F (2015) Natural dyes extraction, stability and application to dye-sensitized solar cells. J Renew Mater 3(4):281–291

Photocatalytic Water Splitting by Suspended Semiconductor Particles

Renato V. Gonçalves, Heberton Wender, Sherdil Khan and Mauricio A. Melo Jr.

1 Introduction

The alarming prospect of climate change and the accelerated global warming urge the humankind to shift the energy focus to sustainable, carbon-free, and renewable energies. Hydrogen with a high-energy yield of 122 kJ/g, which is 2.75 times greater than hydrocarbon-fuels that can be used in fuel cells to generate power, has been considered a powerful and clean fuel without toxic emissions [1–5]. However, the industrial processes to produce hydrogen itself are not environmental friendly, because ca. 96% of the world hydrogen production involve fossil fuels (natural gas, oil and coal) consumption. Within this context, clean hydrogen production through photocatalytic water splitting has been recognized as an exceptional route to produce hydrogen. Photocatalytic hydrogen production via water splitting was reported for the first time in 1972 by Honda and Fujishima using a rutile titanium oxide (TiO_2) photoanode in a photoelectrochemical setup [6]. After the first report, various photocatalytic materials have been developed to enhance the efficiency of the

R.V. Gonçalves (✉)
Instituto de Física de São Carlos, Universidade de São Paulo,
PO Box 369, São Carlos, SP 13560-970, Brazil
e-mail: rgoncalves@ifsc.usp.br

H. Wender
Instituto de Física, Universidade Federal do Mato Grosso do Sul (UFMS),
Cidade Universitária, Campo Grande, MS CEP 79070-900, Brazil

S. Khan
Instituto de Física, Universidade Federal do Rio Grande do Sul (UFRGS),
Av. Bento Gonçalves, 9500, Bairro Agronomia, Porto Alegre, RS CEP 91501-970, Brazil

M.A. Melo Jr.
Department of Chemistry, University of California,
One Shields Avenue, Davis, CA 95616, USA

F.L. de Souza and E. Leite (eds.), *Nanoenergy*,
DOI 10.1007/978-3-319-62800-4_3

hydrogen production from water [2–5, 7–15]. Photocatalytic water splitting can be achieved in two types of configurations

(1) Photoelectrolysis by using semiconducting electrodes in a photoelectrochemical (PEC) cell.
(2) Photolysis by utilizing semiconducting powder suspension in aqueous solutions.

In a photoelectrolysis configuration, two electrodes are immersed in an aqueous electrolyte, one of which is the photoelectrode and the other is the counter electrode that is usually a Pt wire [16, 17]. On the other hand, the photolysis configuration consists of suspended particles; each of that acts as a microelectrode performing oxidation and reduction reaction on its surface. Both configurations are promising for H_2 production from water; however, photolysis is considered much simpler and less expensive as compared to the PEC systems [18–21].

The overall water splitting reaction is an uphill reaction in which Gibbs free energy increases by 237.2 kJ mol^{-1}, as shown in Eq. (1) [22]

$$2H_2O\,(\text{liquid}) \rightarrow H_2(\text{gas}) + \frac{1}{2}O_2\,(\text{gas}), \ldots \Delta G^0 = +237.2 kJmol^{-1} \quad (1)$$

The energy required to drive water splitting can be provided through solar light. In solar water splitting, a photocatalyst immersed in aqueous electrolyte is usually a semiconductor absorbing wide range of solar light. The theoretical minimum semiconductor band gap energy for water splitting is 1.23 eV that corresponds to light of about 1010 nm indicating that visible light possesses enough energy for the water splitting. Thus, suitable photocatalyst materials are required to absorb sunlight with photon energy of 1.23 eV or higher to transfer energy to water molecule and promote two half reactions, as shown in Eqs. (2) and (3).

$$2H^+ + 2e^- \rightarrow H_2(g), \ \ \Delta E^0_{red.} = 0\,V_{NHE}(\text{reduction half reaction}) \quad (2)$$

$$2H_2O(l) \rightarrow O_2(g) + 4H^+ + 4e^-, \ \ \Delta E^0_{ox.} = 1.23\,V_{NHE}\,(\text{oxidation half reaction}) \quad (3)$$

The electronic band structure of a semiconductor material (photocatalyst) plays a key role in the photocatalytic water splitting. The minimum energy required to excite an electron from the valence band into the conduction band level of a semiconductor is called band gap (Eg). To achieve overall water splitting, the valence band level (VB) of the semiconductor should be more positive than the redox potential of O_2/H_2O (1.23 V vs. NHE), while the conduction band level (CB) should be more negative than the redox potential of H^+/H_2 (0 V vs. NHE) [15, 23]. Figure 1 illustrates the basic principle to split water to H_2 and O_2 using sunlight on semiconducting suspended particle photocatalyst.

The overall photocatalytic water splitting reaction is a four-electron process and can be summarized in three fundamental steps

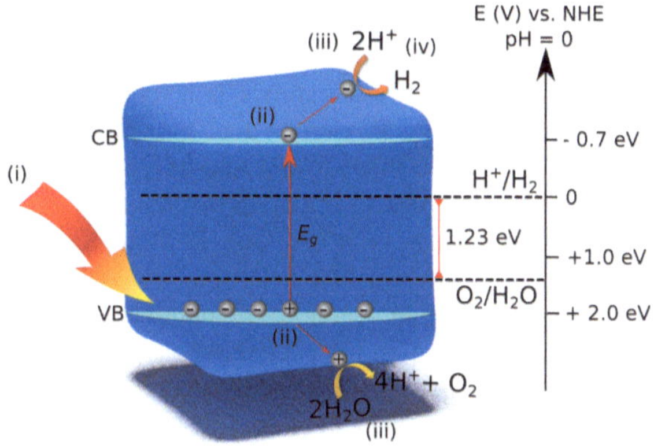

Fig. 1 Schematic steps of water splitting by a semiconductor photocatalyst

(i) Light absorption and the generation of electron–hole pairs;
(ii) Charge separation (electron–hole pairs) and transportation;
(iii) Surface redox reaction.

Upon illumination with energy greater than the band gap (E_g) of the semiconductor, electrons in the valence band (VB) are excited to the conduction band (CB) leaving holes in the VB. The photogenerated electrons and holes that do not exhibit recombination will move toward the photocatalyst surface where they reduce and oxidize the adsorbed water molecules, respectively.

In general, an undesired charge recombination of photogenerated electrons and holes can occur before their migration to the active sites on the surface of the photocatalyst. The charge recombination is often considered as a main limiting process for the loss of efficiency in water splitting reaction [15, 22]. Photogenerated charge carriers can recombine through radiative or nonradiative processes with a recombination time of the order of 10–100 ns (nanosecond), which is around 10 times (or more) faster than the time they take to reach to an adsorbed water molecule on the surface of the photocatalyst [11, 24, 25].

In order to overcome the charge recombination, photocatalyst semiconductor materials can be prepared with high crystallinity to suppressed impurities or crystal defects. Generally, the reduction of impurities and surface crystal defects can inhibit electron–hole recombination and enhance the charge carrier mobility. Another way to suppress the charge recombination and enhance the photocatalytic activity is decreasing the photocatalyst size dimension [11]. The electron–hole diffusion length in nanosized dimension is shorter than in bulkier materials; therefore, the photogenerated charge carriers can easily be separated and reach the surface of the photocatalyst to reduce (H^+/H_2) and oxidize (O_2/H_2O) water molecule adsorbed on actives sites [11, 26].

Another important approach to reduce charge recombination is the deposition of well-dispersed cocatalyst in the form of nanoparticles on the surface of semiconductor photocatalysts, which could improve the charge separation in order to prevent electron–hole recombination, as well as to create active sites to improve the water splitting performance. Cocatalysts materials loaded on the surface of semiconductor photocatalysts can act as active sites for reduction half reactions, oxidation half reactions as well as dual cocatalyst for both reactions [27–30]. Several cocatalysts such as RuO_2, CoPi, and IrO_2 have been reported as materials that trap photogenerated holes to create oxidation sites for O_2 evolution reaction [30–37]. Noble metals, such as Pt, Pd, Au, Rh, and Ru are widely used as cocatalysts to create metal–semiconductor heterojunction to introduce active sites and accelerate the proton reduction reactions in photocatalytic H_2 evolution. Among all these noble metals, platinum has been the most widely used cocatalysts to create a metal–semiconductor heterojunction due to its larger work function, high chemical stability, and lowest overpotential for H_2 evolution [28]. Figure 2 shows a simplified scheme of the semiconductor photocatalyst (heterojunction) loaded with noble metals nanoparticle cocatalyst for H_2 production.

As illustrated in Fig. 2, when noble metal nanoparticles and semiconductors contact each other, a Schottky barrier is formed at the metal–semiconductor interface and charge is transferred so that the Fermi levels equilibrate. Under light irradiation with energy greater than the band gap values of semiconductor, the photoexcited electrons on the CB are transferred toward to metal NPs, which serves as trap centers to reduce the charge recombination and thus improve the proton reduction to form H_2 molecules, while holes on the valence band (VB) migrate to surface of the photocatalyst to oxidize H_2O to form O_2 [27, 28].

A variety of preparation methods for the synthesis of cocatalyst nanoparticles have been developed. The most popular chemical and physical approaches include chemical reduction, impregnation, photodeposition, laser ablation, and magnetron sputtering [38]. An appropriate amount of cocatalysts on the surface of photocatalyst plays an important role to accelerate the photocatalytic water splitting reaction. Figure 3 displays a trend on the loading of cocatalyst and photocatalytic activity.

As shown in Fig. 3, the photocatalytic activity is strongly influenced by the amount of the cocatalyst loaded on the surface of a photocatalyst. The photocatalyst

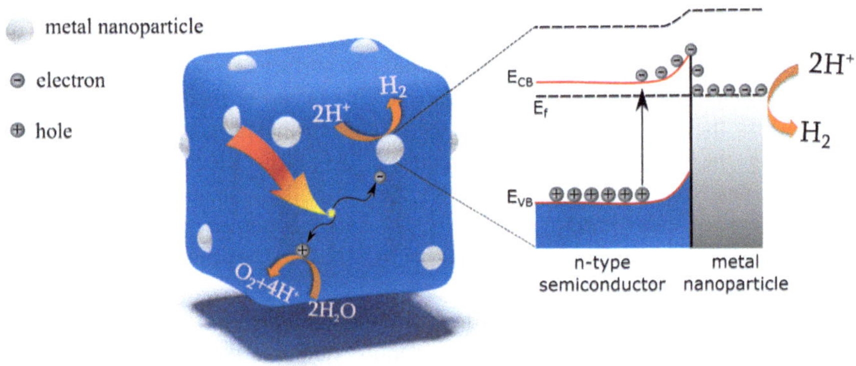

Fig. 2 Mechanism of solar water splitting on metal-semiconductor heterojunction

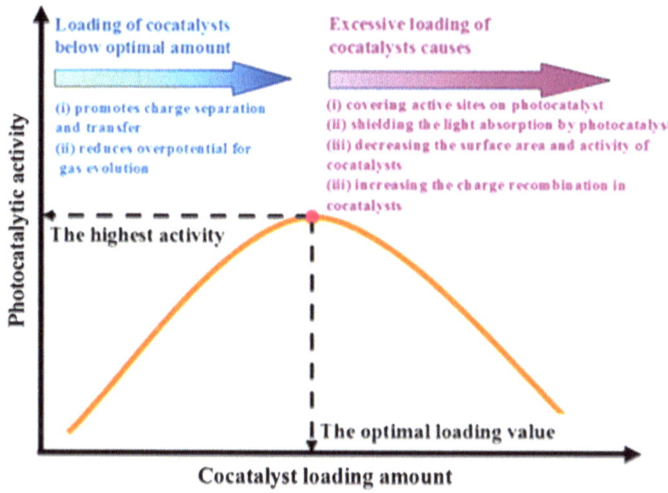

Fig. 3 Influence of cocatalyst loading amount on the photocatalytic activity [28]. Reproduced from Ref. [28] with permission from the Royal Society of Chemistry

can be loaded with an optimal amount of cocatalyst to achieve the highest photocatalytic activity (volcano-type curve). Increasing the loading of cocatalyst on the surface of photocatalyst up to an optimal amount is benefit to promote the separation of photogenerated electrons and holes, leading to an enhanced photocatalytic activity for water splitting reaction. On the other hand, the higher amount of cocatalyst than that of an optimal value can drastically decrease the photocatalytic activity for water splitting reaction. An excessive loading of cocatalyst on the surface of photocatalyst could decrease the photocatalytic activity mainly due to

(i) Cocatalyst covering the active sites for redox reactions;
(ii) Cocatalyst shielding the surface of photocatalyst for light absorption;
(iii) Formation of cocatalyst with large particle size and resulting in the decreasing of surface area;
(iv) High concentration of particles leads to the introduction of charge recombination centers and the decreasing of the photocatalytic activity [28].

2 An Overview of Semiconductor Materials Applied to Solar Water Splitting

The solar water splitting reaction using a semiconductor powder dispersed in water solution is one of the promising pathways to split water and consequently to generate a clean fuel by converting photochemical energy into hydrogen. The semiconductor material used for this purpose is of key importance since all the process begins with photon absorption and every secondary step strongly depends on the semiconductor properties. In this section, we will highlight the most used semiconductor materials for photocatalytic water splitting (PWS), that is, without using bias—only with the powder suspended in solution.

Without a shadow of doubt, the most widely used material for PWS reported up to date is TiO_2. TiO_2 is an n-type semiconductor with a band gap of about 3.2 eV [39]. Since the discovery of water photoelectrolysis using TiO_2 photoelectrodes by Honda and Fujishima in 1971, many materials are being extensively investigated in the search for efficient photocatalyst systems for PWS. According to Kazunari Domen and coworkers, sunlight spectrum can be divided in terms of three main components with distinct wavelengths accounting for 4, 53 and 43% of the total solar energy which are, respectively the ultraviolet (UV) rays ($\lambda < 400$ nm), visible light ($400 < \lambda < 800$ nm) and infrared rays ($\lambda > 800$ nm) [40]. In this way, TiO_2 and many other materials with band gap greater than 3.1 eV such as ZnO, Ta_2O_5, ZrO_2, $NaTaO_3$, $SrTiO_3$, $LiTaO_3$, $K_4N_6O_{17}$ absorb only 4% of the total energy from the sun and, in the case of a technological application contemplating an industrial device or process, their use would not be self-sustaining and economically viable. The infrared radiations do not have sufficient energy to generate electronic transitions of interest for photocatalysis; therefore, we will dedicate the next two sections to highlight UV or visible light active photocatalysts.

2.1 UV Light Active Photocatalysts

UV light accounts for only a small portion of solar energy; however, several works still focus on using wide band gap semiconductors because their ease to prepare, low cost, stability over long period of time and properties that match thermodynamic requirements for water splitting. Some review articles cover the relevant literature until now which are strongly recommended for reading [15, 26, 40–51]. It is important to mention that UV active photocatalysts for PWS are mainly based on transition metal oxides with empty *d* orbitals and consequently low valence band energies, e.g., TiO_2, V_2O_5, and Nb_2O_3 [26]. Due to the high electronegativity of oxygen atoms the metal oxides are very stable, however, absorbs a very narrow portion of the solar light and need high power UV light sources to work [51].

The most commonly explored strategies to improve the photocatalytic activity of UV light active photocatalysts are: (i) doping with an external ion such as N, S, etc., aiming to extend its absorption to the visible (will be discussed further) or to improve light harvesting; (ii) using one or more cocatalyst materials on the photocatalyst surface to facilitate the charge collection and separation; or its association to form a heterojunction with a low band gap semiconductor capable of absorbing visible light [26].

Despite its poor solar light-harvesting ability, Ta_2O_5 has received a prominent place among the most active photocatalysts for water splitting under UV irradiation reported to date [41]. Gonçalves et al. [52] reported Ta_2O_5 nanotubes obtained by anodization of a Ta foil in optimized conditions with performance to produce hydrogen from water/ethanol at an evolution rate of 4.9 mmol h^{-1} g^{-1}. Jun Shen et al. investigated the template free synthesis of fluorine-doped tantalum oxide nanoparticles (F-Ta_2O_5 NPs) and studied their photocatalytic activity for hydrogen evolution from aqueous methanol solution (10 vol.%) under UV irradiation [53]. The as-prepared F-Ta_2O_5 NPs evolved hydrogen at 2.62 mmol g^{-1} h^{-1}. Junko Kondo and coworkers reported crystalline, well-ordered two-dimensional hexagonal mesoporous Ta_2O_5 loaded with 3.0 wt% NiO with efficiency as high as 3360 μmol h^{-1} (11.2 mmol h^{-1} g^{-1}) of hydrogen evolution from pure water using a 450 W high-pressure Hg lamp as excitation source (incident light power was not mentioned) [54]. Hollow tantalum pentoxide (Ta_2O_5) nanoparticles with a band gap of 4.3 eV were recently reported to possess better stability than bulk Ta_2O_5 for photocatalytic hydrogen production from water [55]. Large activities were also reported for carbon nanotubes (CNT)-Ta_2O_5 hybrids loaded with 0.5 wt% of Pt nanoparticles with up to 1600 μmol h^{-1} of hydrogen evolution from sacrificial water splitting under high power UV illumination [56].

TiO_2 is the most widely investigated photocatalyst due to high photoactivity, low cost, low toxicity, and good chemical and thermal stability. It has been reported that pure TiO_2 could not perform overall water splitting and the photocatalytic activities dropped rapidly with reaction time using only the aqueous suspension system, mainly due to problems such as fast and undesired electron–hole recombination [57–59]. Galińska and Walendziewski reported a study on the influence of sacrificial agents on the control of electron–hole recombination on Pt-TiO_2 during photocatalytic water splitting [60]. Authors tested methanol, Na_2S, ethylene diamine tetraacetic acid (EDTA) and I^- and IO_3^- ions as sacrificial reagents to obtain effective water splitting only for EDTA and Na_2S. In the case of methanol, hydrogen evolution was also a result of methanol conversion. Wender and coworkers reported the synthesis of self-organized TiO_2 nanotube arrays produced by anodization of a Ti foil in the presence of an ionic liquid as fluorine source. The photocatalytic activity for hydrogen production using 20 vol.% methanol as sacrificial agent was 0.4 μmol h^{-1} cm^{-2} for 100% anatase photocatalyst [61].

Preparation of metal oxide nanoparticle chemically bonded to reduced graphene oxide (rGO) has been recently reported to enhance photocatalytic activity of semiconductor photocatalysts [62–64]. TiO_2–rGO composite photocatalysts, for example, were first prepared using tetrabutyl titanate and rGO as the starting materials by a sol–gel method [65]. The photocatalytic activity for hydrogen evolution showed that the as-prepared TiO_2/rGO composite exhibited a hydrogen evolution rate of 8.6 μmol h^{-1} under UV–Vis light irradiation, nearly two times larger than that of P25 (4.5 μmol h^{-1}). The material showed small activity for visible light irradiation. A series of TiO_2/rGO composites with different rGO content were subsequently investigated and a maximum hydrogen evolution rate using a 500 W Xe lamp calibrated at 150 mW/cm^2 was achieved for the sample with 5 wt% rGO [66]. The authors interpreted that exceeding contents of rGO decreased the activity of the system by introducing electron–hole recombination centers into the composite. In fact, it was discovered shortly after this work that the oxidized graphene oxide itself could behave as light absorber with appropriated band energy edge positions for hydrogen evolution in aqueous solution or pure water (very slow kinetics) without needing a cocatalyst or a tandem system [67]. The efficient H_2 generation on GO could be attributed to molecule-like behavior of GO sheets and its high dispersity in water. A ternary Cu_2O–TiO_2–rGO nanocomposite was also fabricated by a simple "sonoreduction" with simultaneous reduction of Cu^{2+} to Cu^+, and graphene oxide (GO) to rGO without the addition of any external reducing agent [68]. The photocatalytic activity of the material was enhanced to a maximum H_2 production rate of 110,968 μmol h^{-1} g^{-1} obtained with a 1.0% Cu and 3.0% GO. Authors reported that this is significantly higher than previously reported graphene-based photocatalysts.

Kudo et al. [69] investigated in 2007 the effects of metal cation doping into wide band gap semiconductor showing that doping with lanthanide and alkaline earth ions improved activity of a $NaTaO_3$ photocatalyst for water splitting. Authors also reported that metal cation doping into ZnS, TiO_2, and $SrTiO_3$ created donor energy levels that could transfer electrons to the conduction band of the host photocatalyst resulting in visible light responses for H_2 or O_2 evolution from H_2O/sacrificial reagents. Sol–gel and solid-state methods were applied to synthesize $NaTaO_3$ photocatalysts, followed by loading with NiO cocatalyst to enhance water splitting activity under UV illumination [70]. The NiO/$NaTaO_3$ sol–gel catalysts exhibited a maximum H_2 evolution rate of 9000 μmol h^{-1} g^{-1} at 3 wt% NiO loading, while the NiO/$NaTaO_3$-solid-state catalysts produced a maximum H_2 rate of 147 μmol h^{-1} g^{-1} at 0.7 wt% NiO, using as light source a 400 W high-pressure mercury lamp. To interpret the results, authors proposed that the interdiffusion of cations heavily doped the p-type NiO and n-type $NaTaO_3$, reducing the depletion widths and facilitating charge transfers through the interface barrier.

Metal-loaded strontium titanate ($SrTiO_3$) is reported since early 80s for continuous photochemical generation of both hydrogen and oxygen under UV light by using only pure water without sacrificial agents (overall water splitting) [71, 72]. The synthesis of this semiconductor is simple and its band gap energy has been experimentally determined as 3.2 eV [73]. $SrTiO_3$ was loaded with nickel oxide nanoparticles resulting in improvements of its photocatalytic activity for photochemical overall water splitting without the requirement of rare elements [74]. It was found that reducing the size of titanate particles over 30 nm results in worsening of its photoactivity especially due to effects of charge recombination leading to lower light absorption and slower water oxidation kinetics. The accepted mechanism for this system assumes water oxidation to occur at the titanate and water reduction at NiO_x. However, NiO_x–$SrTiO_3$ system was revisited one year after elucidating that NiO_x–STO is more likely a three component Ni–STO–NiO photocatalyst, in which $SrTiO_3$ absorbs the light, Ni reduces protons, and NiO oxidizes water [75].

Separation of photogenerated charges has been reported between different facets of a $SrTiO_3$ nanocrystal semiconductor [76]. Reduction and oxidation catalytic sites could be separately distributed only on the anisotropic facets of 18-facet $SrTiO_3$

nanocrystals, but randomly distributed on every facet of 6-facet $SrTiO_3$ nanocrystals. It opened possibility of preparing dual-cocatalysts for the anisotropic facets of 18-facet $SrTiO_3$ nanocrystals resulting in a fivefold enhancement of apparent quantum efficiency. Authors attributed the photocatalytic performance to the charge separation between anisotropic facets and the separation of the reduction and oxidation catalytic sites to reduce the charge recombination.

2.2 Visible Light Active Photocatalysts

2.2.1 Native Visible Light Absorber Materials

Fe_2O_3 [77], WO_3 [78, 79], $BiMoO_6$ [80], Bi_2WO_6 [81], $M_3V_2O_8$ (M=Mg, Ni, Zn) [82], $BiFeWO_6$ [83], $MBiSbO_7$ (M=Ga, Fe, Gd) [84], MFe_2O_4 (M=Ca, Zn, Co, Ni) [85, 86], $BiVO_4$, $CuBi_2O_4$, $CuWO_6$, metal nitrides and sulfides, and many other materials have been explored as native visible light absorber semiconductors. However, many of these do not meet all thermodynamic requirements for overall water splitting but can be regarded as important photocatalysts for applications using tandem systems.

Iron is the fourth most common element in the earth's crust and one of its oxide, hematite (α-Fe_2O_3), has emerged as a promising alternative for visible light photoelectrochemical hydrogen evolution since it is a semiconductor with band gap 2.1 eV possessing good chemical stability in water, abundance, low cost, and significant light absorption [16, 77]. The main problem in the use of hematite for water splitting is that its conduction band level is below the redox potential of H^+/H_2 and, consequently, it can only be regarded as an oxygen evolution [87]. In addition, it presents a relatively long visible light absorption depth combined with a very short minority carrier lifetime and mobility; ending up with low charge separation and collection efficiencies [88, 89]. For example, hematite needs to be approximately 375 nm to absorb 95% of incident 550 nm light ($3/\alpha$), whereas the minority carrier collection length is approximately 20 nm. This shortcoming has been addressed via shortening charge collection distance by nanoscale sizing the material while maintaining good light absorption [90, 91]. In addition, different strategies have been recently proposed to overcome these problems, such as, doping, ion irradiation, and surface modification with other oxides. Based on previous reports on substantial band gap, increase of low-dimensional nanomaterials compared to bulk hematite, Wender et al. [77] synthesized hematite nanorings loaded with cobalt hydroxide nanoparticles photocatalysts for efficient hydrogen generation through water photolysis, i.e., without applying external bias. Authors attributed the photoactivity of pure hematite to the size and shape properties of the nanostructures due to possible changes in the conduction and valence band positions. By loading the nanorings with suitable amounts of $Co(OH)_2$ nanoparticle cocatalysts, the photocatalytic activity was significantly increased to a maximum H_2 evolution of about 546 $\mu mol\ h^{-1}\ g^{-1}$ with 7% of $Co(OH)_2$ nanoparticles. In a parallel study,

Mangrulkar et al. [92] prepared nanoferrites for water splitting with much higher activity under visible light compared to commercial iron oxide under similar conditions. By loading Pt cocatalysts, efficiencies as high as 8275 μmol h^{-1} g^{-1} were achieved.

Investigation on synthesis and photocatalytic properties of nanocrystalline $ZnFe_2O_4$ led to a uniform-sized photocatalyst with a quantum yield of H_2 evolution 3.8 times higher than that of conventionally synthesized $ZnFe_2O_4$ [86]. Akihiko Kudo and coworkers [80] reported a comparative study with different bismuth molybdates and showed that aurivillius structured Bi_2MoO_6 with band gap of 2.70 eV showed an intense absorption band in the visible light region and photocatalytic activity for O_2 evolution from an aqueous silver nitrate solution under visible light irradiation compared to Bi_2MoO_6, $Bi_2Mo_2O_9$, and $Bi_2Mo_3O_{12}$.

Materials of general structure $A_2B_2O_7$, known as pyrochlores, where AB_2O_6 is a b-pyrochlore, which is a defective structure, have attracted special attention because their reported visible light driven photocatalytic activity due to the hybridization of Bi 6s-O 2p orbitals [83]. Bismuth-based pyrochlores are reported to possess photocatalytic activity for PWS and broadly includes ferrites ($BiFeO_3$), titanate ($Bi_2Ti_2O_7$) [93, 94], tungstate (Bi_2WO_6), vanadate ($BiVO_4$), and niobate (Bi_2FeNbO_7) [83]. $Bi_2Ti_2O_7$ has been attracting attention because it has a narrower band gap ($\sim$2.8 eV) with suitable band energies for water splitting reactions [95]. $Bi_2Ti_2O_7$ forms a shallow acceptor energy level in the forbidden band and as a result, the holes can be excited into the valence band under irradiation with visible light [94]. In the class of general tungstates, $BiFeWO_6$ with optical band gap energy of 2.35 eV was recently reported to present sunlight photocatalytic activity [83].

$BiVO_4$ has been regarded as an efficient photocatalyst for oxygen evolution under visible light [96, 97]. Due to low energy of its conduction band bulk $BiVO_4$ is not suitable for H_2 evolution. In addition, $BiVO_4$ is characterized by a high recombination rate of photogenerated carriers that have much shorter diffusion length than the thickness required for sufficient light absorption. However, quantum-sized $BiVO_4$ was reported for the first time to decompose pure water into H_2 and O_2 simultaneously under simulated solar light irradiation without any cocatalysts or sacrificial reagents [98]. Authors demonstrated that photocatalytic activity of the quantum-sized $BiVO_4$ arises from the negative shift of conduction band edge by a quantum confinement effect and a decreased overpotential for water reduction.

The nitrides and oxynitrides of d^0-type transition metal elements (Ti^{4+}, Zr^{4+}, Nb^{5+}, Ta^{5+}, etc.) and d^{10}-type (Ge^{3+}, Ga^{4+}, etc.)[99] have attracted attention over the water splitting community since the reported ability to absorb light with wavelength longer than 600 nm, what correspond to more than 35% of the sunlight energy that can be utilized [100]. It is very close to the infrared limit region ($\lambda > 780$ nm). Unfortunately, oxy(nitrides) could not achieve to date overall water splitting and the exact cause is not trivial to point out. Kazunari Domen et al. made a very important discussion about this issues on two recent review articles published in 2013 [100] and 2015 [101].

Besides monoclinic tungsten trioxide WO_3 being recognized as one of the ideal materials for solar-driven water splitting because of its photostability, photosensitivity, good electron transport property and moderate ~150 nm hole diffusion length compared to 2–4 nm of α-Fe_2O_3 and ~100 nm of TiO_2. WO_3 has a relatively high band gap of about 2.7 eV and cannot fully harvest the solar energy for solar-to-hydrogen conversion [42, 102]. In addition, WO_3 conduction band energy is not appropriated for H_2 evolution from water splitting but understanding and manipulating the one half-reaction of photoinduced hole-oxidation to oxygen are of fundamental importance to design and develop an efficient water splitting process. Surface modification, doping, and nanostructure engineering are the most used modification process to address drawbacks of WO_3. For example, the efficiency of O_2 evolution from water splitting was significantly improved over Pt/WO_3 compared to that of pure WO_3 under visible light illumination [103]. WO_3 was associated to CdS to form a CdS/WO_3 nanojunction heterojunction achieving a Z-scheme that could significantly increase the H_2 evolution activity of CdS with lactate as electron donor [104]. The hydrogen evolution rate obtained for 20 wt% CdS/WO_3 photocatalyst was 369 μmol g^{-1} h^{-1}, 5 times higher compared to only CdS with lactic acid as electron donor. By loading Pt, H_2 evolution rates up to 2900 μmol g^{-1} h^{-1} were obtained with visible radiation. Results of transient photovoltage and surface photovoltage indicated that the Z-scheme system of CdS/WO_3 can effectively promote charge separation and depress the charges recombining of photogenerated charge in CdS. WO_2–WO_3 hybrid nanorods were in situ fabricated using a wet-chemistry route and applied for photocatalytic water oxidation using visible and near-infrared light [103].

2.2.2 Metal Doping of Wide Band Gap Semiconductors

Several works proposed doping of wide band gap semiconductor as an efficient strategy to extend photon absorption from the UV to visible light and also to form charge traps to separate electron–hole pairs [44]. Metal doping has been proven to shift the optical absorption edge of semiconductor photocatalysts to the visible range using less than 10 at.% of foreign cations [105]. The interaction of the metal cation states with the valence or conduction band of ZnO, for example, lead to the creation of intra-band gap levels and induced band gap narrowing. As a result, improved photocatalytic activities in the visible light were achieved [106]. The most used dopants for modification of wide band gap oxides properties are transition metals such as Cr, Fe, Ni, V, Mn, Nb, and Cu [107, 108].

For ZnO, several studies revealed an increase in the photocatalytic activity in visible light region with metal doping using Ag, Mn, Cu, Co, Fe, Ce, Eu, and Al and nonmetal doping, including N, C, and S. The literature of metal-doped ZnO case is well covered by a recent review article [106].

Below we will discuss some results achieved for doping TiO_2. Titanium dioxide is the most widely used photocatalyst for hydrogen production from water but it suffers from lack of photocatalytic activity under visible light irradiation. Cr- or

Fe-ion-doped TiO_2 thin films were studied to improve its hydrogen generation by photocatalytic water splitting under visible light irradiation [109]. H_2 production rates were higher for Fe-doped TiO_2 (15.5 μmol h^{-1}) than for Cr-doped TiO_2 (5.3 μmol h^{-1}). Authors explained that Fe ions could trap both electrons and holes, thus avoiding recombination. The material could generate hydrogen for long periods of time due to separated evolution of H_2 and O_2 by eliminating the back-reaction effects. High photocatalytic activities were also achieved for metal and nitrogen co-doped TiO_2 [110], specially the Cu/N co-doped TiO_2 that exhibited a rate of H_2 production of 5.48 mmol h^{-1} under UV light irradiation and 47.6 μmol h^{-1} under visible light irradiation [108]. In another study, Fe and Ni co-doped TiO_2 nanoparticles were prepared and applied for hydrogen production from water/ethanol under visible light irradiation [111]. The average maximum H_2 evolution rate obtained was 361.64 μmol h^{-1} g^{-1} (144.66 μmol h^{-1}) using 5.0% Fe-4.0% Ni/TiO_2 catalyst under 103.9 mW/cm^2 of visible light. Pt-loaded Pt-doped TiO_2 nanocrystals with (001) exposed facets was also recently reported as catalysts for hydrogen production from pure water and methanol–water systems under AM1.5G sunlight illumination (100 mW/cm^2, class AAA solar simulator, 450 W lamp) [112]. The best obtained hydrogen evolution rate was 11.2 mmol h^{-1} g^{-1} (1.12 mmol h^{-1}) under optimized reaction conditions over ca. 1.0 wt% Pt-loaded Pt/TiO_2, which is among the highest hydrogen evolution rates over the noble metal/TiO_2 systems reported to date.

2.2.3 Nonmetal Doping of Wide Band Gap Semiconductors

Doping of high band gap semiconductors is one of the most used strategies for improving their photoactivity for hydrogen evolution. The doping element may introduce a new donor or acceptor energy level into the material, which in some cases is very beneficial for creating thermodynamically required conditions for water oxidation and reduction. Nitrogen, fluorine, sulfur, and carbon are so far the most utilized nonmetal elements for doping semiconductor oxides. A recent article review discusses TiO_2 doping with nonmetal elements up to the year 2012 [49].

A green synergistic approach was recently reported for preparation of C/N co-doped TiO_2 modified with Ag and g-C_3N_4 with enhanced activity under visible light irradiation and high chemical stability [113]. The ternary composite with hierarchical structure and large surface area showed the high photocatalytic activity compared to their single and binary catalysts. Authors interpreted it as a result of the fast generation, separation, and transportation of the photogenerated carriers.

2.2.4 Self-doped Photocatalyst

Self-doping is considered an effective and promising route for obtaining visible-light-responsive photocatalysts. For element doping, self-doping of Ti^{3+} into the crystal lattice of TiO_2, for example, is reported to introduce a new energy level

(0.7–1.2 eV) below the conduction band of TiO_2 [43, 114, 115]. It causes a red shift in the absorption band edge and can increase the carrier concentration. However, it is highly challenging to obtain a stable Ti^{3+}-doped TiO_2 because Ti^{3+} oxidizes easily during the heterogeneous reaction. Fu et al. [116] proposed a simple carbon coating route to stabilize the Ti^{3+}-doped TiO_2 during PWS for hydrogen evolution reaction. The proposed stabilizing mechanism is that the conductive carbon coating layer is a barrier layer and prevents the H_2O and O_2 from diffusing into the surface of the photocatalyst, which can oxidize the surface O vacancies and Ti^{3+} in TiO_2. Recently, a one-step vacuum activation method to achieve the reduction of GO, the self-doping of Ti^{3+} into TiO_2 and the loading of TiO_2 nanoparticles on rGO surface was proposed [117]. The Ti^{3+}–TiO_2/rGO photocatalyst composite showed high hydrogen evolution activity in the visible light irradiation (400 μmol h^{-1} with 0.1 g of catalyst, 0.37 wt% Pt, 20 vol.% methanol solution, 300 W top-irradiated Xe lamp used as a light source and a $\lambda > 400$ nm UV–cutoff filter).

Open-framework V-doped [118], mixed-valence titanium phosphate with intense absorption across the entire visible wavelength range caused by the presence of Ti^{3+} centers in equimolar ratio with respect to Ti^{4+} were recently reported. Controlling V-doping provided an additional charge separation level in the intra-band gap space resulting in fivefold higher hydrogen production rate. A recent review article by Jie-Sheng Chen and coworkers [115] published in 2014 was dedicated to understand the role of self-modification of titanium dioxide materials by Ti^{3+} and/or oxygen vacancies and covers almost all literature until the present date.

2.2.5 Heterojunctions Photocatalysts

Composites comprised of two or more semiconducting materials with suitable band gaps and band positions have been reported to be effective to enhance visible light photocatalytic activity that has been examined in a variety of applications including hydrogen generation and organic pollutants degradation. Some studies focused on the enhancement of UV light active photocatalysts by combining it with visible light active semiconductors. Other studies used composite photocatalysts made of two or more small band gap semiconductors searching for a synergistic ability to produce hydrogen from water through faster charge transfer processes, increased electron–hole pair's lifetime and improved light harvesting.

Heterostructured photocatalyst composite such as CdS/*n*-TiO_2 [119, 120], CdS/Ta_2O_5, CdS/WO_3 [104], CdS/Au/TiO_2 [121], In_2O_3/Ta_2O_5, TiO_2/C_3N_4 [122], TiO_2/$BiVO_4$ [123], Cu_2O/Au/$BiPO_4$ [124], CdS/$CdWO_4$ [125], WO_3-metal-gC_3N_4 (metal=Cu, Ag, Au) [126], BiOI/TiO_2 [127], Si/$MgTiO_3$ [128], RhCrOx/$LaMg_{1/3}Ta_{2/3}O_2N$ [129], $BaZrO_3$/$BaTaO_2N$ [130] and many others were reported to exhibit enhanced visible light photocatalytic activity for hydrogen evolution compared to pure materials [131–133]. The significant advantage of these systems is the separation of electron–hole pairs due to enhanced interfacial charge transfer accompanying the extended light absorption resulting in enhanced photocatalytic activities. Vanadium-doped TiO_2 nanorod nanocomposite anchoring on rGO was successfully synthesized using a simple one-step low temperature hydrothermal method [134]. The visible light activity for H_2 evolution of nanocomposite was higher (over 120 μmol h^{-1}; 20% aqueous methanol solution; 300 W Xe and AM1.5G filter) than that of undoped TiO_2 and non-composited V-TiO_2 nanoparticles. In this noble metal-free nanocomposite material, vanadium could reduce TiO_2 bang gap that in the presence of rGO could effectively trap the photogenerated electrons; thereby, suppressing the electron–hole recombination.

Nitrogen-doped tantalate, $Sr_2Ta_2O_{7-x}N_x$ exhibited 87% more activity for photocatalytic hydrogen production under solar irradiation compared with its undoped counterpart $Sr_2Ta_2O_7$ [135]. A new type of composite containing graphene-Pt and $Sr_2Ta_2O_{7-x}N_x$ demonstrated an additional ~80% increase in hydrogen production and a quantum efficiency of 6.45%, representing ~177% increase from pristine undoped $Sr_2Ta_2O_7$. Graphene worked as an electron transfer highway facilitating charge collection onto Pt cocatalysts. Under optimized conditions, the 5 wt% graphene sample exhibited a hydrogen yield of 293 μmol h^{-1} (±5%) using solar simulated illumination (AM1.5G filter + 300 W Xe lamp).

Bismuth oxide based heterojunctions are reported to possess improved photocatalytic activity under visible light irradiation. Bi_2O_3/Ta_2O_5, Bi_2O_3/TaON, and Bi_2O_3/T_3N_5 composites [136] heterojunctions were reported to show promising conditions for electron–hole separation to generate hydrogen from an aqueous methanol solution. Figure 4 summarizes some of the important properties of the separated materials such as their crystal and electronic information. Best photocatalytic activity for hydrogen evolution under visible light irradiation was obtained from Bi_2O_3/T_3N_5 sample producing 23 μmol h^{-1} g^{-1} in the first 4 h of photoreaction. Mesoporous $Bi_2O_3/TiO_{2-x}N_x$ nanocomposites were synthesized by soft chemical template free homogeneous co-precipitation technique showing hydrogen evolution rate of 198.4 μmol/h (9920 mmol h^{-1} g^{-1}) and an apparent quantum efficiency of 4.3% [137]. The experiments were conducted in the following conditions: 10 vol.% methanol/water solution; 125 W medium pressure Hg lamp as visible light source ($\lambda \geq 400$ nm) and 1 M $NaNO_2$ solution as UV filter.

Compounds		Ta_2O_5	TaON	Ta_3N_5	Bi_2O_3
Crystal unit cell representation					
Color of samples		White	Yellow-green	Red	Yellow
Band gap, eV		3.80	2.29	2.03	2.78
Calculated band positions,	CB	−0.17	−0.17	−0.17	+0.34
eV *vs.* NHE	VB	+3.63	+2.12	+1.86	+3.11

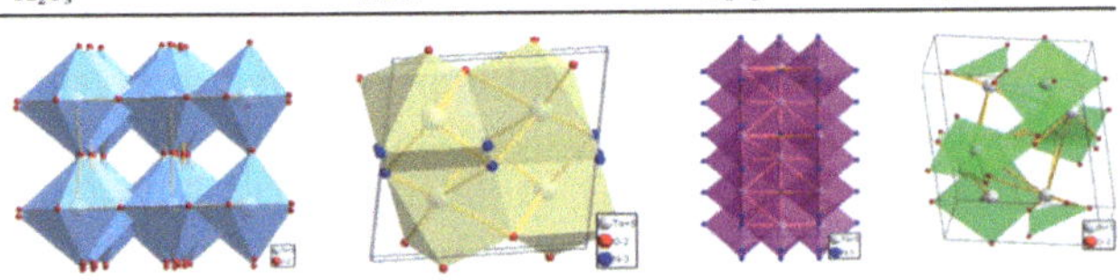

Fig. 4 Crystal structure representation, and electronic structure data of the components of the composites with tetragonal Bi_2O_3 and different phases of tantalum based compounds (orthorhombic Ta_2O_5, monoclinic TaON and monoclinic Ta_3N_5). Adapted from [136]. Reproduced from Ref. [136] with permission from the Royal Society of Chemistry

3 The Tandem Concept for Water Splitting on Particulate Photocatalysts

Even more than 40 years after the first report published by Fujishima and Honda [138], there is still a pursue for the economically competitive implementation of renewable energy provided by sunlight as feedstock for production of fuels in lieu of the largely employed fossil fuels. During these four decades, different approaches have been developed for the improvement of the production yields of the existent photocatalytic methods, mostly focused on the properties of the employed photocatalysts [139, 140].

As stated before, the earliest studies described the hydrogen production from the illumination of semiconductors with band gap energy values higher than 3.2 eV. This fact drove research interest into the improvement of the solar energy absorption for water splitting systems utilizing semiconductors with narrower band gaps and/or modified by doping processes. The use of visible light for the H_2 production from water splitting is vital for the development of efficient systems, as roughly half of the energy provided by the sun is delivered by photons in the visible region [141].

Up to this date, only a limited number of the known photocatalysts can drive the concomitant production of hydrogen and oxygen through the overall water splitting with no addition of sacrificial agents such as methanol, sodium periodate, and others. Moreover, the majority of these semiconductors present large band gap values and are photoactive only under UV excitation, such as the NiO_x-modified strontium titanate (a three component $NiO–SrTiO_3–Ni$ photocatalyst) [75], and the RuO_2-modified $LiNbO_3$ nanowires [142]. This is a constant limitation of water splitting photocatalysts due to the thermodynamic requirements for the photolysis of the water molecule in the presence of semiconductors.

An inventive tactic to circumvent the drawbacks presented by classic single photocatalyst arrangement is the tailored combination of two semiconductors with complementary band energy levels, configuring a tandem action. In this dual-absorber configuration, both semiconductors should absorb light and promote electrons to the respective conduction bands, leaving holes in the valence bands. Each semiconductor component performs one of the half reactions of the overall water splitting, i.e., one photocatalyst should possess a valence band energy level adequate to favor the oxidation of water (VB more positive than the potential of water oxidation), promoting the transfer of four holes per evolved oxygen molecule, as represented by Eq. (3), whereas the other semiconductor should have the CB with suitable energy to make the photoexcited electrons apt to reduce the protons into H_2, according Eq. (2) [140].

The sum of Eqs. (2) and (3) for the half reactions that take place in each components of the tandem system represent the overall water splitting, described by Eq. (1).

One of the most appealing advantages of this two-step photoexcitation mechanism for overall water splitting is that both semiconductors can possess lower band gap energies than those that comprise the requirements to act as photocatalysts in single-component systems. This feature lowers the energy needed to excite the electrons to the CB of the absorbers and allows a more effective absorption of the solar energy, including part of the visible portion. This dual configuration promotes higher quantum efficiencies for the process as higher number of incident photons can be absorbed [143].

In many cases, when the semiconductors have band gap energies that allow them to absorb visible light, the CB edge energy of the oxygen-evolving photocatalyst (OEP) is more positive than the potential of H^+ reduction and the VB edge of the hydrogen-evolving photocatalyst (HEP) is more negative than the potential of water oxidation. Therefore, the excited electrons in the CB of the OEP and the photogenerated holes in the VB of the HEP do not have the proper energy to promote the half reactions defined in Eqs. (2) and (3).

The reaction cycle in the tandem configuration is then completed by the transfer of the excited electrons in the CB of the OEP to the VB of the HEP, quenching the holes formed by the irradiation of the latter. This charge transfer is a crucial step in the mechanism of the overall water splitting performed by the dual-absorber tandem configuration and is often regarded as bottleneck for the achievement of high efficient systems [144]. The first approach used for the shuttle of the excited electrons between the tandem components was the employment of reversible redox mediators in the reaction medium, as depicted in Fig. 5. This two-step photoexcitation mechanism is also called Z-scheme due to its resemblance to the mechanism followed by the natural photosynthesis [145].

When redox pairs, such as Fe^{3+}/Fe^{2+}, $[IO_3]^-/I^-$ and $[Co(bpy)_3]^{3+}/[Co(bpy)_3]^{2+}$ (bpy=bipyridine), are used as electron mediators in Z-scheme water splitting, the holes photogenerated in the VB of the HEP oxide the reducing agent (ex: Fe^{2+}) to

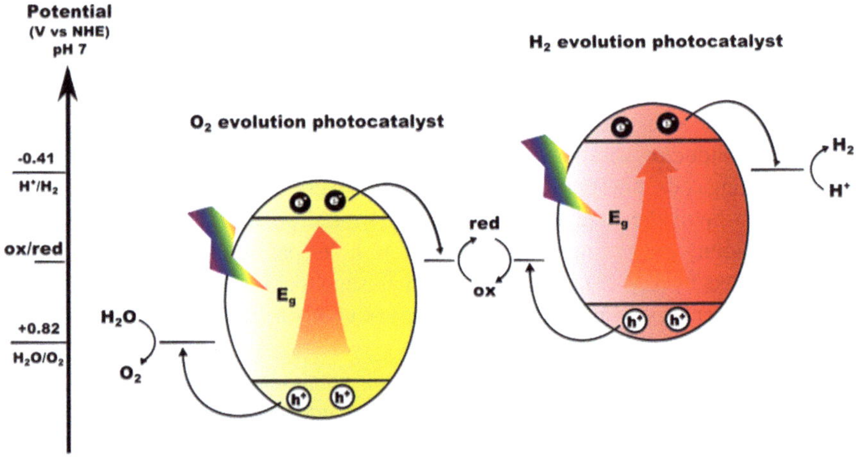

Fig. 5 Reaction mechanism for a dual-absorber water splitting tandem system, also denoted as Z-scheme, using a redox electron mediator for the electron shuttle between the two photocatalysts

the oxidizing agent (ex: Fe^{3+}), which is reduced again by the excited electrons in the CB of the OEP, ensuring the reversibility of the system. This means that the electrochemical redox potential of the redox pairs must lie within the energy edges of the conduction band of the oxygen-evolving photocatalyst and valence band of the hydrogen-evolving photocatalyst [143, 146, 147].

Although effective, the use of redox mediators has been discouraged because of the promotion of backward reactions, the competition of the redox species with water for the photegenerated electrons in the HEP and holes in the OEP, and shading effect by the colored species, which prevents the efficient absorption of the incident photons by the photocatalysts. These are some of the reasons for what redox mediators have been substituted by solid-state mediators and direct inter-particle electron transfer [144, 148, 149].

The first reports based on the Z-scheme concept to develop the photocatalytic water splitting using the ultraviolet radiation date back from the 1990s and early 2000s [145, 147, 150, 151]. In general, these works described inventive ways of combining semiconductors with complementary functionalities, and the search for a reversible redox mediator that would keep the constant production of H_2 and O_2. Some of the mediators that addressed the goals of maintaining the steady production were the Fe^{2+}/Fe^{3+}, $[IO_3]^-/I^-$, and Br_2/Br^- redox couples [145, 147].

Remarkably, in 2001, Abe et al. [147] reported the efficient employment of Pt-loaded TiO_2-anatase as the H_2 evolution photocatalyst, whereas non-loaded TiO_2-rutile acted as the OEP, operating the overall water splitting through a Z-scheme mechanism, with electrons being shuttled by a $[IO_3]^-/I^-$ reversible redox mediator. They observed the simultaneous evolution of H_2 and O_2 in a ratio of 2:1, respectively, in basic conditions, maintaining the production rate for a long period (100 h with the reactor being evacuated in defined intervals to collect the products and avoid the backward reactions on the nanosized Pt cocatalysts), generating a total of 9.6 and 4.8 mmol of H_2 and O_2, respectively.

Another considerable step in the development of tandem systems for photocatalytic water splitting was taken in 2001 when Arakawa and coworkers reported the first Z-scheme system that was functional under visible light irradiation ($\lambda > 420$ nm) [152]. In this study, a mixture of platinum-loaded tungsten oxide (Pt-WO_3) and platinum-loaded strontium titanate (Pt-$SrTiO_3$) co-doped with chromium and tantalum performed the overall water splitting under visible light mediated by the redox cycle of IO_3^- and I^-. These two photocatalysts have shown broad absorption of visible light because of their narrow band gaps and produced H_2 and O_2 at a rate of 1.8 μmol h^{-1} for more than 250 h, representing a quantum efficiency of 0.1% at 420.7 nm [152].

Since then, a vast array of studies based on tandem systems for water splitting in suspension has been tested by employing redox mediators and different visible light absorbers, which led to important improvements in the quantum efficiency of the H_2 production [143, 153, 154]. Nevertheless, as stated before, the use of redox mediators in Z-scheme systems present compelling downsides, such as the absorption of photons by the mediator, creating a shielding effect for the photon absorption by the photocatalysts, the promotion of the backward reactions to form water from the generated H_2 and O_2, and the competition between the redox species and water for the photogenerated electrons and holes.

The circumvention of these issues was efficiently addressed for the first time by Kudo's group in 2009 when they performed the Z-scheme with the HEP and OEP in direct contact with no redox mediator [144]. In this work, they further explored the characteristics of rhodium-doped strontium titanate ($SrTiO_3$:Rh) and bismuth vanadate ($BiVO_4$), which are photocatalysts they have previously applied in Z-scheme studies with the Fe^{2+}/Fe^{3+} couple as electron mediator, under visible light irradiation. This previous system surpassed the quantum efficiency of 0.3% at 440 nm [146]. Nowadays, bismuth vanadate and doped strontium titanate are two of the most promising semiconductors applied in photocatalytic water splitting with suspended particles.

In an inventive approach, under visible light irradiation, they achieved interparticle electron transfer from the conduction band of $BiVO_4$ to the impurity level of $SrTiO_3$:Rh (consisted by the reversible Rh^{3+} state), as the Z-scheme mechanism requires, by adjusting the pH of the reaction medium to 3.5 with H_2SO_4 solution. As the zeta potential of $BiVO_4$ was negative at the pH range of 2–9, the particles possess negatively charged surfaces at pH 3.5. Complementarily, the particles of Ru/$SrTiO_3$:Rh were positively charged at the same pH, as the isoelectric point of this material is $\sim$4. Therefore, the adjustment of the pH medium to 3.5 resulted in the electrostatic attraction between the $BiVO_4$ and Ru/$SrTiO_3$:Rh particles, as depicted in Fig. 6a, and a consequent direct electron transfer through the contact created between the particles of the photocatalysts, as illustrated by the Z-scheme energy diagram in Fig. 6b. The aggregation of the particles was not effective in other pH values, as observed by the authors through optical microscopy, reflecting in a decrease of H_2 and O_2 productions under visible light (Fig. 6c). The enhanced apparent quantum efficiency obtained for this direct-contact system was 1.7% at 420 nm [144].

The same study also pointed out interesting observations regarding this system such as: [1] ruthenium is a more effective cocatalyst for the hydrogen-evolving photocatalyst than platinum because the former does not favor the back reaction; and [2] the doping Rh level in the forbidden band not only favors the visible light absorption but also has a pivotal role in the electron transfer between the particles due to its reversible nature.

The aptness of the $BiVO_4$-Ru/$SrTiO_3$:Rh tandem system with no electron mediator led to further investments in this configuration to the achievement of higher quantum yields. The same authors have synthesized composites formed by these photocatalysts through a so-called liquid–solid-state reaction, which resulted in well-crystallized $BiVO_4$ particles connected to Ru/$SrTiO_3$:Rh and a quantum efficiency of 1.6% at 420 nm [155]. In this approach, $BiVO_4$ was synthesized by stirring Ru/$SrTiO_3$:Rh particles and V_2O_5 in an aqueous nitric acid solution containing $Bi(NO_3)_3$, for 5 days. The resulting composite presented a tightly connected $BiVO_4$–Ru/$SrTiO_3$:Rh interface and could perform the photocatalytic water splitting under neutral conditions, which is an important improvement from the previous system.

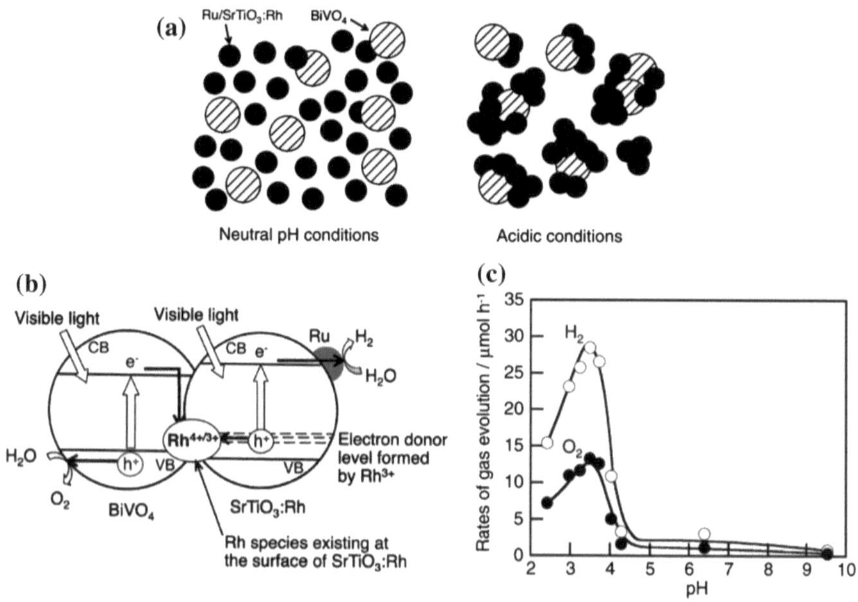

Fig. 6 **a** Representation of the suspension of Ru/$SrTiO_3$:Rh and $BiVO_4$ at neutral and acidic conditions, **b** the energy diagram for the Z-scheme showing the direct contact between the semiconducting components, and **c** pH dependence photocatalytic activity of the tandem formed by Ru/$SrTiO_3$:Rh and $BiVO_4$ [144]. Reproduced from Ref. [144] with permission from the American Chemical Society

A breakthrough in the field of particulate photocatalysts operating through a Z-scheme mechanism was achieved in 2016 by Domen and colleagues [148]. The group developed a more elaborated assemblage to achieve an optimized charge carriers mobility between the HEP and OEP employed. In this case, the hydrogen-evolving photocatalyst was $SrTiO_3$ co-doped with rhodium and lanthanum and loaded with ruthenium nanoparticles as cocatalysts, whereas the oxygen-evolving photocatalyst was $BiVO_4$ doped with molybdenum and loaded with ruthenium oxide cocatalysts. Such modifications in the structures of the pristine components were performed to concomitantly maximize the absorption of the visible portion of the solar spectrum and the separation of the photogenerated charge carriers.

A mixture of the particles of the photocatalysts was assembled in sheets through the particle transfer method (Fig. 7a) and embedded in a gold layer, which served as a conductive solid-state mediator. The photocatalyst layers were annealed at 300 °C in air to reduce the contact resistance between the semiconductors and the gold layer and further modified with deposits of Cr_2O_3, which is known to suppress the backward reactions on the surfaces of the catalysts. This tandem system gave a record apparent quantum yield of 33% at 419 nm and solar-to-hydrogen (STH) conversion efficiency of 1.1% at 331 K and 10 kPa in pure water (pH 6.8) with no electron mediator, pH adjustment, or applied voltage. The Z-scheme mechanism through which this system operates and the roles of each component represented by the energy level diagram is shown in Fig. 7b.

Although a very high quantum efficiency was achieved in this system, further examination is needed to make the production of such device more economically interesting. Studies based on alternatives for this arrangement have already been reported by substituting gold with carbon conductor layer, as solid-state mediator, and by substituting Ru/$SrTiO_3$:La,Rh with $RhCrO_x/LaMg_{1/3}Ta_{2/3}O_2N$ as HEP due to the wider visible light absorption reach of the latter (up to 600 nm) [156–158].

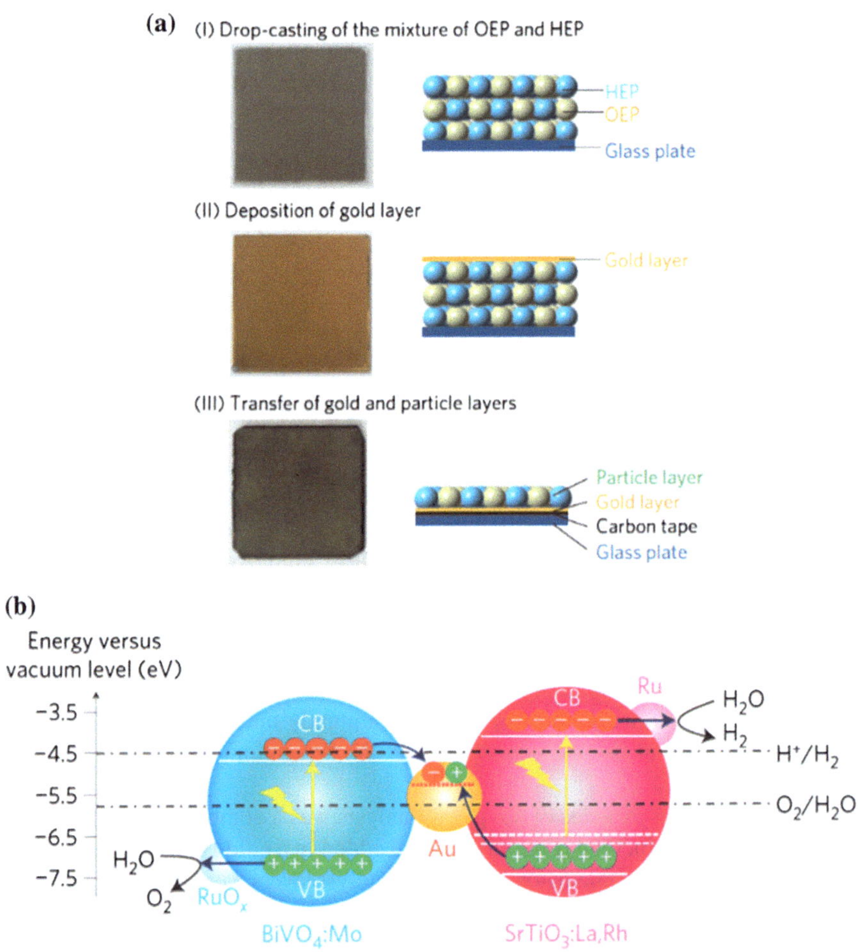

Fig. 7 Schematic representation of **a** the assembled photocatalyst sheets through the particle transfer method, and **b** the Z-scheme mechanism using the solid-state gold layer as electron mediator between $RuO_x/BiVO_4$:Mo and Ru/$SrTiO_3$:La, Rh in the photocatalyst sheets [148]. Reproduced from Ref. [148] with permission from the Nature Materials

4 Efficiency Determination in Solar Water Splitting

In solar water splitting, ranking the device performances is very important. There are some standard criteria defining that performance. The widely accepted benchmark for semiconducting materials is solar-to-hydrogen (STH) efficiency [40]. STH efficiency can be determined in powder suspension as well as photoelectrochemical systems. Generally, in powder suspension systems, it is customary to determine the apparent quantum efficiency (AQE). The objective of the following section is to overview proper experimental protocols and challenges to determine these efficiencies correctly.

4.1 STH (Solar-to-Hydrogen) Efficiency

STH efficiency defines the ratio of the chemical energy of the evolved hydrogen to the solar energy input from broadband solar irradiance (AM 1.5 G) during the water splitting reaction and can be calculated from Eq. (4).

$$\mathrm{STH} = \left[\frac{\text{Output energy as } \mathrm{H_2}}{\text{Energy of incident solar light}} = \frac{r_{\mathrm{H_2}} \times \Delta G}{P_{\mathrm{total}} \times S}\right]_{\mathrm{AM1.5G}}, \quad (4)$$

where r_{H_2} (mmol H_2/s) is the rate of hydrogen production, ΔG = 237 kJ/mol at 25 °C is gain in Gibbs free energy, P_{total} (mW/cm^2) is incident illumination power density and S (cm^2) is illuminating area of the reactor. In photocatalytic systems for overall water splitting, both H_2 and O_2 should evolve in a stoichiometric amount (2:1) in the absence of any oxidation or reduction agent [40]. While determining STH efficiency in the lab instead of using radiation from the sun, different types of white light sources are used. Care must be taken while calibrating and adjusting the light-sources profile to AM 1.5 conditions set forth by the American Society of Testing and Materials [159–161]. Otherwise, the STH efficiency values may be over or underestimated. STH efficiency of some photocatalysts in suspended particulate systems as well as in sheets obtained from powder samples are given in references [162–166].

In the experimental setup of powder suspension systems performing STH is sometimes challenging. There exist film-coating techniques to form sheets of the powders on appropriate substrates that can simplify these measurements [165, 166]. Alternatively, if hydrogen evolution cannot be directly measured, the STH can be determined in photoelectrochemical setup, from the photocurrent, solar power input, and the faradaic efficiency for hydrogen evolution (η_F) from Eq. (5).

$$\mathrm{STH} = \left[\frac{\left|J_{\mathrm{sc}}(\mathrm{mA\,cm^{-2}})\right| \times (1.23\,\mathrm{V}) \times \eta_{\mathrm{F}}}{P_{\mathrm{total}}(\mathrm{mW\,cm^{-2}})}\right]_{\mathrm{AM1.5G}}, \quad (5)$$

where J_{sc} is the photocurrent density, 1.23 V is the thermodynamic water splitting potential and η_F can be obtained from the current density versus time curve [167]. Generally, a true STH efficiency is measured in a two-electrode cell [168]. The working electrode (WE) and the counter electrode (CE) are immersed in the same pH solution (without any sacrificial reagent) and are operated under short circuit condition. Hence, no external voltage is applied across the electrodes and the reaction response is overall water splitting; purely energized from the sunlight. Along with the light source calibration to AM 1.5 G stated above another important factor while STH efficiency determination is the water film thickness, i.e., transmission of light through water before hitting the semiconducting electrode; for more detail on this subject reader is referred elsewhere [169]. It is important to mention, Eqs. (4) and (5) are valid only for the stoichiometric H_2 and O_2 evolution in the absence of any sacrificial agent. The experimentally determined STH efficiencies of some of the photocatalysts are given in Table 1.

Theoretically the band gap of a material plays an important role to calculate maximum STH efficiency considering thermodynamic energy conservation. Figure 8 shows some materials with their maximum theoretical STH efficiency and their photocurrent as a function of their band gaps.

Higher STH efficiency from photoelectrochemical devices necessitates the fact, the ideal materials would be those absorbing wide range of visible light. Within this context, the UV absorbing materials such as TiO_2, Ta_2O_5, ZrO_2 are ruled out for their lower STH efficiency (Table 1). On the other hand, visible light absorbing and also the tandem devices present a great potential for superior STH efficiencies [159, 178]. Theoretical STH efficiency values serve a good guide for the researcher to find materials to maximize light absorption; however, the band gap of the materials alone is not the complete picture to describe STH efficiency. The semiconductor

Table 1 List of some promising photocatalysts for water splitting and their reported solar to hydrogen (STH) efficiency and apparent quantum efficiency (AQE)

	Electrolyte	STH	AQE	
CoO	Pure water	5%	–	[170]
C_3N_4/CDots	Pure water	2%	4.42% at 600 nm (H_2 evolution)	[171]
Pt/TiO_2	C_2H_5OH (pH 10)	–	5.7% at 340–410 nm (H_2 evolution)	[172]
Pt/Ta_2O_5	C_2H_6O (aq)	–	22% at 254 nm (H_2 evolution)	[173]
CoO_x/MgO/Ta_3N_5	La_2O_3/$AgNO_3$ (aq)	–	11.3% 500–600 nm (O_2 evolution)	[174]
Pt/ZrO_2/TaON	NaI (aq)	–	6.3% at 420.5 nm (O_2 evolution)	[175]
g-C_3N_4	–	–	26.5% (400 nm) H_2 evolution	[176]
$SrTiO_3$:La, Rh/Au/$BiVO_4$:Mo	Pure water (pH 6.8)	1.1%	33% at 419 nm (H_2 evolution)	

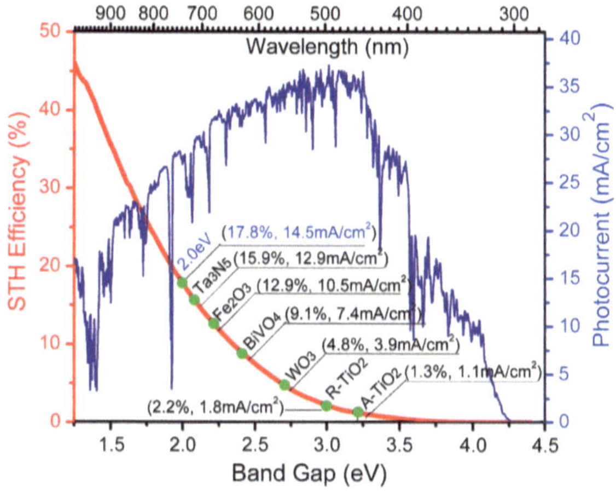

Fig. 8 Theoretical maximum solar-to-hydrogen (STH) efficiency and the photocurrent density as a function of band gap under AM 1.5 G irradiation (100 mW cm^{-2}) [177]. Reproduced from Ref. [177] with permission from the Royal Society of Chemistry

should overcome the overpotential losses and thermodynamic barriers for overall water splitting. In addition, electronic structure should be such that the recombination and trapping of the photogenerated charge carriers should be avoided [179]. Efforts are required to fine tune the materials electronic structure to achieve their highest STH efficiency.

4.2 *Apparent Quantum Efficiency (AQE)*

Photocatalytic water splitting activity in powder suspension systems can be directly evaluated by the H_2 or O_2 gas evolution rate (μmol h^{-1}) using a gas chromatograph. The gas evolution rate strongly depends on the experimental setup including light source and type of a reaction cell [15, 168]. To evaluate the photocatalytic performance, typically apparent quantum efficiency (AQE) is determined from the ratio of the number of reacted electrons involved in the reaction to the number of injected photons [177, 180, 172]

$$\mathrm{AQE} = \frac{\text{Number of reacted electrons}}{\text{Number of incident photons}} \times 100 \tag{6}$$

Since direct determination of the number of reacted electrons via experimental methods is difficult, that is why the word apparent is used in AQE. For a single particle system for H_2 production, two electrons are involved whereas O_2 production costs four electrons. Therefore, AQE for H_2 and O_2 evolution can be determined using Eqs. (7) and (8), respectively.

$$\mathrm{AQE} = \frac{2 \times \text{number of } \mathrm{H_2} \text{ molecules}}{\text{Number of incident photons}} \tag{7}$$

$$\mathrm{AQE} = \frac{4 \times \text{number of } \mathrm{O_2} \text{ molecules}}{\text{Number of incident photons}} \tag{8}$$

In this experiment, time course of radiation of the photocatalytic activity is an important factor. The number of incident photons can be estimated from Eq. (9).

$$\text{Number of incident photon} = \frac{P \times \lambda \times A}{hc} \times t, \tag{9}$$

where P (W m^{-2}) is the optical power input measured by power meter (calibrated photodiode), λ (m) is the wavelength of the monochromatic light, A (m^2) is the irradiation area of the reactor, h (J s) is Planck's constant, c (m s^{-1}) is the speed of light, and t (s) is the illumination time. Number of H_2 or O_2 molecules can be determined from the gas evolution rate using the data from the gas chromatograph from Eq. (10).

$$\text{Number of evolved molecules} = n_{(\mathrm{H_2\,or\,O_2}),t} \times N_\mathrm{A}, \tag{10}$$

where n (mol) is the amount of H_2 or O_2 evolved during time t of the incident light exposure and N_A (mol^{-1}) is Avogadro's constant. Wavelength has a great effect on AQE; therefore, while reporting AQE it is important to mention the wavelength of the incident light for which the AQE is determined. AQEs of some of the photocatalysts are compared in Table 1.

5 Conclusion

Photocatalytic hydrogen production from water splitting utilizing semiconducting suspended particles is an important approach for future clean energy generation. The band gap of the semiconductor should be small enough so that it can absorb wide range of visible light. The water redox potential should be sandwiched in between the band gap of the semiconductor for overall water splitting. The semiconductor should be stable in aqueous media and the electronic structure should be such that the photogenerated charge carriers are produced without recombination and transported efficiently to the reactive sites for PWS. There are few semiconductors that are native visible light absorbers and are utilized for water splitting, and several promising efforts have been made to metal/nonmetal doping of UV light absorbing materials so that they can absorb visible light. In addition to use single semiconductor arrangement for overall water splitting, another important and efficient system utilized for PSW is called Z-scheme or Tandem configuration. In this setup tailored combination of two semiconductors with adequate band energy levels are utilized performing a tandem action. Depending on their energy levels, each semiconductor component performs one of the half reactions of the overall water splitting, i.e., one initiates water oxidation and the other promotes the water reduction. Once PWS is performed, it is very important for the researchers to rank their photocatalysts against other existing photocatalysts. Within this regard, performing STH efficiency and AQE measurements are important and standard protocols should be followed for these measurements. Despite the fact, that there are many photocatalysts approached higher STH efficiencies; yet, there exists a plenty of room to find efficient and stable photocatalysts for PSW for industrial point of view.

References

1. Kapdan IK, Kargi F (2006) Bio-hydrogen production from waste materials. Enzyme Microb Technol 38(5):569–582
2. Shi Z, Wen X, Guan Z, Cao D, Luo W, Zou Z (2015) Recent progress in photoelectrochemical water splitting for solar hydrogen production. Ann Phys 358:236–247
3. Tachibana Y, Vayssieres L, Durrant JR (2012) Artificial photosynthesis for solar water-splitting. Nat Photon 6(8):511–518
4. Frank Osterloh BAP (2011) Recent developments in solar water-splitting photocatalysis. MRS Bull 36(1):17–22
5. Fan Y, Li D, Deng M, Luo Y, Meng Q (2009) An overview on water splitting photocatalysts. Front Chem China 4(4):343–351
6. Fujishima A, Honda K (1972) Electrochemical photolysis of water at a semiconductor electrode. Nature 238(5358):37–38
7. Khan S, Zapata MJM, Baptista DL, Gonçalves RV, Fernandes JA, Dupont J, Santos MJL, Teixeira SR (2015) Effect of oxygen content on the photoelectrochemical activity of crystallographically preferred oriented porous Ta N nanotubes. J Phys Chem C 119(34): 19906–19914
8. Khan S, Zapata MJM, Pereira MB, Gonçalves RV, Strizik L, Dupont J, Santos MJL, Teixeira SR (2015) Structural, optical and photoelectrochemical characterizations of monoclinic Ta N thin films. Phys Chem Chem Phys 17(37):23952–23962
9. Tee SY, Win KY, Teo WS, Koh L-D, Liu S, Teng CP, et al (2017) Recent progress in energy-driven water splitting. Adv Sci 4(5):1600337
10. Acar C, Dincer I, Naterer GF (2016) Review of photocatalytic water-splitting methods for sustainable hydrogen production. Int J Energ Res 40(11):1449–1473
11. Osterloh FE (2013) Inorganic nanostructures for photoelectrochemical and photocatalytic water splitting. Chem Soc Rev 42(6):2294–2320
12. Barber J, Tran PD. From natural to artificial photosynthesis (2013). J Royal Soc Interface 10(81)
13. Teoh WY, Scott JA, Amal R (2012) Progress in heterogeneous photocatalysis: from classical radical chemistry to engineering nanomaterials and solar reactors. J Phys Chem Lett 3(5): 629–639
14. Navarro Yerga RM, Álvarez Galván MC, del Valle F, Villoria de la Mano JA, Fierro JLG (2009) Water splitting on semiconductor catalysts under visible-light irradiation. Chemsuschem 2(6):471–485
15. Kudo A, Miseki Y (2009) Heterogeneous photocatalyst materials for water splitting. Chem Soc Rev 38(1):253–278
16. Sivula K, Le Formal F, Grätzel M (2011) Solar water splitting: progress using hematite (α-Fe2O3) photoelectrodes. Chemsuschem 4(4):432–449

17. Grimes CA, Varghese OK, Ranjan S (2008) Light, water, hydrogen the solar generation of hydrogen by water photoelectrolysis. Springer, New York, p 517. ISBN: 978-0-387-33198-0 (Print) 978-0-387-68238-9 (Online)
18. Pinaud BA, Benck JD, Seitz LC, Forman AJ, Chen Z, Deutsch TG, et al (2013) Technical and economic feasibility of centralized facilities for solar hydrogen production via photocatalysis and photoelectrochemistry. Energ Environ Sci 6(7):1983–2002
19. Yuan Y-P, Ruan L-W, Barber J, Joachim Loo SC, Xue C (2014) Hetero-nanostructured suspended photocatalysts for solar-to-fuel conversion. Energ Environ Sci 7(12):3934–3951
20. Osterloh FE (2014) Boosting the efficiency of suspended photocatalysts for overall water splitting. J Phys Chem Lett 5(15):2510–2511
21. Wang Q, Hisatomi T, Jia Q, Tokudome H, Zhong M, Wang C, et al (2016) Scalable water splitting on particulate photocatalyst sheets with a solar-to-hydrogen energy conversion efficiency exceeding 1%. Nat Mater 15(6):611–615 (advance online publication)
22. Walter MG, Warren EL, McKone JR, Boettcher SW, Mi QX, Santori EA, et al (2011) Solar water splitting cells (vol 110, pg 6446, 2010). Chem Rev 111(9):5815
23. Chen X, Li C, Gratzel M, Kostecki R, Mao SS (2012) Nanomaterials for renewable energy production and storage. Chem Soc Rev 41(23):7909–7937
24. Serpone N, Lawless D, Khairutdinov R (1995) Size effects on the photophysical properties of colloidal anatase TiO_2 particles: size quantization versus direct transitions in this indirect semiconductor? J Phys Chem 99(45):16646–16654
25. Fan W, Zhang Q, Wang Y (2013) Semiconductor-based nanocomposites for photocatalytic H_2 production and CO_2 conversion. Phys Chem Chem Phys 15(8):2632–2649
26. Jafari T, Moharreri E, Amin A, Miao R, Song W, Suib S (2016) Photocatalytic water splitting—the untamed dream: a review of recent advances. Molecules 21(7):900
27. Yang J, Yan H, Zong X, Wen F, Liu M, Li C (2013) Roles of cocatalysts in semiconductor-based photocatalytic hydrogen production. Phil Trans Royal Soc A: Math, Phys Eng Sci 371(1996)
28. Ran J, Zhang J, Yu J, Jaroniec M, Qiao SZ (2014) Earth-abundant cocatalysts for semiconductor-based photocatalytic water splitting. Chem Soc Rev 43(22):7787–7812
29. Wang D, Hisatomi T, Takata T, Pan C, Katayama M, Kubota J et al (2013) Core/shell photocatalyst with spatially separated co-catalysts for efficient reduction and oxidation of water. Angew Chem Int Ed 52(43):11252–11256
30. Maeda K, Xiong A, Yoshinaga T, Ikeda T, Sakamoto N, Hisatomi T et al (2010) Photocatalytic overall water splitting promoted by two different cocatalysts for hydrogen and oxygen evolution under visible light. Angew Chem Int Ed 49(24):4096–4099
31. Wang D, Li R, Zhu J, Shi J, Han J, Zong X et al (2012) Photocatalytic water oxidation on $BiVO_4$ with the electrocatalyst as an oxidation cocatalyst: essential relations between electrocatalyst and photocatalyst. J Phys Chem C 116(8):5082–5089
32. Breault TM, Brancho JJ, Guo P, Bartlett BM (2013) Visible light water oxidation using a co-catalyst loaded anatase-structured ti1–(5x/4)Nb_xO_2–y–δNy Compound. Inorg Chem 52(16):9363–9368
33. Ma SSK, Maeda K, Abe R, Domen K (2012) Visible-light-driven nonsacrificial water oxidation over tungsten trioxide powder modified with two different cocatalysts. Energ Environ Sci 5(8):8390–8397
34. Meekins BH, Kamat PV (2011) Role of water oxidation catalyst IrO_2 in shuttling photogenerated holes across TiO_2 interface. J Phys Chem Lett 2(18):2304–2310
35. Asai R, Nemoto H, Jia Q, Saito K, Iwase A, Kudo A (2014) A visible light responsive rhodium and antimony-codoped $SrTiO_3$ powdered photocatalyst loaded with an IrO_2 cocatalyst for solar water splitting. Chem Commun 50(19):2543–2546
36. Ma B, Yang J, Han H, Wang J, Zhang X, Li C (2010) Enhancement of photocatalytic water oxidation activity on IrO_x–ZnO/Zn_2–xGeO_4–x–3yN_2y catalyst with the solid solution phase junction. J Phys Chem C 114(29):12818–12822
37. Ai G, Mo R, Li H, Zhong J (2015) Cobalt phosphate modified TiO_2 nanowire arrays as co-catalysts for solar water splitting. Nanoscale 7(15):6722–6728

38. Kraeutler B, Bard AJ (1978) Heterogeneous photocatalytic preparation of supported catalysts. Photodeposition of platinum on titanium dioxide powder and other substrates. J Am Chem Soc 100(13):4317–4318
39. López R, Gómez R (2011) Band-gap energy estimation from diffuse reflectance measurements on sol–gel and commercial TiO_2: a comparative study. J Sol-Gel Sci Technol 61(1):1–7
40. Hisatomi T, Kubota J, Domen K (2014) Recent advances in semiconductors for photocatalytic and photoelectrochemical water splitting. Chem Soc Rev 43(22):7520–7535
41. Takata T, Tanaka A, Hara M, Kondo JN, Domen K (1998) Recent progress of photocatalysts for overall water splitting. Catal Today 44(1–4):17–26
42. Bak T, Nowotny J, Rekas M, Sorrell CC (2002) Photo-electrochemical hydrogen generation from water using solar energy. materials-related aspects. Int J Hydrogen Energy 27(10): 991–1022
43. Chen X, Mao SS (2007) Titanium dioxide nanomaterials: synthesis, properties, modifications, and applications. Chem Rev 107(7):2891–2959
44. Ni M, Leung MKH, Leung DYC, Sumathy K (2007) A review and recent developments in photocatalytic water-splitting using TiO_2 for hydrogen production. Renew Sustain Energy Rev 11(3):401–425
45. Hernandez-Alonso MD, Fresno F, Suarez S, Coronado JM (2009) Development of alternative photocatalysts to TiO_2: challenges and opportunities. Energy Environ Sci 2(12): 1231–1257
46. Chen X, Shen S, Guo L, Mao SS (2010) Semiconductor-based photocatalytic hydrogen generation. Chem Rev 110(11):6503–6570
47. Maeda K, Domen K (2010) Photocatalytic water splitting: recent progress and future challenges. J Phys Chem Lett 1(18):2655–2661
48. Walter MG, Warren EL, McKone JR, Boettcher SW, Mi Q, Santori EA et al (2010) Solar water splitting cells. Chem Rev 110(11):6446–6473
49. Pelaez M, Nolan NT, Pillai SC, Seery MK, Falaras P, Kontos AG et al (2012) A review on the visible light active titanium dioxide photocatalysts for environmental applications. Appl Catal B 125:331–349
50. Osterloh FE (2013) Inorganic nanostructures for photoelectrochemical and photocatalytic water splitting. Chem Soc Rev 42(6):2294–2320
51. Sivula K, van de Krol R. Semiconducting materials for photoelectrochemical energy conversion. Nat Rev Mat 16010
52. Gonçalves RV, Migowski P, Wender H, Eberhardt D, Weibel DE, Sonaglio FC et al (2012) Ta_2O_5 nanotubes obtained by anodization: effect of thermal treatment on the photocatalytic activity for hydrogen production. J Phys Chem C 116(26):14022–14030
53. Li Z, Liu J, Li J, Shen J (2012) Template free synthesis of crystallized nanoporous F-Ta_2O_5 spheres for effective photocatalytic hydrogen production. Nanoscale 4(13):3867–3870
54. Noda Y, Lee B, Domen K, Kondo JN (2008) Synthesis of crystallized mesoporous tantalum oxide and its photocatalytic activity for overall water splitting under ultraviolet light irradiation. Chem Mater 20(16):5361–5367
55. Lin S, Shi L, Yoshida H, Li M, Zou X (2013) Synthesis of hollow spherical tantalum oxide nanoparticles and their photocatalytic activity for hydrogen production. J Solid State Chem 199:15–20
56. Cherevan AS, Gebhardt P, Shearer CJ, Matsukawa M, Domen K, Eder D (2014) Interface engineering in nanocarbon-Ta_2O_5 hybrid photocatalysts. Energy Environ Sci 7(2):791–796
57. Sato S, White JM (1980) Photodecomposition of water over Pt/TiO_2 catalysts. Chem Phys Lett 72(1):83–86
58. Munuera G, Gonzalez-Elipe AR, Fernandez A, Malet P, Espinos JP (1989) Spectroscopic characterisation and photochemical behaviour of a titanium hydroxyperoxo compound. J Chem Soc, Faraday Trans 1: Phys Chem Condens Phases 85(6):1279–1290
59. Duonghong D, Borgarello E, Graetzel M (1981) Dynamics of light-induced water cleavage in colloidal systems. J Am Chem Soc 103(16):4685–4690

60. Galińska A, Walendziewski J (2005) Photocatalytic water splitting over Pt–TiO_2 in the presence of sacrificial reagents. Energy Fuels 19(3):1143–1147
61. Wender H, Feil AF, Diaz LB, Ribeiro CS, Machado GJ, Migowski P et al (2011) Self-organized TiO_2 nanotube arrays: synthesis by anodization in an ionic liquid and assessment of photocatalytic properties. ACS Appl Mater Interfaces 3(4):1359–1365
62. Kamat PV (2010) Graphene-based nanoarchitectures. anchoring semiconductor and metal nanoparticles on a two-dimensional carbon support. J Phys Chem Lett 1(2):520–527
63. Zhang H, Lv X, Li Y, Wang Y, Li J (2010) P25-graphene composite as a high performance photocatalyst. ACS Nano 4(1):380–386
64. Xiang Q, Yu J, Jaroniec M (2011) Enhanced photocatalytic H_2-production activity of graphene-modified titania nanosheets. Nanoscale 3(9):3670–3678
65. Zhang XY, Li HP, Cui XL (2009) Preparation and photocatalytic activity for hydrogen evolution of TiO_2/graphene sheets composite. Chin J Inorg Chem 25(11):1903–1907
66. Zhang X-Y, Li H-P, Cui X-L, Lin Y (2010) Graphene/TiO_2 nanocomposites: synthesis, characterization and application in hydrogen evolution from water photocatalytic splitting. J Mater Chem 20(14):2801
67. Yeh T-F, Syu J-M, Cheng C, Chang T-H, Teng H (2010) Graphite oxide as a photocatalyst for hydrogen production from water. Adv Func Mater 20(14):2255–2262
68. Babu SG, Vinoth R, Praveen Kumar D, Shankar MV, Chou H-L, Vinodgopal K et al (2015) Influence of electron storing, transferring and shuttling assets of reduced graphene oxide at the interfacial copper doped TiO_2 p–n heterojunction for increased hydrogen production. Nanoscale 7(17):7849–7857
69. Kudo A, Niishiro R, Iwase A, Kato H (2007) Effects of doping of metal cations on morphology, activity, and visible light response of photocatalysts. Chem Phys 339(1–3): 104–110
70. Hu C-C, Teng H (2010) Structural features of p-type semiconducting NiO as a co-catalyst for photocatalytic water splitting. J Catal 272(1):1–8
71. Zhu Y, Salvador PA, Rohrer GS (2016) Controlling the relative areas of photocathodic and photoanodic terraces on the $SrTiO_3$(111) Surface. Chem Mater 28(14):5155–5162
72. Wagner FT, Somorjai GA (1980) Photocatalytic and photoelectrochemical hydrogen production on strontium titanate single crystals. J Am Chem Soc 102(17):5494–5502
73. van Benthem K, Elsässer C, French RH (2001) Bulk electronic structure of $SrTiO_3$: experiment and theory. J Appl Phys 90(12):6156–6164
74. Townsend TK, Browning ND, Osterloh FE (2012) Nanoscale strontium titanate photocatalysts for overall water splitting. ACS Nano 6(8):7420–7426
75. Townsend TK, Browning ND, Osterloh FE (2012) Overall photocatalytic water splitting with NiOx–$SrTiO_3$—a revised mechanism. Energy Environ Sci 5(11):9543
76. Mu L, Zhao Y, Li A, Wang S, Wang Z, Yang J et al (2016) Enhancing charge separation on high symmetry $SrTiO_3$exposed with anisotropic facets for photocatalytic water splitting. Energy Environ Sci 9(7):2463–2469
77. Wender H, Gonçalves RV, Dias CSB, Zapata MJM, Zagonel LF, Mendonça EC et al (2013) Photocatalytic hydrogen production of $Co(OH)_2$ nanoparticle-coated α-Fe_2O_3 nanorings. Nanoscale 5(19):9310
78. Darwent JR, Mills A. Photo-oxidation of water sensitized by WO_3 powder (1982) J Chem Soc, Faraday Trans 2 78(2):359
79. Erbs W, Desilvestro J, Borgarello E, Graetzel M (1984) Visible-light-induced oxygen generation from aqueous dispersions of tungsten(VI) oxide. J Phys Chem 88(18):4001–4006
80. Shimodaira Y, Kato H, Kobayashi H, Kudo A (2006) Photophysical properties and photocatalytic activities of bismuth molybdates under visible light irradiation. J Phys Chem B 110(36):17790–17797
81. Kudo A, Hijii S (1999) H_2 or O_2 evolution from aqueous solutions on layered oxide photocatalysts consisting of Bi3+ with 6s2 configuration and d0 transition metal ions. Chem Lett 28(10):1103–1104

82. Wang D, Tang J, Zou Z, Ye J (2005) Photophysical and photocatalytic properties of a new series of visible-light-driven photocatalysts $M_3V_2O_8$ (M=Mg, Ni, Zn). Chem Mater 17(20): 5177–5182
83. Radha R, Srinivasan A, Manimuthu P, Balakumar S (2015) Tailored sunlight driven nano-photocatalyst: bismuth iron tungstate ($BiFeWO_6$). J Mater Chem C 3(39):10285–10292
84. Luan J, Guo N, Chen B (2014) Hydrogen production with Ga_2BiSbO_7, Fe_2BiSbO_7 and Gd_2BiSbO_7 as photocatalysts under visible light irradiation. Int J Hydrogen Energy 39 (3):1228–1236
85. Cao J, Kako T, Li P, Ouyang S, Ye J (2011) Fabrication of p-type $CaFe_2O_4$ nanofilms for photoelectrochemical hydrogen generation. Electrochem Commun 13(3):275–278
86. Dom R, Subasri R, Hebalkar NY, Chary AS, Borse PH (2012) Synthesis of a hydrogen producing nanocrystalline $ZnFe_2O_4$ visible light photocatalyst using a rapid microwave irradiation method. RSC Adv 2(33):12782
87. Kudo A, Miseki Y (2009) Heterogeneous photocatalyst materials for water splitting. Chem Soc Rev 38(1):253–278
88. Klahr BM, Hamann TW (2011) Current and voltage limiting processes in thin film hematite electrodes. J Phys Chem C 115(16):8393–8399
89. Klahr B, Gimenez S, Fabregat-Santiago F, Bisquert J, Hamann TW (2012) Electrochemical and photoelectrochemical investigation of water oxidation with hematite electrodes. Energ Environ Sci. 5(6):7626–7636
90. Klahr BM, Martinson ABF, Hamann TW (2011) Photoelectrochemical investigation of ultrathin film iron oxide solar cells prepared by atomic layer deposition. Langmuir 27(1): 461–468
91. Kay A, Cesar I, Grätzel M (2006) New benchmark for water photooxidation by nanostructured α-Fe_2O_3 films. J Am Chem Soc 128(49):15714–15721
92. Mangrulkar PA, Polshettiwar V, Labhsetwar NK, Varma RS, Rayalu SS (2012) Nano-ferrites for water splitting: unprecedented high photocatalytic hydrogen production under visible light. Nanoscale. 4(16):5202
93. Murugesan S, Huda MN, Yan Y, Al-Jassim MM, Subramanian V (2010) Band-engineered bismuth titanate pyrochlores for visible light photocatalysis. J Phys Chem C 114(23):10598–10605
94. Zhou D, Yang H, Tu Y, Tian Y, Cai Y, Hu Z, et al (2016) In situ fabrication of $Bi_2Ti_2O_7$/TiO_2 heterostructure submicron fibers for enhanced photocatalytic activity. Nanoscale Res Lett 11(1)
95. McInnes A, Sagu JS, Wijayantha KGU (2014) Fabrication and photoelectrochemical studies of $Bi_2Ti_2O_7$ pyrochlore thin films by aerosol assisted chemical vapour deposition. Mater Lett 137:214–217
96. Kudo A, Omori K, Kato H (1999) A novel aqueous process for preparation of crystal form-controlled and highly crystalline $BiVO_4$ powder from layered vanadates at room temperature and its photocatalytic and photophysical properties. J Am Chem Soc 121 (49):11459–11467
97. Seabold JA, Choi K-S (2012) Efficient and stable photo-oxidation of water by a bismuth vanadate photoanode coupled with an iron oxyhydroxide oxygen evolution catalyst. J Am Chem Soc 134(4):2186–2192
98. Sun S, Wang W, Li D, Zhang L, Jiang D (2014) Solar light driven pure water splitting on quantum sized $BiVO_4$ without any cocatalyst. ACS Catal 4(10):3498–3503
99. Maeda K, Teramura K, Saito N, Inoue Y, Kobayashi H, Domen K (2006) Overall water splitting using (oxy)nitride photocatalysts. Pure Appl Chem 78(12)
100. Moriya Y, Takata T, Domen K (2013) Recent progress in the development of (oxy)nitride photocatalysts for water splitting under visible-light irradiation. Coord Chem Rev 257(13–14):1957–1969
101. Takata T, Pan C, Domen K (2016) Recent progress in oxynitride photocatalysts for visible-light-driven water splitting. Sci Technol Adv Mater 16(3):033506

102. Bamwenda GR, Sayama K, Arakawa H (1999) The effect of selected reaction parameters on the photoproduction of oxygen and hydrogen from a WO_3–Fe^{2+}–Fe^{3+} aqueous suspension. J Photochem Photobiol, A 122(3):175–183
103. Tanaka A, Hashimoto K, Kominami H (2014) Visible-light-induced hydrogen and oxygen formation over Pt/Au/WO_3 photocatalyst utilizing two types of photoabsorption due to surface plasmon resonance and band-gap excitation. J Am Chem Soc 136(2):586–589
104. Zhang LJ, Li S, Liu BK, Wang DJ, Xie TF (2014) Highly efficient CdS/WO_3 photocatalysts: Z-scheme photocatalytic mechanism for their enhanced photocatalytic H_2 evolution under visible light. ACS Catal 4(10):3724–3729
105. Juan M, Coronado FF, Hernández-Alonso MD, Portela R (2013) Design of advanced photocatalytic materials for energy and environmental applications. Springer, Berlin
106. Samadi M, Zirak M, Naseri A, Khorashadizade E, Moshfegh AZ (2016) Recent progress on doped ZnO nanostructures for visible-light photocatalysis. Thin Solid Films 605:2–19
107. Choi W, Termin A, Hoffmann MR (1994) The role of metal ion dopants in quantum-sized TiO_2: correlation between photoreactivity and charge carrier recombination dynamics. J Phys Chem 98(51):13669–13679
108. H-y Lin, C-y Shih (2016) Efficient one-pot microwave-assisted hydrothermal synthesis of M (M=Cr, Ni, Cu, Nb) and nitrogen co-doped TiO_2 for hydrogen production by photocatalytic water splitting. J Mol Catal A: Chem 411:128–137
109. Dholam R, Patel N, Adami M, Miotello A (2009) Hydrogen production by photocatalytic water-splitting using Cr- or Fe-doped TiO_2 composite thin films photocatalyst. Int J Hydrogen Energy 34(13):5337–5346
110. Selcuk MZ, Boroglu MS, Boz I (2012) Hydrogen production by photocatalytic water-splitting using nitrogen and metal co-doped TiO_2 powder photocatalyst. React Kinet, Mech Catal 106(2):313–324
111. Sun T, Fan J, Liu E, Liu L, Wang Y, Dai H et al (2012) Fe and Ni co-doped TiO_2 nanoparticles prepared by alcohol-thermal method: application in hydrogen evolution by water splitting under visible light irradiation. Powder Technol 228:210–218
112. Banerjee B, Amoli V, Maurya A, Sinha AK, Bhaumik A (2015) Green synthesis of Pt-doped TiO_2 nanocrystals with exposed (001) facets and mesoscopic void space for photo-splitting of water under solar irradiation. Nanoscale. 7(23):10504–10512
113. Jiang Z, Liu D, Jiang D, Wei W, Qian K, Chen M et al (2014) Bamboo leaf-assisted formation of carbon/nitrogen co-doped anatase TiO_2 modified with silver and graphitic carbon nitride: novel and green synthesis and cooperative photocatalytic activity. Dalton Trans 43(36):13792
114. Serpone N (2006) Is the band gap of pristine TiO_2 narrowed by anion- and cation-doping of titanium dioxide in second-generation photocatalysts? J Phys Chem B 110(48):24287–24293
115. Su J, Zou X, Chen J-S (2014) Self-modification of titanium dioxide materials by Ti^{3+} and/or oxygen vacancies: new insights into defect chemistry of metal oxides. RSC Adv 4 (27):13979
116. Fu G, Zhou P, Zhao M, Zhu W, Yan S, Yu T et al (2015) Carbon coating stabilized Ti^{3+}-doped TiO_2 for photocatalytic hydrogen generation under visible light irradiation. Dalton Trans 44(28):12812–12817
117. Qiu B, Zhou Y, Ma Y, Yang X, Sheng W, Xing M et al (2015) Facile synthesis of the Ti^{3+} self-doped TiO_2-graphene nanosheet composites with enhanced photocatalysis. Sci Rep 5:8591
118. Serra M, Baldovi HG, Albarracin F, Garcia H (2016) Visible light photocatalytic activity for hydrogen production from water–methanol mixtures of open-framework V-doped mixed-valence titanium phosphate. Appl Catal B 183:159–167
119. Strataki N, Antoniadou M, Dracopoulos V, Lianos P (2010) Visible-light photocatalytic hydrogen production from ethanol–water mixtures using a Pt–CdS–TiO_2 photocatalyst. Catal Today 151(1–2):53–57

120. Jang J, Kim H, Joshi U, Jang J, Lee J (2008) Fabrication of CdS nanowires decorated with TiO_2 nanoparticles for photocatalytic hydrogen production under visible light irradiation. Int J Hydrogen Energy 33(21):5975–5980
121. Yun HJ, Lee H, Kim ND, Lee DM, Yu S, Yi J (2011) A combination of two visible-light responsive photocatalysts for achieving the z-scheme in the solid state. ACS Nano 5 (5):4084–4090
122. Yan J, Wu H, Chen H, Zhang Y, Zhang F, Liu SF (2016) Fabrication of TiO_2/C_3N_4 heterostructure for enhanced photocatalytic Z-scheme overall water splitting. Appl Catal B 191:130–137
123. Jian Z, Huang S, Cao Y, Zhang Y (2016) Hydrothermal preparation and characterization of $TiO_2/BiVO_4$ composite catalyst and its photolysis of water to produce hydrogen. Photochem Photobiol 92(3):363–370
124. Li J, Yuan H, Zhu Z (2015) Fabrication of $Cu_2O/Au/BiPO_4$ Z-scheme photocatalyst to improve the photocatalytic activity under solar light. J Mol Catal A: Chem 410:133–139
125. Jia X, Tahir M, Pan L, Huang Z-F, Zhang X, Wang L et al (2016) Direct Z-scheme composite of CdS and oxygen-defected $CdWO_4$: an efficient visible-light-driven photocatalyst for hydrogen evolution. Appl Catal B 198:154–161
126. Li H, Yu H, Quan X, Chen S, Zhang Y (2016) Uncovering the key role of the fermi level of the electron mediator in a Z-Scheme photocatalyst by detecting the charge transfer process of WO_3-metal-gC_3N_4 (Metal=Cu, Ag, Au). ACS Appl Mater Interfaces 8(3):2111–2119
127. Zhang X, Zhang L, Xie T, Wang D (2009) Low-temperature synthesis and high visible-light-induced photocatalytic activity of $BiOI/TiO_2$ heterostructures. J Phys Chem C 113(17):7371–7378
128. Zhu W, Han D, Niu L, Wu T, Guan H (2016) Z-scheme $Si/MgTiO_3$ porous heterostructures: noble metal and sacrificial agent free photocatalytic hydrogen evolution. Int J Hydrogen Energy 41(33):14713–14720
129. Pan Z, Hisatomi T, Wang Q, Chen S, Nakabayashi M, Shibata N et al (2016) Photocatalyst sheets composed of particulate $LaMg1/3Ta2/3O_2N$ and Mo-Doped $BiVO_4$ for Z-scheme water splitting under visible light. ACS Catalysis 6(10):7188–7196
130. Maeda K, Lu D, Domen K (2013) Solar-driven z-scheme water splitting using modified $BaZrO_3$–$BaTaO_2N$ solid solutions as photocatalysts. ACS Catalysis. 3(5):1026–1033
131. Xu L, Shi W, Guan J (2012) Preparation of crystallized mesoporous CdS/Ta_2O_5 composite assisted by silica reinforcement for visible light photocatalytic hydrogen evolution. Catal Commun 25:54–58
132. Xu L, Guan J, Gao L, Sun Z (2011) Preparation of heterostructured mesoporous In_2O_3/Ta_2O_5 nanocomposites with enhanced photocatalytic activity for hydrogen evolution. Catal Commun 12(6):548–552
133. Xu L, Guan J, Shi W (2012) Enhanced interfacial charge transfer and visible photocatalytic activity for hydrogen evolution from a Ta_2O_5-based mesoporous composite by the incorporation of quantum-sized CdS. ChemCatChem 4(9):1353–1359
134. Agegnehu AK, Pan C-J, Tsai M-C, Rick J, Su W-N, Lee J-F et al (2016) Visible light responsive noble metal-free nanocomposite of V-doped TiO_2 nanorod with highly reduced graphene oxide for enhanced solar H2 production. Int J Hydrogen Energy 41(16):6752–6762
135. Mukherji A, Seger B, Lu GQ, Wang L (2011) Nitrogen doped $Sr_2Ta_2O_7$ coupled with graphene sheets as photocatalysts for increased photocatalytic hydrogen production. ACS Nano 5(5):3483–3492
136. Adhikari SP, Hood ZD, More KL, Ivanov I, Zhang L, Gross M et al (2015) Visible light assisted photocatalytic hydrogen generation by Ta_2O_5/Bi_2O_3, $TaON/Bi_2O_3$, and Ta_3N_5/Bi_2O_3composites. RSC Adv 5(68):54998–55005
137. Naik B, Martha S, Parida KM (2011) Facile fabrication of $Bi2O_3/TiO_{2-x}N_x$ nanocomposites for excellent visible light driven photocatalytic hydrogen evolution. Int J Hydrogen Energy 36(4):2794–2802

138. Fujishima A, Honda K (1972) Electrochemical photolysis of water at a semiconductor electrode. Nature 238:37
139. Osterloh FE (2013) Inorganic nanostructures for photoelectrochemical and photocatalytic water splitting. Chem Soc Rev 42(6):2294–2320
140. Maeda K (2013) Z-scheme water splitting using two different semiconductor photocatalysts. ACS Catal 3:1486–1503
141. Kudo A, Miseki Y (2009) Heterogeneous photocatalyst materials for water splitting. Chem Soc Rev 38(1):253–278
142. Radhakrishnan AN, Rao PP, Linsa KS, Deepa M, Koshy P (2011) Influence of disorder-to-order transition on lattice thermal expansion and oxide ion conductivity in (Ca (x)Gd(1−x))(2)(Zr(1−x)M(x))$_2$O$_7$ pyrochlore solid solutions. Dalton Trans 40(15):3839–3848
143. Sasaki Y, Kato H, Kudo A (2013) [Co(bpy)3](3+/2+) and [Co(phen)3](3+/2+) electron mediators for overall water splitting under sunlight irradiation using Z-scheme photocatalyst system. J Am Chem Soc 135:5441–5449
144. Sasaki Y, Nemoto H, Saito K, Kudo A (2009) Solar water splitting using powdered photocatalysts driven by Z-schematic interparticle electron transfer without an electron mediator. J Phys Chem C 113
145. Fujihara K (1998) TO, M. Matsumura. Splitting of water by electrochemical combination of two photocatalytic reactions on particles. J Chem Soc, Faraday Trans 94:3705–3709
146. Kato H, Hori M, Konta R, Shimodaira Y, Kudo A (2004) Construction of Z-scheme type heterogeneous photocatalysis systems for water splitting into H_2 and O_2 under visible light irradia. Chem Lett 33:1348–1349
147. Abe R, Sayama K, Domen K, Arakawa H (2001) A new type of water splitting system composed of two different TiO_2 photocatalysts (anatase, rutile) and IO^{3-}/I^- shuttle redox mediator. Chem Phys Lett 344:339–344
148. Wang Q, Hisatomi T, Jia Q, Tokudome H, Zhong M, Wang C (2016) Scalable water splitting on particulate photocatalyst sheets with a solar-to-hydrogen energy conversion efficiency exceeding 1. Nat Mater 15(6):611–615
149. Iwase A, Ng YH, Ishiguro Y, Kudo A, Amal R (2011) Reduced graphene oxide as a solid-state electron mediator in Z-scheme photocatalytic water splitting under visible light. J Am Chem Soc 133:11054–11057
150. Sayama K, Yoshida R, Kusama H, Okabe K, Abe Y, Arakawa H (1997) Photocatalytic decomposition of water into H_2 and O_2 by a two-step photoexcitation reaction using a WO_3 suspension catalyst and an Fe^{3+}:Fe^{2+} redox system. Chem Phys Lett 277:387–391
151. Tennakone K, Tantrigoda R, Abeysinghe S, Punchihewa S, Fernando CA (1990) Water photolysis via reversible oxidation and reduction between MnO_2 and $MnO_4{}^{2-}$. J Photochem Photobiol, A-Chem 52:43–46
152. Sayama K, Mukasa K, Abe R, Abe Y, Arakawa H (2001) Arakawa. Stoichiometric water splitting into H_2 and O_2 using a mixture of two different photocatalysts and an IO^{3-}/I^- shuttle redox mediator under visible light irradiation. Chem Commun 2416–2417
153. Maeda K, Higashi M, Lu D, Abe R, Domen K (2010) Efficient nonsacrificial water splitting through two-step photoexcitation by visible light using a modified oxynitride as a hydrogen evolution photocatalyst. J Am Chem Soc 132:5858–5868
154. Yan J, Wu H, Chen H, Zhang Y, Zhang F, Liu SF (2016) Fabrication of TiO_2/C_3N_4 heterostructure for enhanced photocatalytic Z-scheme overall water splitting. Appl Catal B-Environ 191:130–137
155. Jia Q, Iwase A, Kudo A (2014) $BiVO_4$–Ru/$SrTiO_3$:Rh composite Z-scheme photocatalyst for solar water splitting. Chem Sci 5:1513–1519
156. Wang Q, Hisatomi T, Suzuki Y, Pan Z, Seo J, Katayama M et al (2017) Particulate photocatalyst sheets based on carbon conductor layer for efficient Z-scheme pure-water splitting at ambient pressure. J Am Chem Soc 139(4):1675–1683
157. Pan Z, Hisatomi T, Wang Q, Chen S, Iwase A, Nakabayashi M, Shibata N, Takata T, Katayama M, Minegishi T, Kudo A, Domen K (2016) Photoreduced graphene oxide as a

conductive binder to improve the water splitting activity of photocatalyst sheets. Adv Funct Mater 26:7011–7019
158. Pan Z, Hisatomi T, Wang Q, Chen S, Iwase A, Nakabayashi M, Shibata N, Takata T, Katayama M, Minegishi T, Kudo A, Domen K (2016) Photocatalyst sheets composed of particulate LaMg1:3Ta$_2$:3O$_2$N and Mo-doped $BiVO_4$ for Z-scheme water splitting under visible light. ACS Catal 6:7188–7196
159. Doscher H, Young JL, Geisz JF, Turner JA, Deutsch TG (2016) Solar-to-hydrogen efficiency: shining light on photoelectrochemical device performance. Energy Environ Sci 9 (1):74–80
160. Murphy AB, Barnes PRF, Randeniya LK, Plumb IC, Grey IE, Horne MD et al (2006) Efficiency of solar water splitting using semiconductor electrodes. Int J Hydrogen Energy 31 (14):1999–2017
161. Harald M, Andreas I, Robert K, Willem Z, Heinz AO, Ewan DD (2005) Spectral mismatch in calibration of photovoltaic reference devices by global sunlight method. Meas Sci Technol 16(6):1250
162. Kato H, Sasaki Y, Shirakura N, Kudo A (2013) Synthesis of highly active rhodium-doped $SrTiO_3$ powders in Z-scheme systems for visible-light-driven photocatalytic overall water splitting. J Mater Chem A. 1(39):12327–12333
163. Sasaki Y, Kato H, Kudo A (2013) [Co(bpy)3]3+/2+ and [Co(phen)3]3+/2+ electron mediators for overall water splitting under sunlight irradiation using Z-scheme photocatalyst system. J Am Chem Soc 135(14):5441–5449
164. Wang Q, Hisatomi T, Ma SSK, Li Y, Domen K (2014) Core/shell structured La- and Rh-codoped $SrTiO_3$ as a hydrogen evolution photocatalyst in Z-scheme overall water splitting under visible light irradiation. Chem Mater 26(14):4144–4150
165. Wang Q, Hisatomi T, Suzuki Y, Pan Z, Seo J, Katayama M et al (2017) Particulate photocatalyst sheets based on carbon conductor layer for efficient Z-scheme pure-water splitting at ambient pressure. J Am Chem Soc 139(4):1675–1683
166. Youn DH, Jang J-W, Kim JY, Jang JS, Choi SH, Lee JS (2014) Fabrication of graphene-based electrode in less than a minute through hybrid microwave annealing. Sci Rep 4:5492
167. Li Y, Zhang L, Torres-Pardo A, González-Calbet JM, Ma Y, Oleynikov P et al (2013) Cobalt phosphate-modified barium-doped tantalum nitride nanorod photoanode with 1.5% solar energy conversion efficiency. Nat Commun 4:2566
168. Li X, Yu J, Low J, Fang Y, Xiao J, Chen X (2015) Engineering heterogeneous semiconductors for solar water splitting. J Mat Chem A 3(6):2485–2534
169. Doscher H, Geisz JF, Deutsch TG, Turner JA (2014) Sunlight absorption in water—efficiency and design implications for photoelectrochemical devices. Energy Environ Sci 7 (9):2951–2956
170. Liao L, Zhang Q, Su Z, Zhao Z, Wang Y, Li Y et al (2014) Efficient solar water-splitting using a nanocrystalline CoO photocatalyst. Nat Nano. 9(1):69–73
171. Liu J, Liu Y, Liu N, Han Y, Zhang X, Huang H et al (2015) Metal-free efficient photocatalyst for stable visible water splitting via a two-electron pathway. Science 347 (6225):970–974
172. Escobedo Salas S, Serrano Rosales B, de Lasa H (2013) Quantum yield with platinum modified TiO_2 photocatalyst for hydrogen production. Appl Catal B: Environ 140–141:523–536
173. Souza VS, Scholten JD, Weibel DE, Eberhardt D, Baptista DL, Teixeira SR et al (2016) Hybrid tantalum oxide nanoparticles from the hydrolysis of imidazolium tantalate ionic liquids: efficient catalysts for hydrogen generation from ethanol/water solutions. J Mater Chem A. 4(19):7469–7475
174. Chen S, Shen S, Liu G, Qi Y, Zhang F, Li C (2015) Interface engineering of a CoO_x/Ta_3N_5 photocatalyst for unprecedented water oxidation performance under visible-light-irradiation. Angew Chem Int Ed 54(10):3047–3051

175. Maeda K, Higashi M, Lu D, Abe R, Domen K (2010) Efficient nonsacrificial water splitting through two-step photoexcitation by visible light using a modified oxynitride as a hydrogen evolution photocatalyst. J Am Chem Soc 132(16):5858–5868
176. Martin DJ, Qiu K, Shevlin SA, Handoko AD, Chen X, Guo Z et al (2014) Highly efficient photocatalytic H_2 evolution from water using visible light and structure-controlled graphitic carbon nitride. Angew Chem Int Ed 53(35):9240–9245
177. Li J, Wu N (2015) Semiconductor-based photocatalysts and photoelectrochemical cells for solar fuel generation: a review. Catal Sci Technol 5(3):1360–1384
178. May MM, Lewerenz H-J, Lackner D, Dimroth F, Hannappel T (2015) Efficient direct solar-to-hydrogen conversion by in situ interface transformation of a tandem structure. Nat Commun 6:8286
179. Khan S, Santos MJL, Malfatti CF, Dupont J, Teixeira SR (2016) Pristine Ta_3N_5 nanotubes: trap-driven high external biasing perspective in semiconductor/electrolyte interfaces. Chem Eur J 22(51):18501–18511
180. Chen XB, Shen SH, Guo LJ, Mao SS (2010) Semiconductor-based photocatalytic hydrogen generation. Chem Rev 110(11):6503–6570

Latest Advances on the Columnar Nanostructure for Solar Water Splitting

Flavio L Souza and Edson R Leite

1 Introduction

Energy from the sun can provide sufficient power for all our energy needs, and a potentially efficient route to storing this energy is to convert sunlight into chemical energy in the form of chemical bonds which is a form of an artificial photosynthesis process. Considering the abundance of H_2O on the planet, water splitting is a natural pathway for artificial photosynthesis. The pioneer work of Fujishima and Honda [1] disseminated the worldwide research focused on the sunlight-induced water-splitting process into hydrogen as a clean and renewable energy source. In their classical report, the authors demonstrated that the water splitting can be induced by light using TiO_2 semiconductor as a photoanode. As in conventional water electrolysis, O_2 evolution occurs at the anode, H_2 evolution occurs at the cathode, and an aqueous electrolyte completes the current loop. One or both electrodes can be photoactive semiconductors, where a space charge (depletion) layer is formed at the semiconductor/liquid junction (SCLJ). Under light irradiation a pair of electron–holes is generated. The photogenerated carriers are separated by the space charge field, and the minority carriers (holes for an n-type photoanode) travel to the SCLJ to perform one-half of the water-splitting reaction. Figure 1a

F.L. Souza (✉)
Centro de Ciências Naturais e Humanas, Universidade Federal do ABC—UFABC,
Santo André, São Paulo, Brazil
e-mail: flavio.souza@ufabc.edu.br

E.R. Leite (✉)
Departamento de Química, Universidade Federal de São Carlos,
São Carlos, São Paulo, Brazil
e-mail: leite@pq.cnpq.br

F.L. Souza and E.R. Leite (eds.), *Nanoenergy*,
https://doi.org/10.1007/978-3-319-62800-4_4

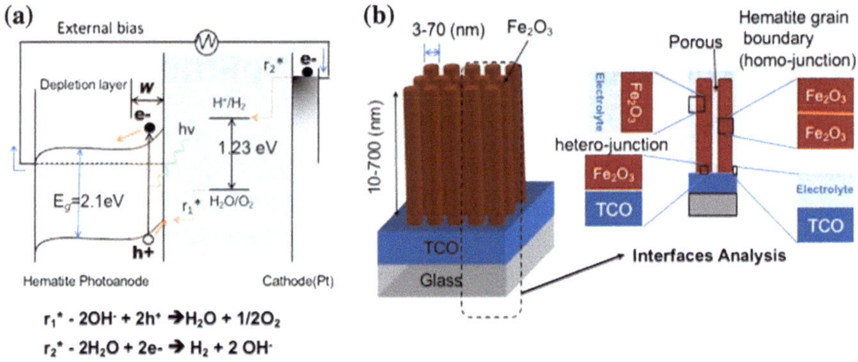

Fig. 1 **a** Schematic representation of the n-type photoanode (hematite) energy diagram and reactions related to water splitting; **b** illustration of porous hematite thin films where we can observe several kinds of interfaces (solid–solid and solid–liquid). Reprinted from Ref. [15]. Copyright 2016, with permission from Royal Society Chemistry

illustrates the energy diagram of an n-type photoanode and the reactions related to the water splitting. We can write that the photocurrent density (J_{ph}) is given by

$$J_{ph} = J_{abs} \cdot \eta_{sep} \cdot \eta_{cat}, \tag{1}$$

where J_{abs} is the photoabsorption rate expressed as current density, η_{sep} is the carrier separation efficiency and η_{Cat} is the catalytic efficiency for water oxidation. Based on Eq. (1), we can notice that to improve the photocurrent density it is necessary to increase J_{abs}, η_{sep}, and η_{cat}. However, this is not a simple task. For instance, increasing the film thickness lead to an increase in the J_{abs}. However, an increase in the film thickness can result in a decrease of the η_{sep}, in special for semiconductors where the hole diffusion length (L_P) is much shorter than the depletion layer width (W_{SC}) ($L_p \ll W_{SC}$). Under this condition, the photocurrent is primarily due to the carriers generated in the depletion layer. One alternative to increase the η_{sep} is improving the electronic conductivity of the material, which allows to improve the charge separation process. Again, this is not that simple because one typical strategy to improve the electronic conductivity of n-type semiconductor is to increase the ionized donor density (N_D) through the doping with donor impurities. This can improve the conductivity; however, increasing N_D will decrease the W_{SC} and consequently can decrease the $\eta_{sep.}$ We can describe W_{SC} as

$$W_{SC} = W_O(V_{SC})^{1/2}, \tag{2}$$

where $W_O = (2\varepsilon_r\varepsilon_o/eN_D)^{1/2}$, e is the charge of the electron, ε_r is the semiconductor dielectric constant, ε_o is the vacuum permittivity, and V_{SC} is the potential at the surface space charge layer.

As we can see the improvement of the J_{ph} is a very complex task and demand the use of strategies that allow the improvement of the J_{abs} without modification of the η_{sep}. One way to do that is the development of nanomaterials with controlled morphology [2].

Hematite (α-Fe_2O_3) is considered the most promising material to be used as a photoanode (PA) for water splitting, with a potential to convert 16.8% of the sun's energy into hydrogen and very high photoelectrochemical stability in water [2]. However, this ionic semiconductor presents poor intrinsics electronic and optical properties, such as (i) a relatively low absorption coefficient, requiring thick films for efficient light absorption; (ii) a poor majority carrier (electrons) conductivity; and (iii) a short minority carrier diffusion length (holes) [2–4]. In addition, hematite requires a large overpotential for water oxidation. These properties of the hematite hinder the photoelectrochemical cell (PEC) performance by increasing the electron/hole recombination. In general, there are three main electron/hole recombination pathways that can occur in bulk, and in solid–solid and solid–liquid (electrolyte) interfaces [2, 4]. The bulk electron/hole recombination process is directly related to the poor electronic conductivity and short hole collection depth of the hematite. Besides, the recombination can also occur in two different solid–solid interfaces: in the hematite/transparent conducting oxide (TCO) electrode [5] and in the hematite/hematite grain boundary [6]. Finally, there are significant recombination processes in the solid–liquid (electrolyte) interfaces; recombination losses due to the back-injection of electrons, from the TCO exposed area, into the electrolyte [7]; and a recombination that occurs in the hematite–electrolyte interface, as a result of surface states and traps that delay the hole transfer from the valence band of the semiconductor to the electrolyte [8–11]. It is clear that the junction of these several electron/hole recombination pathways lead to a hematite PA with a low photocurrent and large overpotential for water oxidation.

In recent years, an impressive effort has been done to reduce the recombination processes that occur in each pathway. For instance, to reduce bulk recombination, researchers are mainly focused on producing nanostructured hematite with different morphologies and/or doping the ionic semiconductor with different donors elements.[1, 2] These strategies decrease bulk recombination by shortening the hole transport distance and by increasing its electrical conductivity, respectively. Recombination losses due to the back-injection of electrons from areas exposed to TCO into the electrolyte can be minimized by the addition of an underlayer [7, 12]; the recombination in the hematite–electrolyte interface can be controlled by using the passivating layer [13], depositing good electrocatalystic materials for oxygen evolution reaction [11], and surface etching [14]. It is clear that the variety of electron/hole combination pathways and interfaces is related to the thin film morphology of hematite PA. Hematite PA morphology is based on polycrystalline (with grain sizes in the nanometer scale) and mesoporous films, normally textured along the [110] direction and grown over a TCO substrate.[14] Thus, this material is characterized by (i) nonuniform grain sizes; (ii) a columnar structure; (iii) porosity;

(iv) impurity segregation; and (v) grain–grain misorientation, forming homo-junctions and hetero-junctions. Figure 1b illustrates, in a simplified way, microstructural features and interfaces found in a typical nanostructured hematite PA.

The development of high-performance hematite photoanodes depends greatly on the thin film deposition technique. Thus, the understanding of how the nanostructure develops, as well as find new film deposition routes, is a strategic and fundamental area for mastering water photoelectrolysis technology. We will describe in this chapter recent progress regarding the design of hematite nanostructure, surface activation by combining high annealing temperature and dopant impregnation and their impact on the photoelectrochemical performance for water splitting. Herein, the discussion will be conducted based on two chemical approaches to manufacture hematite photoanode that provide columnar morphology: colloidal deposition and aqueous solution hydrothermal-assisted process.

2 Colloidal Deposition

Colloidal nanocrystals (NCs) are among the most versatile nanomaterial platforms to prepare thin films due to their superb compositional and morphological tunability [16]. Among several methodologies listed in the literature to produce hematite photoanodes, colloidal nanoparticles deposition (CND) process has shown promising results [16, 17]. The CND process presents the following parameters as critical variables: the surface chemical composition of the inorganic particle; the amount of organic material in the colloidal dispersion; the nanoparticle or nanocrystals concentration; the crystallinity and the morphology of nanoparticles; colloidal dispersion viscosity; and sintering procedure (heating rate, temperature, atmosphere, and soaking time). These parameters have strong influences in the formation of single-phase hematite film that consequently affects the performance of the photoanode to promote the water splitting. The CND process can by divided basically into three steps as displayed schematically in Fig. 2. The first step consists of the attainment of a colloidal solution of the nanocrystal to be deposited. This solution is then deposited by traditional techniques such as spin or dip coating (step I). After deposition, the substrate is drying to promote the solvent elimination (step II). The final step is film sintering, where several physical–chemical transformations can occur, such as crystallization, phase transformation, grain growth, and others.

Sivula et al. [17] were the first to prepare mesoporous hematite photoanodes by a CND method. In their pioneer work, they reported a yield water-splitting current of 0.56 mA/cm^2 under standard condition (AM 1.5G 100 mW/cm^2, 1.23 V vs. RHE). They reported also a strong dependence of the photocurrent in relation to the sintering temperature. The best photocurrent result was reported for a sintering temperature of 800 °C. Sivula et al. associated the high photoactivity to the unintentional doping by Sn^{+4}, originated by Sn^{+4} diffusion from the TCO film (F-doped SnO_2 film). This unintentional doping was confirmed to occur by

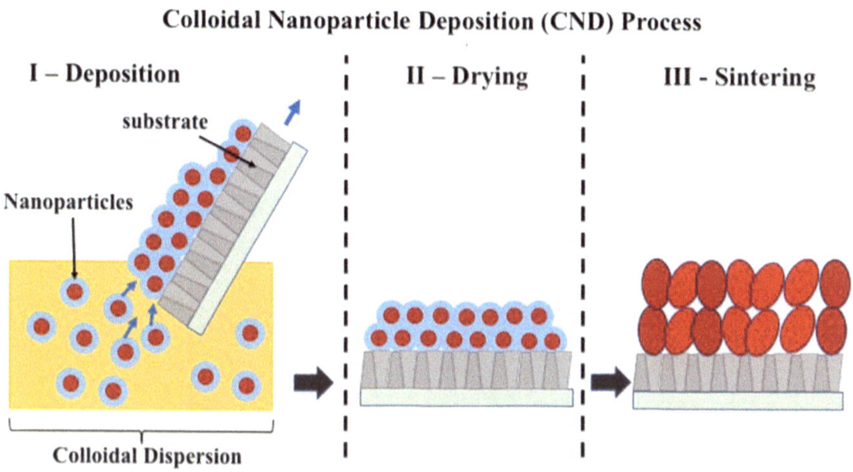

Fig. 2 Schematic representation of the colloidal nanoparticle deposition process

Gonçalves and Leite [18] by using scanning transmission electron microscopy (STEM) and high-resolution transmission electron microscopy (HRTEM). However, they showed that this unintentional doping occurs in a limited region close to the TCO film, called as a sharp interdiffusion zone ($\sim$25 nm of thickness). Gonçalves and Leite [18] also showed by HRTEM that at low sintering temperature, the hematite/TCO interface presents several small gaps (in the range of few nanometers), making a very poor electric contact between the hematite photoanode film and the electron collector film (TCO film). Figure 3 shows HRTEM images, where we observed the presence of gaps at the interface for sintered films at low temperature [18]. Increasing the sintering temperature, these gaps are no longer observed. It was also observed the creation of semi-coherent interfaces between hematite and TCO films.

Gonçalves et al. [19] developed an interesting route to prepare hematite photoanode by using the colloidal deposition approach. In that research, the authors used a magnetite nanocrystal as the precursor for hematite. During the sintering process, the magnetite (Fe_3O_4) is transformed into hematite; the Fe_3O_4 transforms into gamma-maghemite γ-Fe_2O_3 at 200 °C, and γ-Fe_2O_3 is transformed into α-Fe_2O_3 (hematite) at 450 °C. This last reaction is critical because pure hematite is obtained just above the Neel point (725 °C) with a minimum number of defects [20]. Thus, a sintering temperature that assures the complete elimination of the organic layer and oxidation of magnetite must be selected.

The magnetite colloidal approach produced orange-red transparent hematite films after sintering at 820 °C. Figure 4 illustrates cross-sectional (see Fig. 4a) and top-view (see Fig. 4b) analyses of the film morphology characterized by SEM. Figure 4a shows a film with a thickness around 930 nm and a typical columnar grain structure. Also, a good interface between the TCO and hematite film is apparent as well as the presence of an elongated porous which promotes the

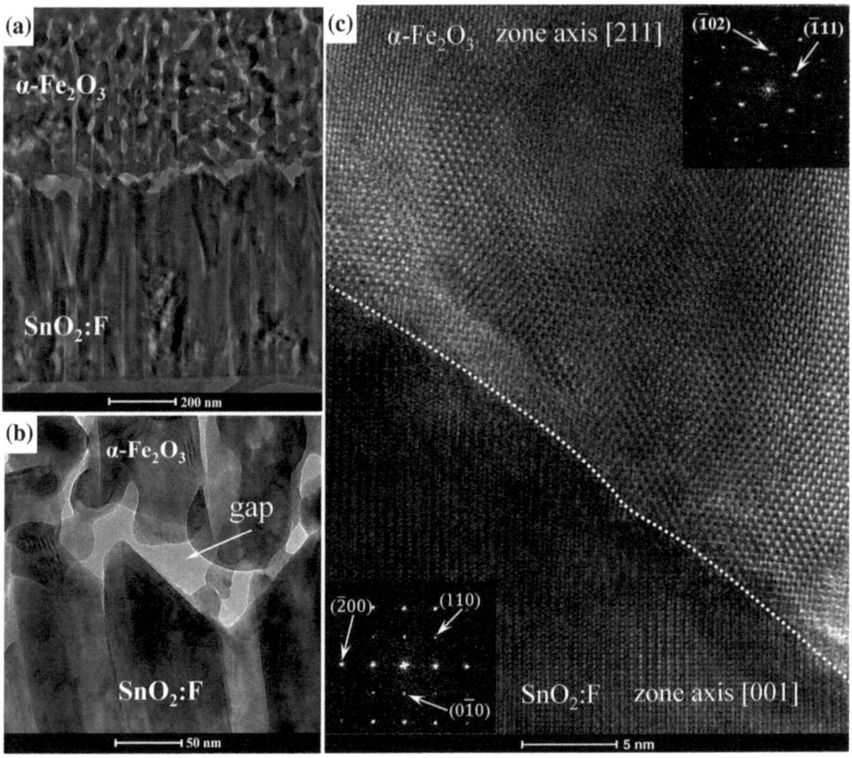

Fig. 3 Cross-sectional TEM images of the hematite thin film sintered at 550 °C. **a** Low-magnification BF-TEM image, **b** high magnification BF-TEM image, and **c** HRTEM image of the TCO/hematite interface. A *dashed line* has been added at the interface as a guide to the eyes. *Insets* fast Fourier transform (FFT) analysis of the hematite and SnO_2, showing the zone axis of each phase. Reprinted from Ref. [18]. Copyright 2013, with permission from Cambridge University press

connection between the top of the hematite film and the TCO surface. Figure 4b (top-view analysis) exhibits the nanostructured nature of the film with elongated grains (a typical size of 50 nm) and open porosity, which characterizes a mesoporous film. The XRD analysis of the film sintered at 820 °C (see Fig. 4c) exhibits the formation of a hematite phase with preferential orientation in the [110] axis vertical to the substrate. The crystallite size for the thicker film is calculated from XRD data and Scherrer's equation and considers the (110) plane at 36 nm which supports the FE-SEM top-view analysis.

Photoelectrochemical characterization shows that the approach developed by Gonçalves and coauthors produced a high-performance hematite electrode for water photooxidation with an onset potential as low as 0.8 V_{RHE}. This value is comparable to the best results reported in the literature for a hematite photoanode modified without catalytic materials [21]. A high plateau of photocurrent (photocurrent

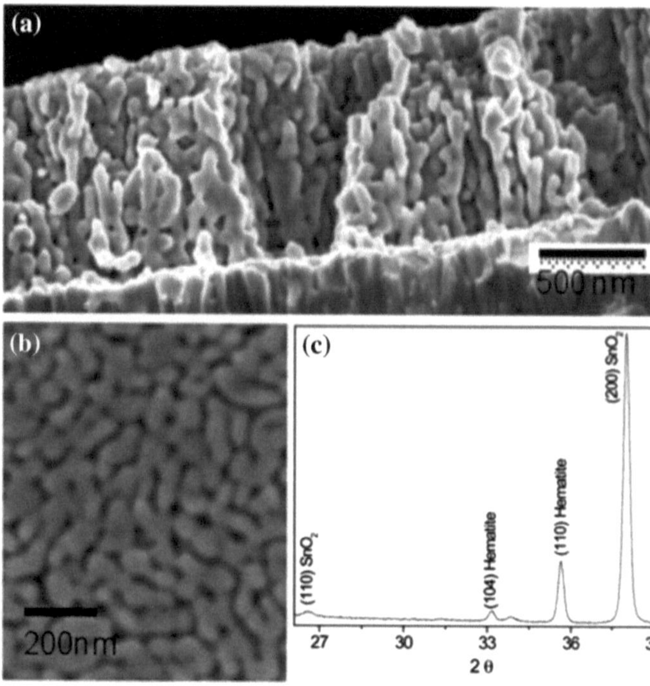

Fig. 4 SEM and XRD analyses of the hematite thin film prepared from the colloidal deposition of magnetite nanocrystals and sintered at 820 °C: **a** SEM cross-sectional analysis; **b** SEM top-view analysis; **c** XRD pattern. Reprinted from Ref. [19]. Copyright 2011, with permission from American Chemical Society

density of 1.1 mA cm^{-2} at 1.23 V_{RHE}) is also observed, even for a film with a grain size in the range of 50 nm. The Mott–Schottky analysis of the hematite film clearly indicates this material possesses a favorable surface rather than an improvement of bulk properties. Lower overpotential and high photocurrent values are attributed to favorable surface properties of the hematite grains obtained by using a high temperature and an oxygen atmosphere during the sintering process. Trevellick et al. [22] addressed two origins of unfavorable surface properties that contribute to the low photocurrent in a hematite photoanode: (1) induced surface reconstruction (at least partially) for a structure that has electrochemical properties very similar to Fe_3O_4 properties. It is well established that Fe_3O_4 is not a good catalytic material to promote water oxidation; [3, 23] and (2) a small intrinsic faradaic rate constant for the oxidation of water on Fe_2O_3 associated with trap levels at or near the surface. The extensive grain growth process promoted by the high sintering temperature (higher than the Neel point) and oxygen must contribute to the stabilization of the surface, avoiding any structure similar to Fe_3O_4. Those treatments must also decrease the surface state recombination as well as the recombination in the depletion region. The reason for this modification is not clear and is under investigation; however, we can propose that the high sintering temperature under oxygen flow leads to a passivation of the surface trapping state due to the formation of an elongated textured hematite grain with a high fraction of (001) surface. In general, in relation to other hematite surfaces, it is accepted that the (001) surface is relatively inert (in water) for the protonation and deprotonation reactions needed for charge accumulation [24]. This behavior can be favorable in decreasing the hematite surface trapping state in a water-based electrolyte. Another important contribution of the Gonçalves et al. [19] works was the morphology of the photoanode. A columnar grain structure oriented in a favorable crystallographic orientation for electron collection and a mesoporous structure was obtained which permits the penetration of the electrolyte up to the TCO interface which must also contribute to the high photocurrent.

One of the problems associated with CND process is the thin film thickness control. In general, it is necessary to perform several deposition and sintering cycles to obtain films with a thickness of 200–400 nm. The need for multiple deposition/sintering cycles increases the cost of this photoanode deposition technology, as well as promoting the formation of layers with different thermal histories, which may affect the performance and reproducibility of the device. Recently, Gonçalves et al. [25] developed a process in which a colloidal suspension of superparamagnetic nanoparticles is deposited by dip coating in the presence of magnetic field. These authors believe the influence of the magnetic field in the CND process is associated with the magnetorheological (MR) fluid behavior of the superparamagnetic colloidal dispersion used in their work. MR fluids are materials, which show a reversible transition from a liquid to a nearly solid state under the presence of external magnetic fields. In the absence of a magnetic field, MR fluids behave as Newtonian fluids. However, when a magnetic field is applied (transverse to the direction of flow), a yielding, shear thinning and viscoelastic behavior is observed [26]. These modifications induced by the magnetic field in the rheological behavior of the superparamagnetic colloidal dispersion have a direct impact in the dip-coating process, allowing to prepare films with thickness ranging from 150 to 250 nm, in a single step. Figure 5 describes in a schematic way the CND process assisted by magnetic field. The photoanode prepared by the CND process in the presence of magnetic field showed a significant improvement of the photoelectrochemistry performance regarding the water electrolysis. The authors reported that for undoped hematite photoanode deposited in the presence of a magnetic field, a photocurrent of 1.4 mA cm^{-2} at 1.23 V_{RHE} and a plateau of 1.9 mA cm^{-2} at 1.5 V_{RHE} under standard AM 1.5 G illumination. Considering now the Sn-doped hematite photoanode, deposited in the presence of magnetic field, the photocurrent achieved was of 2.7 mA cm^{-2} at 1.23 V_{RHE} and reached a plateau of 3.8 mA cm^{-2} at 1.5 V_{RHE}. This is one of the best results reported in the literature so far. It is important to point out that, even for a film formed by several deposition circles and similar thickness, the photoelectrochemical performance is not the same, when compared with a film prepared in the presence of magnetic field. Figure 6 shows current–potential curve for hematite photoanodes

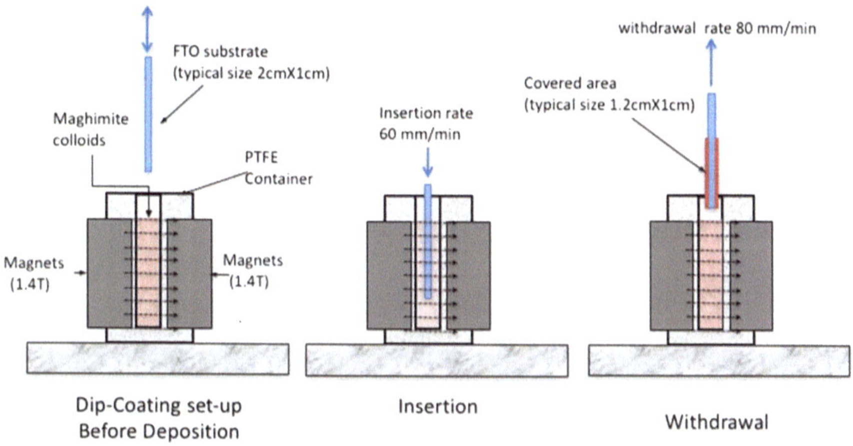

Fig. 5 Detail about the CND magnetic field assisted used by Gonçalves and Leite. Reprinted from Ref. [25]. Copyright 2014, with permission from Royal Society of Chemistry

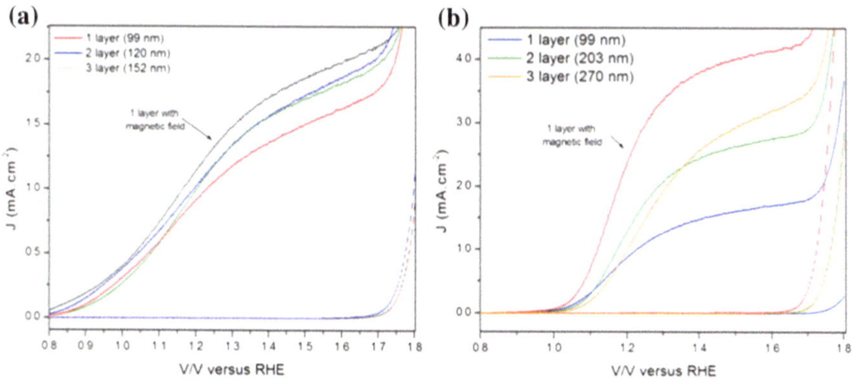

Fig. 6 Comparison of the current potential curve for hematite produced with and without a magnetic field applied during the dip-coating deposition process: **a** Undoped hematite; and **b** Sn-doped hematite. Reprinted from Ref. [25]. Copyright 2014, with permission from Royal Society of Chemistry

produced with and without a magnetic field applied during the dip-coating deposition process. As we can notice, the performance of the photoanodes processed in the presence of magnetic field is superior.

Note that the magnetic field improves the photoelectrochemistry performance during the deposition process. Even with multiple deposition circles (without a magnetic field), photoanodes did not achieve the same performance as photoanodes processed with an external magnetic film.

The undoped film deposited in the presence of magnetic field showed a film thickness of 180 nm and Sn-doped film deposited in the presence of magnetic field showed a film thickness of 250 nm [25]. The film thickness is indicated in Fig. 4a legend for the films prepared without the magnetic field.

3 Hydrothermal-Assisted Synthesis

As discussed in the previous sections, the latest advances on hematite n-type semiconductor are related to design nanoarchitecture with morphology highly active to split water into hydrogen and oxygen gas under sunlight irradiation. An alternative method to the *CND* [19] that has also gained great attention is the Purpose-Built Material (*PBM*) first proposed by Vayssieres [27]. The fundamental concept behind of the *PBM* process is to control experimentally the interfacial tension (surface free energy) based on a Gibbs adsorption equation model. In other words, by monitoring the kinetics and thermodynamics of nucleation, growth and aging process of metal oxide might be possible to minimize the interfacial tension by leading the system stabilization. Such control allows to design nanostructured materials with one-dimensional morphology and suitable characteristic on various substrates [28]. To design this low-dimensional and anisotropic nanomaterial onto several kinds of substrates a simple apparatus was proposed, which consisted in a preparation of a precursor solution of metal salts or complexes with a specific ionic pH and ionic strength [27, 28]. The precursor solution in the presence of commercial substrates in a closed bottle is heated at different times at temperature below 100 °C, as illustrated in Fig. 7 (1st step). Few hours later or days the substrate is completely covered by well-distributed one-dimensional (1D) morphology, which the nanometric dimensions are controlled by time, precursor concentration and temperature, as illustrated in Fig. 7 (second and third steps).

In fact, *PBM* can be the useful chemical route to synthesize transition metal oxide with anisotropic, normal, or oriented crystal along to the substrate by manipulating the precursor solution to control the interfacial adsorptions of ions. Vayssieres et al. [31, 32] using the *PBM* method investigated the photoelectrochemical performance of a vertical-aligned hematite nanorod electrode.

Figure 8 illustrates the top-view and transversal scanning electron microscopy (SEM) images of the hematite samples designed by using the *PBM* method. Top-view SEM images (Fig. 8a, b) exhibited large covered area with homogenous deposition, while the transversal SEM image (Inset Fig. 8b) confirms the formation of vertical rods onto the commercial substrate. In these seminar works, they conducted a systematic study considering several parameters (morphology, film thickness, electrolyte composition, and dye sensitization) that could directly influence the overall photoelectrochemical performance. The results were discussed in terms of incident-to-photo-conversion efficiency (IPCE) showing an enhancement of 100 and 7 times better when compared with spherical particles thin films (Fig. 9a, b), earlier investigated [31, 32]. The nanorod morphology exhibited

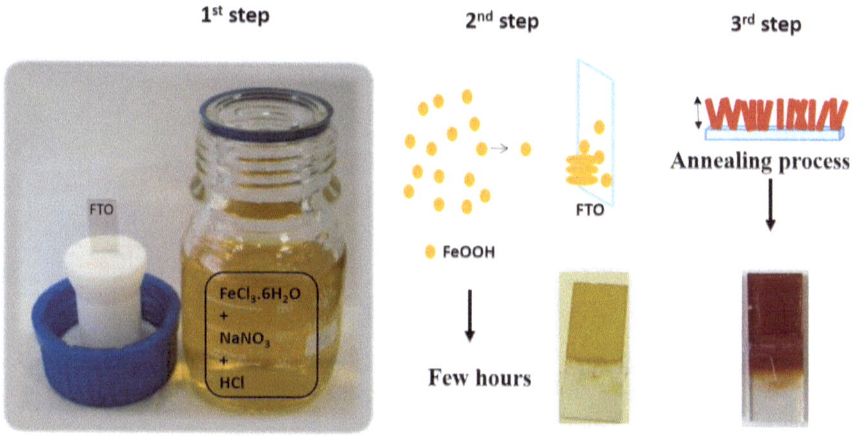

Fig. 7 Schematic representation of *PBM* process: first, precursor solution and substrate immersion in autoclavable bottle, second, heating at temperature below 100 °C nucleation process into substrate surface and homogeneous layer deposition, and third, some cases the desired oxide phase is obtained after annealing process or can conducted to increase substrate–layer interface [29, 30]

Fig. 8 Top-view SEM images of hematite films synthesized by *PBM*. **a** Low magnification, **b** High magnification. *Inset* in (**b**) is the cross-sectional SEM images from the cleaved film

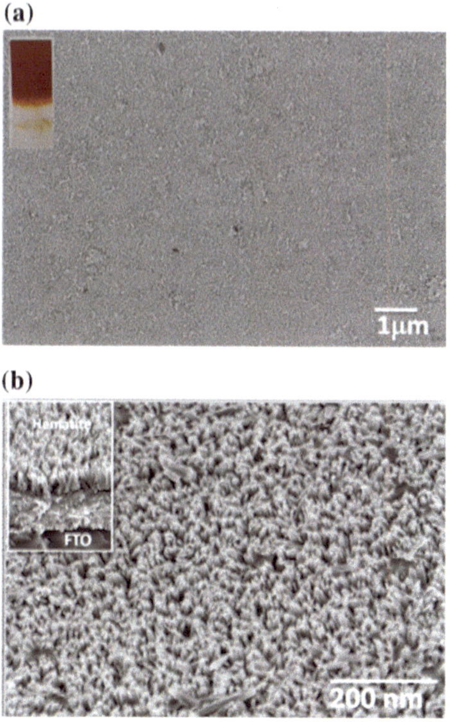

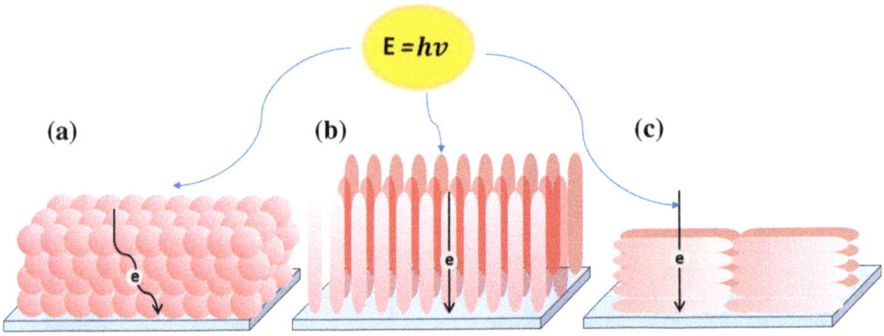

Fig. 9 Schematic representation of electron transport through hematite morphology grown, **a** spherical particles parallel, **b** vertical rods orthogonal and **c** rods parallel to the conductor substrate (Figure adapted from Ref. [31])

superior performance against spherical particles varying the electrolyte pH from 6.8 to 12 and even the sunlight irradiation was conducted through the electrolyte or electrode (front or back side).

The better IPCE values were attributed to the vertical nanorod morphology, which favor a better electron transport path through the back contact (substrate) with the presence of a fewer grain boundary in contrast with sintered spherical morphology. In addition, investigation films composed by vertical and parallel nanorod (Fig. 9b, c) was conducted under sunlight irradiation (Back and front side). At front side illumination, the vertical nanorods provide a slight better performance, whereas illuminated from substrate side (back side) no difference was observed in IPCE performance showing that the orientation not always matters for photoelectrochemical application. Since this pioneer work, significant progress worldwide has been achieved using this hierarchical nanoarchitecture based on aqueous solution heated at low temperature under hydrothermal conditions [29, 30]. To boost the yield of water oxidation reaction, through the hematite with vertical nanorod morphology several modifications on synthesis parameters, addition of dopants, surface modification, and high annealing temperatures were proposed along these two decades. Some of the best results achieved on undoped and modified hematite with vertical nanorod morphology were compiled in Table 1, which will be used to guide the discussion regarding of the evolution in performance and comprehension. The authors know that the selected works in this table do not represent the real number of published paper regarding of the PBM method; however, these reports give a good idea about what has been proposed to boost the vertical nanorod hematite performance. In the case of undoped hematite nanorod (HHN) photoelectrochemical performance, the best values achieved up to this date, values in italic in Table 1, are from 1.12 to 1.48 mA cm^{-2} at 1.23 V_{RHE} (RHE, reversible hydrogen reference electrode).

Even some changes were done in the original synthesis parameters (such as time (reduction), different precursors), pretreatment, or different substrates, the use of

Table 1 Photoelectrochemical performance of undoped (UHN) and modified hematite nanorod MHN) under sunlight irradiation at 1.23 V versus reversible hydrogen reference electrode

Undoped hematite nanorod—UHN	Modified hematite nanorod—MHN	Annealing temperature (°C)	UHN-*J* mA cm^{-2} at 1.23 V versus RHE	MHN-*J* mA cm^{-2} at 1.23 V versus RHE	Year	Reference
X	(N3-dye)	550	0.05	<0.15	2000	[32]
X	–	550	0.01	–	2002	[31]
X	Sn-HN	800	*1.24*	**1.86**	2011	[33]
X	Ti-HN	550–750	0.87	**1.91**	2012	[34]
X	Cr-HN	750	0.03	0.05	2012	[35]
X	–	390	0.9	–	2012	[29]
X	N-HN	550	–	**1.8**	2012	[36]
X	Co_3O_4-HN	550–800	0.72	1.2	2012	[37]
X	Ti-HN	550	0.01	0.15	2012	[38]
X	Sn-HN	750	*1.24*	**2.25**	2013	[39]
X	Ti-HN with Co (ions)	650	0.3	1.2	2013	[40]
X	Ti-HN	750	0.45	0.66	2013	[41]
X	Zr-HN	750	0.03	0.33	2013	[42]
X	Mn-HN	550–750	0.45	**1.4**	2014	[43]
X	HN surface passivation with Au	750	0.03	0.25	2014	[44]
X	Ag-HN	750	0.05	0.18	2014	[45]
X	N-HN (layer < 100 nm)	750	0.39	0.95	2015	[46]
X	Ti-HN	650	–	1.9	2015	[47]
X	ATO-HN	500–800	0.60	**2.12**	2015	[48]
X	Nb-HN	750	0.15	0.65	2016	[49]
X	Sn-HN	750	*1.12*	**1.62**	2016	[50]

high annealing temperature in between 700 and 800 °C seems to be an important role to improve the UHN performance. In fact, the positive effect of high annealing temperature has been extensively addressed as a key factor to improve the light-driven water oxidation performance of undoped hematite processed for all different synthetic methods [51–53]. Yat Li and coworkers in 2011 investigated the influence of temperature on the UHN properties showing photocurrent response around 1.24 mA cm^{-2} at 1.23 V_{RHE} when the temperature at 800 °C was used [33]. They demonstrated that the high annealing temperature promoted increases in rod length and diameter, probably due to the aggregation, and by XPS was evidenced the presence of Sn that come from the substrate by diffusion. At higher temperature, more percentage of Sn from the substrate was found; however, the photocurrent response sensibly dropped due to the substrate damage (glass distortion and decreased the FTO conductivity). This unintentionally doping, first observed by

Sivula et al. [17] studying hematite prepared by CND, seems to increase electrical conductivity acting as an electron donor and being responsible to enhance the UHN current response under light irradiation. Wong et al. [39] achieved a similar photocurrent response at 1.23 V_{RHE} by altering the chemical precursor and using high annealing temperature at 750 °C per 30 min. Carvalho et al. [50] reported a systematic investigation regarding of the influence of synthesis time and annealing temperature on the UHN performance. The optimized synthesis conditions, which exhibited a high photocurrent response around 1.12 mA cm^{-2} at 1.23 V_{RHE}, was found to be 2 h followed by annealing at 750 °C per 30 min.

The high annealing temperature was found to increase the UHN surface wettability, which means that the high temperature improves the solid–liquid interface favoring a more efficient water oxidation reaction (as illustrated in Fig. 10 the effect of two different annealing temperatures). According to the literature, the electron transport on bare hematite is given through the bulk; the high annealing temperature allied with the nanorod morphology vertically aligned to the substrate seems to increase the electron transfer toward the substrate reducing the recombination rate [54]. Indeed, high annealing temperature combined with vertical rod morphology might favor a better solid–solid interface; the crystal growth at [110] direction is known from literature that direction could increase up to four orders of magnitude of the electrical conductivity. Moreover, in some case, diffusion of Sn from the substrate FTO layer was observed after high annealing temperature leading an increase in the number of donor charge density and its transfer toward the substrate [33, 55, 56]. These abovementioned effects related to high annealing temperature have been reported on literature known only for rod morphology fabricated by PBM route, but to several other morphologies and methods [17, 18]. For all those studies listed in Table 1 that reveal photocurrent response for UHN below 1 mA

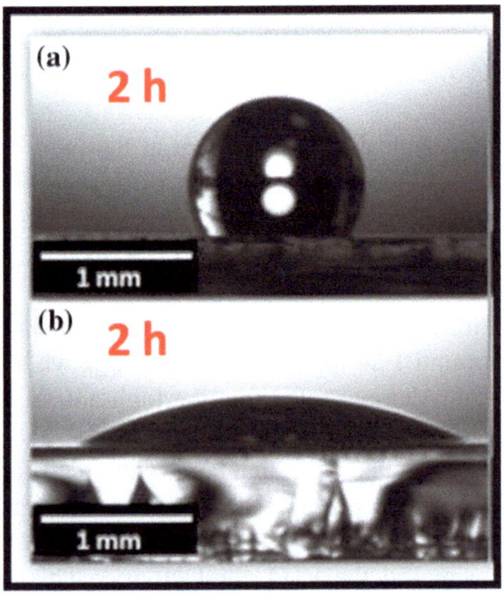

Fig. 10 The contact angle images for UHN prepared by PBM process at 2 h followed by annealing at **a** 550 °C 1 h and **b** 750 °C per 30 min

cm^{-2} at 1.23 V_{RHE}, a similar explanation above was given and the critical parameter was found to be the high recombination rate due to a poor electron transport toward the substrate or hole diffusion through the solid–liquid interface.

To keep booting, the vertical nanorod hematite (UHN) performance for water oxidation reaction assisted by light several modifications has been proposed to increase the solid–liquid interface (hole diffusion), as summarized in Table 1. The most common and successful modification applied for the vertical nanorod hematite is a deposition of solution containing metal transition ions (such as Sn, Co, Ti, Mn, etc., highlighted in bold, Table 1) over the UHN layer before or after the annealing process. The deposition of a solution with Ti or Sn combined with high annealing temperature revealed a better way to increase the hole diffusion toward to the hematite surface (see values highlighted in bold, Table 1). In special, the addition of Sn on the surface of different hematite morphologies [25, 57] including UHN (Table 1) had led the overall efficiency rising to almost double in comparison with the bare material. Yat Li et al. [33] investigated the intentional addition of Sn on UHN surface and combined with high annealing temperature showed an enhancement in relation to the UHN from 1.24 to 1.86 mA cm^{-2} at 1.23V_{RHE}, respectively (Fig. 11). The addition of Sn over UHN was done before the annealing treatment, where after this process at high temperature a significant change in morphology was observed. The SEM images revealed that the Sn-HN change from nanorod to coral-like morphology with smaller features sizes, which increased the surface area available for chemical reaction (see Fig. 11).

They also investigated the dynamic of charge by ultrafast spectroscopy showing that the hole–electron combination rate within the first picosecond remained high even with Sn addition. Then, the improvement in the photocurrent response was attributed a higher surface area promoted by coral-like morphology and a possible increase in electrical conductivity (Fig. 11). Therefore, this significant change in morphology by the addition of Sn followed by annealing process has been commonly reported in literature independent of the manufacturing method.

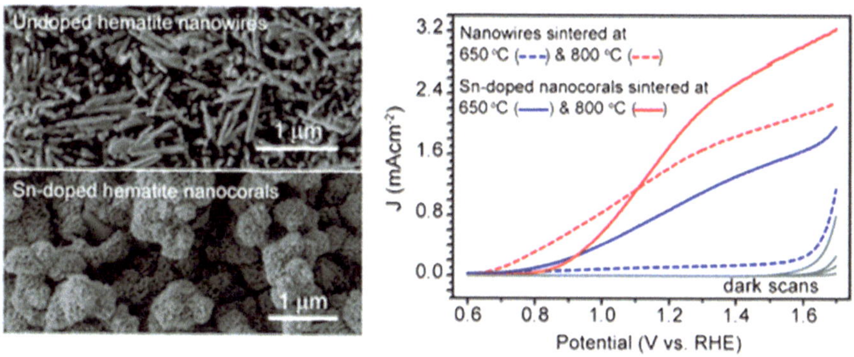

Fig. 11 *Top-left* SEM images of UHN, *Bottom-left* SEM images of UNH modified with addition of Sn, *Right* Photocurrent response for UHN and Sn-UHN annealed at 650 °C represented by *dashed-blue dot* and *solid blue*, and at 800 °C *dashed-red dot* and *red-solid curves*, respectively. Reprinted from Ref. [33]. Copyright 2011, with permission from American Chemical Society

Nevertheless, Wong and coworkers [39] achieved the highest photocurrent response by adding Sn over the HN layer followed by annealing at 750 °C per 30 min, which shift from 1.24 to 2.25 mA cm^{-2} at 1.23 V_{RHE}. Contrary to commonly reported, no substantial change in nanorod morphology was observed in the SEM images after addition of Sn combined with high annealing temperature [38]. This pronounced improvement was related to the efficient surface passivation by the formation of $Fe_xSn_{1-x}O_4$ at the HN surface that revealed in the TEM images [38], which possibly reduces the recombination rate at electrode–electrolyte interface. Besides of the photocurrent improvement, a shift of 100 mV on the onset potential toward a cathodic direction was observed with the addition of Sn, which increases the necessity of external bias application.

This cathodic shift in the onset potential has been investigated in literature due to the addition of Sn, but the reasons remain unclear yet. As discussed above, Carvalho and Souza [50] studied the addition of Sn on HN surface followed by high annealing temperature; no substantial change in the morphology was observed, as illustrated in Fig. 12. In fact, the morphology was preserved with a slight reduction in the rod length and diameter. As the added Sn was not detected by X-ray diffraction technique and overall rod dimension decreases, it is possibly segregated at nanorod surface. The SEM and TEM images exhibit the formation of a layer covering the nanorod surface (as observed by Wong et al. [39]); however, after

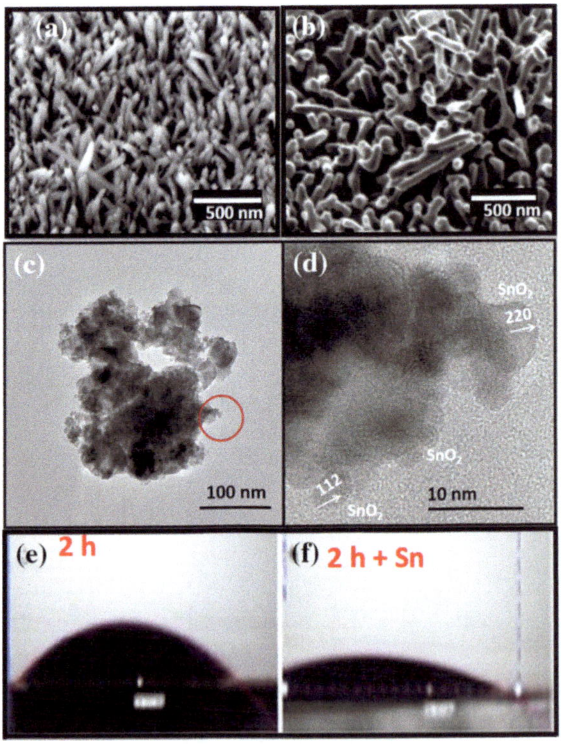

Fig. 12 Top-view SEM images after high annealing temperature at 750 °C per 30 min, **a** UHN, **b** Sn-UHN. HRTEM images of Sn-UHN powder gentle removed from the substrate, **c** low magnification, and **f** high magnification. Contact angle images of **e** UHN, and **f** Sn-UHN prepared at same conditions in (**a**) and (**b**)

analyzing the chemical composition of this layer by energy dispersive spectroscopy (EDS) using the TEM images (Fig. 12c, d), it was found to be SnO_2 phase instead of $Fe_xSn_{1-x}O_4$. To shed some light by the influence of addition of Sn, the donor charge density was estimated from Mott–Schottky plot showing no significant changes. On the other hand, the Sn-HN–electrolyte interface has substantially increased, as illustrated by the contact angle images (Fig. 12e, f), that reveal high surface water affinity. The final contact angle measured at around 18° is considered to have a super-hydrophilic surface (at contact angle below 20°).

The addition of Sn combined with high annealing temperature was found to preserve the morphology promoting a slight reduction in the rod dimensions, increasing the surface–electrolyte interface, but not substantially affects the electrical conductivity. Then, the improved photocurrent response from 1.12 to 1.62 mA cm^{-2} at 1.23 V_{RHE} was exclusively attributed to a better solid–liquid interface that favored a more efficient hole diffusion. Since it is accepted that the columnar morphology combined with high annealing temperature favors the crystal orientation at higher conductivity direction of hematite and better substrate–layer interface, an enhancement in efficiency of electron transfer through the back contact is expected. The addition of catalysts over the hematite surface was expected to improve the solid–liquid interface favoring a more efficient hole diffusion. As consequence, this "free" electron (due to a reduction in the recombination rate) could also be injected through the substrate increasing the material efficiency.

4 Conclusion and Perspectives

Latest advances in hematite electrodes manufactured by *CND* and *PBM* process are an excellent choice to address and overcome the most significant drawback of this material, which is the poor electronic properties. The current discussion addressed in this chapter leads us to believe that the columnar morphology and preferential crystal orientation at specific direction onto the substrate seems to be essential to obtain highly active bare hematite for water oxidation assisted by sunlight irradiation. On the other hand, the addition of catalysts (e.g., Ti and Sn) over the columnar surface, which are responsible for the highest performance in hematite so far, looks like to be dominated by different parameters besides columnar morphology. In fact, the presence of catalysts combined with high annealing temperature strongly affects the columnar morphology causing a substantial change in it, which lead us to attribute the highest performance essentially due to an overall improvement in interfaces. By combining morphology, high annealing temperature, and surface changes, a significant reduction in the overall hole–electron recombination rate and a pronounced values of photocurrent response is expected due to (i) solid–solid interface improved by high temperature, which will facilitate the electron transfer through the back contact, and (ii) solid–liquid interface modified by the presence of catalyst will favor the chemical reactions by a more efficient hole diffusion. Essentially from the results discussed here we learned that hematite in the presence of catalyst followed by high annealing temperature become a polycrystalline material. In contrast of the semiconductor technologies that are fundamentally supported in single crystal, the polycrystalline catalyst–hematite modified exhibited the highest performance in photoelectrochemical application. In fact, for pure hematite electrodes, the presence of grain boundary had been killed the electron transfer and hole diffusion efficiency, but the addition of catalyst seems to promote a significant reduction in the grain boundary energy favoring a highly photoactivity. In other words, it seems to be essential for bare hematite morphologies with few numbers of grain boundary, columnar aspect with preferential crystal orientation, or defect that will help to drop down the recombination; while for catalyst modifying hematite, the remaining challenges are identified and understood, in polycrystalline morphology, in which a kind of grain boundary is essential to promote a free electron path through the back contact and how the catalyst affect the grain boundary energy that will possibly allow us to control them. Over 35 years of investigation of hematite, a plenty of rooms for optimization, fundamental investigation, and catalytic and electronic mechanisms remain in this field to make this abundant material true for a commercial application. Since our planet is essentially abundant of seawater and constantly graced by the presence of inexhaustive source of light, to find a way to directly convert and store solar energy into chemical energy (hydrogen and oxygen) will be a cheapest manner to provide clean and sustainable energy to supply the world demand.

References

1. Fujishima A, Honda K (1972) Nature 238:37–38
2. Sivula K, Le Formal F, Grätzel M (2011) Chemsuschem 4:432–449
3. Walter MG, Warren EL, McKone JR, Boettcher SW, Mi Q, Santori EA, Lewis NS (2010) Chem Rev 110:6446–6473
4. Cho IS, Han HS, Logar M, Park J, Zheng X (2016) Adv Energy Mat 6
5. Pendlebury SR, Cowan AJ, Barroso M, Sivula K, Ye J, Grätzel M, Klug DR, Tang J, Durrant JR (2012) Energy Environ Sci 5:6304–6312
6. Warren SC, Voïtchovsky K, Dotan H, Leroy CM, Cornuz M, Stellacci F, Hébert C, Rothschild A, Grätzel M (2013) Nat Mater 12:842–849
7. Steier L, Herraiz-Cardona I, Gimenez S, Fabregat-Santiago F, Bisquert J, Tilley SD, Grätzel M (2014) Adv Func Mater 24:7681–7688
8. Bertoluzzi L, Bisquert J (2012) J phys Chem Lett 3:2517–2522
9. Barroso M, Mesa CA, Pendlebury SR, Cowan AJ, Hisatomi T, Sivula K, Grätzel M, Klug DR, Durrant JR (2012) Proc Natl Acad Sci 109:15640–15645
10. Young KM, Klahr BM, Zandi O, Hamann TW (2013) Catal Sci Technol 3:1660–1671
11. Nellist MR, Laskowski FA, Lin F, Mills TJ, Boettcher SW (2016) Acc Chem Res 49:733–740
12. Tamirat AG, Rick J, Dubale AA, Su W-N, Hwang B-J (2016) Nanoscale Horizons 1:243–267
13. Shinde PS, Annamalai A, Kim JH, Choi SH, Lee JS, Jang JS (2015) Sol Energy Mater Sol Cells 141:71–79
14. Shinde PS, Choi SH, Kim Y, Ryu J, Jang JS (2016) Phys Chemistry Chem Phys 18:2495–2509

15. Soares MR, Gonçalves RH, Nogueira IC, Bettini J, Chiquito AJ, Leite ER (2016) Phys Chemistry Chem Phys 18:21780–21788
16. Gadiyar C, Loiudice A, Buonsanti R (2017) J Phys D Appl Phys 50:074006
17. Sivula K, Zboril R, Le Formal F, Robert R, Weidenkaff A, Tucek J, Frydrych J, Gratzel M (2010) J Am Chem Soc 132:7436–7444
18. Goncalves RH, Leite ER (2014) J Mater Res 29:47–54
19. Gonçalves RH, Lima BH, Leite ER (2011) J Am Chem Soc 133:6012–6019
20. Bjoerksten U, Moser J, Graetzel M (1994) Chem Mater 6:858–863
21. Tilley SD, Cornuz M, Sivula K, Grätzel M (2010) Angew Chem 122:6549–6552
22. Dare-Edwards MP, Goodenough JB, Hamnett A, Trevellick PR (1983) J Chem Soc, Faraday Trans 1: Phys Chem Condens Phases 79:2027–2041
23. Trasatti S (1980) J Electroanal Chem Interfacial Electrochem 111:125–131
24. Yanina SV, Rosso KM (2008) Science 320:218–222
25. Gonçalves RH, Leite ER (2014) Energy Environ Sci 7:2250–2254
26. de Vicente J, Klingenberg DJ, Hidalgo-Alvarez R (2011) Soft Matter 7:3701–3710
27. Vayssieres L, Beermann N, Lindquist S-E, Hagfeldt A (2001) Chem Mater 13:233–235
28. Vayssieres L, Guo J, Nordgren J (2000) Purpose-built anisotropic metal oxide nanomaterials, MRS Proceedings. Cambridge University Press, USA, p C7. 8
29. de Carvalho VAN, Luz RAS, Lima BH, Crespilho FN, Leite ER, Souza FL (2012) Journal of Power Sources 205 (2012) 525–529
30. Ferraz LC, Carvalho WM Jr, Criado D, Souza FL (2012) ACS Appl Mater Interfaces 4: 5515–5523
31. Lindgren T, Wang H, Beermann N, Vayssieres L, Hagfeldt A, Lindquist S-E (2002) Sol Energy Mater Sol Cells 71:231–243
32. Beermann N, Vayssieres L, Lindquist SE, Hagfeldt A (2000) J Electrochem Soc 147: 2456–2461
33. Ling Y, Wang G, Wheeler DA, Zhang JZ, Li Y (2011) Nano Lett 11:2119–2125
34. Deng J, Zhong J, Pu A, Zhang D, Li M, Sun X, Lee S-T (2012) J Appl Phys 112:084312
35. Shen S, Jiang J, Guo P, Kronawitter CX, Mao SS, Guo L (2012) Nano Energy 1:732–741
36. Ling Y, Wang G, Reddy J, Wang C, Zhang JZ, Li Y (2012) Angew Chem Int Ed 51: 4074–4079
37. Xi L, Tran PD, Chiam SY, Bassi PS, Mak WF, Mulmudi HK, Batabyal SK, Barber J, Loo JSC, Wong LH (2012) J Phys Chem C 116:13884–13889
38. Miao C, Ji S, Xu G, Liu G, Zhang L, Ye C (2012) ACS Appl Mater Interfaces 4:4428–4433
39. Xi L, Chiam SY, Mak WF, Tran PD, Barber J, Loo SCJ, Wong LH (2013) Chemi Sci 4: 164–169
40. Miao C, Shi T, Xu G, Ji S, Ye C (2013) ACS Appl Mater Interfaces 5:1310–1316
41. Shen S, Kronawitter CX, Wheeler DA, Guo P, Lindley SA, Jiang J, Zhang JZ, Guo L, Mao SS (2013) J Mat Chem A 1:14498–14506
42. Shen S, Guo P, Wheeler DA, Jiang J, Lindley SA, Kronawitter CX, Zhang JZ, Guo L, Mao SS (2013) Nanoscale 5:9867–9874
43. Chiam SY, Kumar MH, Bassi PS, Seng HL, Barber J, Wong LH (2014) ACS Appl Mater Interfaces 6:5852–5859
44. Shen S, Li M, Guo L, Jiang J, Mao SS (2014) J Colloid Interface Sci 427:20–24
45. Shen S, Zhou J, Dong C-L, Hu Y, Tseng EN, Guo P, Guo L, Mao SS (2014) Scientific Reports 4:6627
46. Freitas AL, Carvalho WM, Souza FL (2015) J Mater Res 30:3595–3604
47. Li X, Bassi PS, Boix PP, Fang Y, Wong LH (2015) ACS Appl Mater Interfaces 7:16960–16966
48. Wang D, Zhang Y, Peng C, Wang J, Huang Q, Su S, Wang L, Huang W, Fan C (2015) Adv Sci 2
49. Fu Y, Dong CL, Lee WY, Chen J, Guo P, Zhao L, Shen S (2016) ChemNanoMat 2:704–711
50. Carvalho WM, Souza FL (2016) ChemPhysChem 17:2710–2717

51. Wickman B, Fanta AB, Burrows A, Hellman A, Wagner JB, Iandolo B (2017) Scientific reports 7
52. Zhou F, Kotru S, Pandey R (2002) Thin Solid Films 408:33–36
53. Kaouk A, Ruoko T-P, Pyeon M, Gönüllü Y, Kaunisto K, Lemmetyinen H, Mathur S (2016) J Phys Chem C 120:28345–28353
54. Kronawitter C, Zegkinoglou I, Rogero C, Guo J-H, Mao S, Himpsel F, Vayssieres L (2012) J Phys Chem C 116:22780–22785
55. Annamalai A, Kannan AG, Lee SY, Kim D-W, Choi SH, Jang JS (2015) J Phys Chem C 119:19996–20002
56. Li M, Yang Y, Ling Y, Qiu W, Wang F, Liu T, Song Y, Liu X, Fang P, Tong Y (2017) Nano Lett 17:2490–2495
57. Ling Y, Li Y (2014) Part Part Syst Charact 31:1113–1121

Biofuel Cells

Roberto A.S. Luz, Andressa R. Pereira, Rodrigo M. Iost and Frank N. Crespilho

1 Introduction

The search for clean and sustainable energy supplies arises from the last century with the continuous depletion of fossil fuels and natural source of gases. Nowadays, the continuous development of new sustainable and pollutant-free technologies rather than the continuous combustion of nonrenewable fossil fuels like petroleum and/or natural gases for energy production is still a big challenge, with which are necessary to be developed to supply the global energy demand [1, 2]. Focus has been done on the study of electrochemical conversion of fuels to produce electrical energy by electrochemical reactions. The electrochemical energy produced by oxidation/reduction reactions has been extensively studied during the last decades as an alternative and environmentally favorable energy source [3]. The redox reactions are intrinsic to batteries, fuel cells, electrochemical capacitors and biofuel cells, and use different principles of energy storage and energy conversion [4–9]. Although batteries are capable to store and discharge electrical energy, fuel cells have made remarkable progress and many applications are envisioned for clean and more efficient energy conversion without loss of mass [10]. A fuel cell consists of two electrodes, anode and cathode, which are immersed in an electrolyte solution shuttling electrons across the interface electrode/electrolyte from an input fuel on the anode driven to an external electrical circuit to the cathode electrode. In general, for a large range of organic compounds used as fuels, the products generated are H_2O and CO_2. The efficiency is significantly higher, e.g. alkaline fuel cells, and

A.R. Pereira · R.M. Iost · F.N. Crespilho (✉)
Instituto de Química de São Carlos, Universidade de São Paulo,
São Carlos, SP 13560-970, Brazil
e-mail: frankcrespilho@iqsc.usp.br

R.A.S. Luz
Departamento de Química, Universidade Estadual do Piauí,
Teresina, PI 64002-150, Brazil

F.L. de Souza and E. Leite (eds.), *Nanoenergy*,
DOI 10.1007/978-3-319-62800-4_5

highlights an advantage over other electrochemical energy generation systems. In addition, the development of electrochemical devices has resulted in great breakthroughs over conventional batteries [8–11]. However, the operation of fuel cells is based on non-selective and expensive metals as catalysts for the conversion of energy. At the same time, another strategy has been drawn for clean energy generation, such as the study of bioelectrochemical systems [12, 13].

The study of bioelectrochemical reactions originated from biological organisms was revealed in the 1790s when Galvani demonstrated nerves contracting in a frog's leg upon application of external energy produced by a static electricity generator [14]. The interest in fuel cell development, however, stemmed from the USA space program, started in the 1950s, which explored the possibility of using microorganisms or biomolecules to generate power [4, 15–19]. This so-called biological fuel cell, or biofuel cell (BFC), is a particular kind of biodevice that utilizes biological components, such as enzymes or microorganisms, as selective recognition elements towards a specific molecule capable of producing electrical energy by electrochemical redox processes [15–23]. The BFC system is similar to traditional fuel cells and consists of an anode and cathode. On the other hand, the electrodes are modified with biological molecules separated or not by a semi-permeable membrane. When a specific fuel is introduced into the anodic compartment, the oxidation reaction takes place and the electrons shuttle at the anode are used to reduce other chemical species, like molecular oxygen, at the cathode. However, the amount of electrical energy generated by biological molecules is usually very low, and among other features, is still a limitation in the use on biological fuel cells.

The use of microorganisms for energy conversion is not new. Recent researches on BFCs have been arised since the utilization of the first microbial fuel cell in 1912 by Potter and co-workers [24]. The study showed a simple example of energy generation by fermentation of microorganisms by bacteria in glucose media. These electrochemical biodevices based on biological components offer many advantages, such as their use of renewable [25, 26], non-pollutant [27] fuels and their ability to operate at low temperatures [28]. Besides these advantages, the performance of BFCs in terms of power energy density generated by biochemical reactions and their stability is all responsible for the broad research in the area in the last decade. Moreover, the main focus is to utilize biological components, as opposed to metal catalysts, to catalyze biochemical reactions, and generate energy with implantable biodevices in live organisms, like in the human body [26]. Since the first experiment reported by Yahiro in 1964 [29], which used the enzyme glucose oxidase (GOx) as the biological component for the anode operating with glucose as fuel [29–40], many advances have been made which focus on their various applications, such as portable implantable biodevices [26] and also recently in bioelectrochemical Haber–Bosch process in a H_2/N_2 fuel cell [41]. In addition to their use of renewable fuels, the favorable and mild temperature operation conditions make BFCs a promising route for future energy conversion devices.

In this chapter, we present an overview of the fundamental aspects in enzyme- and microbial-based BFCs for biological energy conversion devices, and we explore their main operation characteristics, enzymatic immobilization on electrodes, and nanostructured materials for BFC applications. The focus on this chapter is to establish the correlation between enzymatic thermodynamic and kinetic processes at solid conducting interfaces, describing the utilization in modified electrodes for obtention of BFCs for in vitro and in vivo studies. Some perspectives for miniaturized biodevices will also be addressed in the next topics.

2 Fundamental Aspects in BFCs Development

The investigation of the attachment of biomolecules to solid conducting substrates is well-known and arises owing the wide interest for biosensing [42, 43] and more recently in energy conversion biodevices [26]. The fundamental principle in BFCs is based on a biological transducing element (enzymes and/or microorganisms) for the conversion of chemical energy from the fuel into electrical current by the electrode [7, 44, 45]. BFCs consist of two electrodes that can be separated or not by a semi-permeable membrane [46–54], as can be seen in Fig. 1. First, species are oxidized on the anode and the electrons produced are driven throughout an external

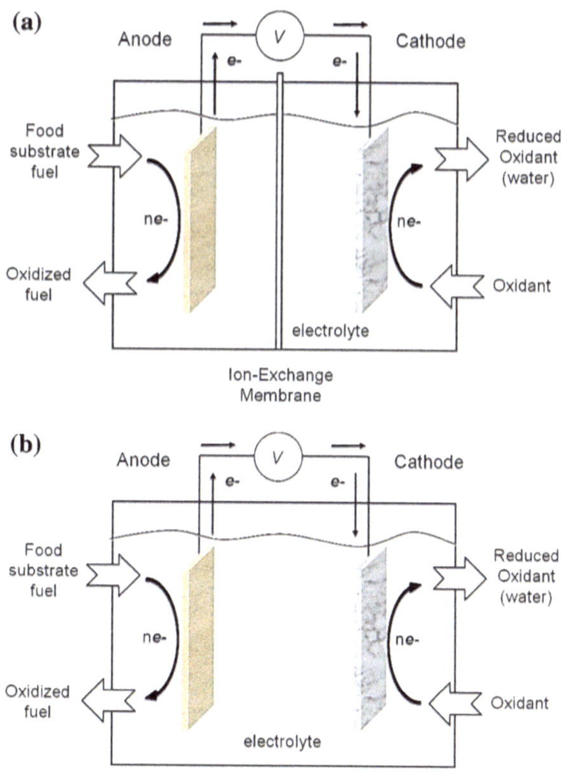

Fig. 1 Scheme of a general representation of two BFC configurations, **a** with a semi-permeable membrane, and **b** in the absence of a membrane. The fuel is oxidized on the anode by a biological component, and the electrons released are transferred throughout an external circuit, which is associated with an oxidant element (usually O_2), to generate H_2O

circuit reaching the cathode [55] and, generally, water is produced when the redox reaction takes place due to the presence of molecular oxygen in the environment [48].

The early development of BFCs consists in the investigation of an efficient collector material for the electrodes at the same time to improve the shuttle of electrons from the biomolecule to the electrode surface, such as the use of efficient architectures for the obtention of the electrodes [56–60]. For the last, the immobilization procedure of the biomolecule onto the solid support is an important feature [61–68]. The investigation of the immobilization procedure has been explored by cross-linking agents [69], physical adsorption [70, 71], self-assemble monolayers [72, 73], bilipidic membranes [74, 75], and also the entrapment in polymeric matrices [60, 76, 77], as schematized in Fig. 2.

As one of the simpler pathways for the obtention of bioelectrodes, the use of organic polymers plus inorganic materials can result in a composite or hybrid materials with unique properties [50, 78–82]. These methodologies are usually used for the obtention of the biocathodes and also bioanodes in BFCs. As an example, Scodeller and co-workers [81] reported the use of purified *Trametes trogii* laccase as a biocatalyst for an oxygen redox reaction on biocathodes composed of layer-by-layer self-assembled laccase and osmium complexes on mercaptopropanesulfonate modified gold electrodes. This work shows a relevant operating system based on the detection of H_2O_2 as a result of O_2 electroreduction and generated by

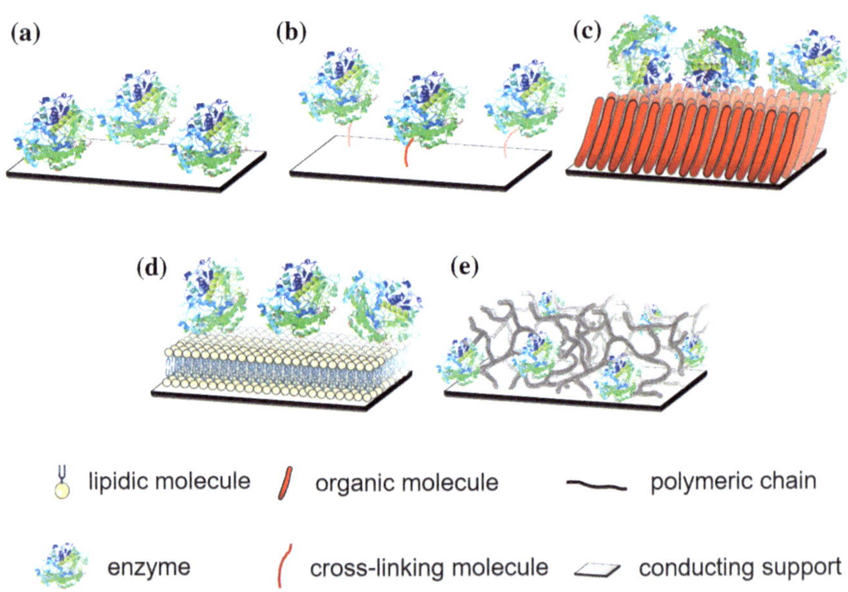

Fig. 2 Scheme of enzymatic immobilization onto a conducting solid support. **a** Physical adsorption. **b** Cross-linking. **c** Self-assemble monolayer. **d** Bilipidic membrane and **e** Entrapment in polymeric matrix

laccase biocatalysis mediated by an osmium complex. On the other hand, bioanodes have also been investigated using these polymeric platforms. Raitman et al. [82] electropolymerized polyaniline (PAni) and polyacrylic acid (PAA) on electrode surfaces resulting in PAni/PAA films. Then, an amino-FAD cofactor was covalently linked to the PAA functional group chains, and apo-glucose oxidase was reconstituted on the FAD cofactor. The bioelectrocatalytic activation of the reconstituted enzyme was attributed to the appropriate alignment of the enzyme with respect to the electroactive PAni/PAA films, which is facilitated the electron transfer between the cofactor site embedded in the enzyme and the electrode surface [82]. Carbon materials have also been shown promising for biodevices mainly owing to the chemical and physical stability [83–87], also in order of favoring the immobilization of the enzymes onto the electrode [88, 89]. Some studies have shown promising in the development of carbon-based platforms for BFCs. Cosnier and co-workers [90] reported a porous carbon matrix for the immobilization of enzymes onto the electrode surface, resulting in bioelectrodes with high current density for both anode and cathode, highlights for the high current and potential stability over one month for the biodevice. Another strategy has also been reported about the use of entrapment of redox enzymes onto biscrolled carbon nanotube yarns [91], and also hydrogels for improvement of bioelectrodes performance in H_2/O_2 BFCs [92]. In addition to the use of enzymes, microorganisms like bacteria's have also been explored for the obtention of efficient BFCs [15–17, 93–101].

Microbial BFCs provides some advantages over enzymes such as higher current densities originated from the magnitude of oxidation processes [23] and a decreased susceptibility to interfering agents or reaction inhibitors [102]. However, the improvement and understanding of operational characteristics in microbial BFCs is not a simple task. Further knowledge about some experimental factors such the continuous growth due to microorganism multiplication into the electrochemical compartment and also the electron shuttle through the biological membrane, which influences the final cell voltage, are some of the greatest complication in this type of biodevice [16, 17, 96, 103]. At the same time, it is more difficult to access the energy produced inside this cell during the metabolic process.

Conventional microbial-based BFCs are similar to other cell types with two electrodes (anode and cathode) separated, in general, by a proton exchange membrane [50, 104–107]. The oxidation of redox-active microorganisms occurs at the anode, while the cathode can be composed of a simple metal electrode, a bacterial biocathode or an oxidoreductase enzyme. The functional activity of microorganisms facilitates the oxidation of organic compounds, which produces electrical energy driven throughout an external circuit and generates water from the reduction of oxygen, while the proton exchange occurs via a semi-permeable membrane. A schematic configuration of a microbial BFC is presented in Fig. 3.

There is no doubt about that the development of microbial BFCs is a promising route for obtention of clean energy conversion devices. As in enzymatic biodevices, many operational difficulties associated with BFCs in general such as lifetime and

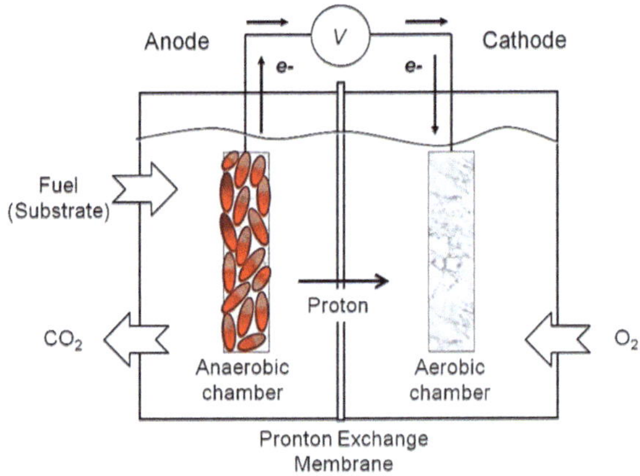

Fig. 3 A general scheme of a microbial-based BFC. The principle of power energy generation is similar to an enzymatic-based BFC

low operation stability present big challenges associated with the large-scale application of this type of bioelectrochemical system [7, 20, 21, 45, 108]. One of the main challenges today is to improve the interface for an optimum biorecognition; to achieve this goal, focus has been placed on search for different electrode configurations based on nanostructured materials, which have provided several advances in BFC development. The next section examines these two kinds of BFCs in detail and summarizes some of the most important studies reported concerning the utilization of nanostructured materials for BFCs development, which will also be detailed further. For the last feature the understanding of the BFCs thermodynamics is an important factor for future an efficient bioconversion.

3 Power Energy and Thermodynamics in BFCs

The electrical energy from biological cells can be generated by anodic/cathodic reactions from enzymatic biocatalytic processes or energetically favorable metabolic reactions on microorganisms [4, 7, 15, 16, 19–22]. The evaluation of an electrochemical system's power density, which is commonly measured as the power generated per area or volume of electrochemical cell is currently evaluated, also in the case of many different kinds of electrode configurations [4, 5, 7]. The energy released in an electrochemical reaction is described by Gibb´s energy for a reversible electrochemical reaction. To describe the reversible cell potential generated in an electrochemical reaction, the Nernst equation is used (1) [109]

$$E = E^0 + \frac{RT}{nF} \log Q \tag{1}$$

in which E^0 is the standard potential, Q is the reaction quotient (activity quotient), n is the number of electrons involved in the reaction. R, T, and F are the universal gas constant, the absolute temperature and the Faraday constant, respectively. In a biological fuel cell, electrons are shuttle from the biomolecule to the electrode and the maximum current (I) achievable and, consequently, the maximum power, will depend on efficiency of electrical communication between the enzyme and the electrode. In general, the power output for a fuel cell can be represented by (2)

$$P_{\mathrm{Cell}} = E_{\mathrm{Cell}} \int I \mathrm{d}t, \tag{2}$$

where the overall cell voltage (E_{Cell}) is given by (3) [4]

$$E_{\mathrm{Cell}} = E_{\mathrm{C}} - E_{\mathrm{A}} - \Sigma IR_{\mathrm{e}} \tag{3}$$

The term ΣIR_{e} represents the sum of the internal potential losses of the electrochemical cell. E_{C} and E_{A} are the experimental cathodic and anodic potentials, respectively, and may be related to the equilibrium potential (E_{e}) by the overpotential (η) of the cell according Eq. (4)

$$\eta = E - E_{\mathrm{e}} \tag{4}$$

One important factor for the optimization of the current density and the cell voltage in an electrochemical cell can be obtained when the ohmic resistance losses (ΣIR_{e}) is minimized and the driving force ($E_{\mathrm{C}} - E_{\mathrm{A}}$) is maximized. The configuration of electrodes on an electrochemical cell and their operation at low current densities are examples of how power generation can be optimized in BFCs. Although the low current operation generated by biomolecules is a substantial problem in BFCs, the increase in current generated by electrochemical reactions can be used to improve the electrochemical response on the electrode. As the BFCs are applied for different purposes such as portable power supplies, implantable devices, drug delivery systems and microchips, the size of the biodevice should be considered [110–114]. Thus, it is more common to calculate the geometric and volumetric power density, according to the following Eqs. (5) and (6) [4]

$$\text{Power density} = \frac{P_{\mathrm{Cell}}}{A} \tag{5}$$

$$\text{Volumetric power} = \frac{P_{\mathrm{Cell}}}{V_{\mathrm{e}}} = \frac{A_{\mathrm{e}}}{A} P_{\mathrm{Cell}} \tag{6}$$

in which A, A_{e}, and V_{e} are respectively the geometric electrode area, the electrode area per unit of volume and electrode volume. For a complete biofuel cell system, it is also needed to consider the fuel as adding volume. The power energy produced in a fuel cell can be determined by an experimental polarization curve [115], also accomplished for BFCs [116]. For the last, the efficiency is also evaluated (7)

$$\text{Efficiency} = \frac{-nF\Delta E}{\Delta H} \tag{7}$$

As discussed earlier, the performance of biofuel cells is influenced by many operation factors, such by the use of different architectures [117, 118]. Then, many efforts has also been made in order to understand the fundamentals of electron transfer reactions in biological systems, also when in contact with electrodes, such by the use of Marcus theory [61]. Further fundamental studies is also been investigated by the use of nanostructured materials together with biomolecules as a successful strategy for more efficient biodevices.

4 Nanostructured Materials Applied to Modified Electrodes and BFCs

Recent advances in bioelectrochemistry are responsible for the continuous development of nanostructured materials and interfaces [5]. During the past few decades, the intense search for new materials with different properties has been the main focus in the emerging fields of nanoscience and nanotechnology [119]. The development of nanostructures shows the possibility to control physical and chemical properties at molecular level [120]. Materials at nanometer scale are characterized by size, shape, and large surface area, which greatly influences their electrochemical properties; that is, electrochemical properties are influenced by the size/surface energy and can improve electron transfer kinetics in bioelectrodes [121]. Several nanomaterials have been applied to modified electrodes, such as carbon nanostructures [70, 122–125], nanoparticles of different compositions and sizes [122, 126–128] among others.

Carbon nanotubes represent one of the allotropic forms of carbon, and several works have reported their use in modified electrodes. Since their discovery in 1991 by Iijima and co-workers [129], carbon nanotubes have received major attention due to their unique electronic, electrochemical, and mechanic properties, as well as their potential application in many areas [130–132]. The electrochemical properties of carbon nanotubes have been extensively investigated in the last decade by Compton and co-workers [133–137]. In general, their electroactivity was initially associated with their unique tubular structure and also possibly by edge effects which were later related to metallic impurities [134]. On the other hand, carbon nanotubes have been extensively reported in modified electrodes in order to improve significantly the electrochemical detection of dopamine [138]. Another important approach was reported by the use of carbon nanotubes for the amperometric detection of glucose [139]. Some other fundamental studies have also concerned protein film voltammetry studies of proteins in order to obtain the direct communication with electrodes, or direct electron transfer reactions (DET) for development of efficient BFCs [33, 35, 88, 130, 139–142]. Gao and co-workers

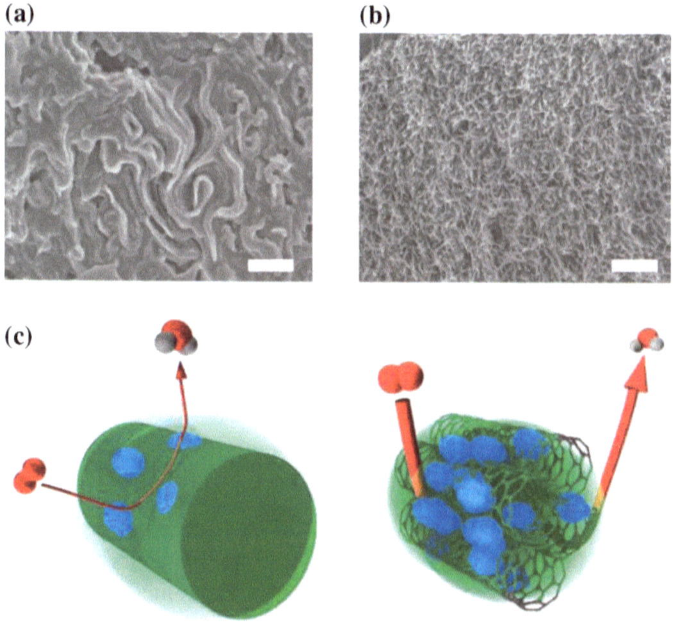

Fig. 4 SEM images of **a** CF and **b** CNT fiber. Schematic representation of oxygen reduction on **c** CF and **d** CNT fiber (the *green shade* represents the redox polymer and the *blue color* represents the enzymes). Reproduced with permission of Macmillan Publishing Group

[88] developed an electrode composite material based on carbon fibers (CFs) and oriented carbon nanotube (CNT) microwires for the electroreduction of oxygen (Fig. 4). These results show that CFs have a relatively nonporous structure. Further, the obtention of CNT microwires showed a large porous surface which allows for the permeation of both electrolyte solutions and oxygen, which are important factors for efficient bioelectrodes. In addition, the resistance of mass transfer along the porous materials can be reduced, generating a current density of 750 $\mu A\ cm^{-2}$ and only a 20% loss of their power density.

Graphene—a monolayer of carbon atoms—is a promising nanomaterial first reported in 2007 [143, 144], although such interesting electronic properties was already observed in 2004 in thin layer carbon films [145]. The electrochemical properties of graphene have also been explored during the past decade, such as in graphene oxide and reduced graphene oxide. Usually, graphene oxide is converted to the graphene form and utilized in modified electrodes for electrochemical devices and biodevices. Some studies have shown that heterogeneous electron transfer reactions is enormously influenced [146, 147], also for sensing devices [148]. For this purpose, several studies have reported the preparation of modified electrode-based graphene. Moreover, the electrochemical properties of graphene allow for their inclusion in interesting applications such as biosensing [149–151], sensing [152, 153], and BFC development [154, 155] such by the use of sol-gel method for immobilization of graphene nanosheets [154] in comparison with carbon nanotubes towards the development of a bioanode and biocathode, with a maximum power density of 24.3 μA cm^{-2} and a lifetime of 7 days. Recently, an improved electron transfer of the GOx enzyme immobilized on a flexible carbon fiber (FCF) modified with graphene oxide have been demonstrated [70]. The presence of graphene at the enzyme/electrode interface diminishes the activation energy by decreasing the distance between the electrode surface and the flavin adenine dinucleotide cofactor ($FAD/FADH_2$) of GOx, which gives an estimated electron-transfer rate constant twice the achieved with the pristine carbon electrode.

5 Enzyme-Based Modified Electrodes Applied to BFCs

Enzymes are currently used in electrodes because of the ability to convert chemical energy into electrical energy [156]. Here, it will be described some characteristics and advantages of the utilization of enzyme-based modified electrodes, while in the next topic will be described microbial-based modified electrodes. One of the advantages of enzyme-based modified electrodes in biofuel cells is the advantage that biodevices can be easily miniaturized, providing BFCs much smaller than equivalently powered microbial cells; besides, this type of cell allows operation closer to the redox potential of the enzyme itself [157].

Biological redox reactions are known to be specific and occur in live organisms to accelerate several biochemical processes associated with metabolism [20, 21, 23]. Most of these reactions need to be catalyzed to occur at the necessary rate to sustain life and in this case, the catalysts are enzymes. Once that a catalyst is not consumed or modified during the reaction, it can be used, in principle, indefinitely to convert substrate into product. As any other catalyst, enzymes act to decrease the activation energy barrier in biochemical reactions without, however, undergoing structural modifications at the end of the reaction [61, 158]. Such biochemical reactions, or catalyzed enzymatic reactions, take place with specific enzymes of a particular nature being especially suitable for a specific reaction to produce energy by coupled redox reactions that can be utilized in specific applications [21–23, 26, 32, 159–166]. For this purpose, enzymes with a metallic redox center are very interesting due to the possibility to be part of redox couple reactions as cofactors by a charge transfer with the electrode surface. Cofactors are, in some cases, organic molecules that transfer electrical charge between enzymes and the electrode surface, thereby allowing to generate a flow of current in the electrochemical system. The utilization of biocatalysts are very attractive and they present several advantages over metal catalysts, such environmentally friendly conditions (e.g., mild temperature and pH), advanced operation lifetime [167], and the possibility to use inexpensive biological fuel cell components, besides being renewable [168]. From another point of view, such biological systems allow energy generation from more complex fuels [15, 16, 19–21, 23]. Figure 5 shows a conventional configuration of

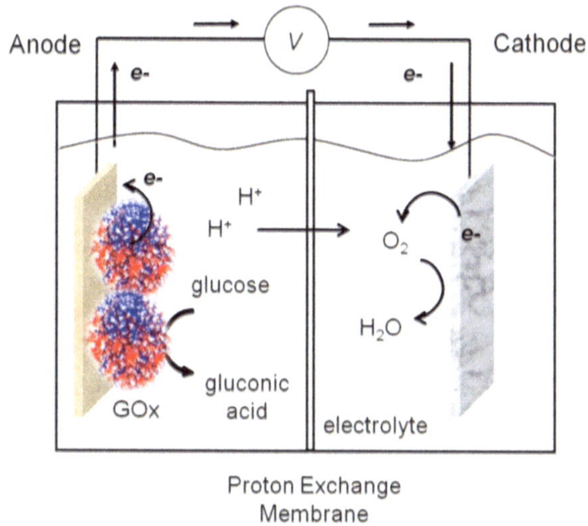

Fig. 5 Schematic representation of an enzyme-based BFC configuration. As an example of operation, glucose is oxidized with the enzyme GOx immobilized on the anode to generate H_2O on the cathode

an enzyme-based BFC operating with the oxidation of glucose on the anode by GOx.

There are two types of BFCs according to this classification. The first is based on the type of product generated by the enzymatic reactions. In this type of BFCs, enzymes do not directly produce the energy originated by the biological processes. The products catalyzed by such enzymes are used as fuel for fuel cells, such as the generation of hydrogen as fuel by hydrogenase enzymes [27]. In the second type of BFCs, biological components are directly involved in energy production, which originates from biochemical reactions. The enzymes present on the anodic electrodes catalyze the oxidation of an organic species to generate a current that depends directly on the enzymatic performance [169–171]. The key to this type of cell is to establish better communication between the enzymes and electrodes for an efficient electron transfer and a higher-than-usual current generated by the biocatalytic processes [169–171]. Enzymes are large biological molecules with redox properties that can be divided into three groups based on the localization of the enzymatic cofactors in the protein structure. The first group is classified by the presence of nicotinamide adenine nucleotide (NADH/NAD^+) or nicotinamide adenine dinucleotide phosphate (NADPH/$NADP^+$) redox centers, which are weakly bound to the proteic structure of the enzyme that actuate directly on cellular electron transfer reactions. The second group of enzymes is classified as having redox centers at or near the periphery of their protein shell structure. Peroxidases and other copper enzymes belong to this group. Enzymes of the third group have redox centers that are strongly bound to the proteic structure and are located deep inside the enzyme or glycoprotein structure.

The activity density and, consequently, the current densities at enzyme functionalized electrodes depends on the loading of enzyme immobilized on the electrode material per unit of area and the rates of two reactions: the turnover of the substrate by the enzyme working at the electrode and the transfer of electrons from active site of the enzyme to the electrode surface [172]. Thus, the major difficulty with these enzymatic redox catalysts is immobilizing the enzymes on solid electrodes and accessing the redox center localized deep inside the protein structure without the loss of their bioactivity properties. The use of these specific enzymes on electrode surfaces began many decades ago, when the first biodevices were studied to develop a biosensor with high selectivity and specificity [173]. The development of such devices brings about the possibility of studying electrochemical processes when biomolecules actuate out of their natural environments [21, 23, 174]. Although the redox enzymatic reactions can occur when enzymes are in contact with the solid electrode surfaces, they partly lose some of their bioactivity properties due to denaturation by the immobilization processes [21, 23, 174]. In fact, the main interest associated with the use of immobilization methods and developing biodevices that maximize the enzymatic biocatalysis is to better maintain the integrity of the enzyme structures by optimization of the electrode materials and immobilization methods. Biomolecules such as enzymes can be immobilized on a substrate surface by physical or chemical adsorption [61, 175–178]. Physical adsorption occurs as a result of a residual charge present on certain chemical groups of the enzymes, which causes spontaneous adsorption onto a substrate or molecule with the opposite residual structural charge. On the other hand, chemical adsorption is caused by the cross-linking of chemical groups on an enzyme's structure with other molecules on the substrate, or by the presence of bifunctional groups [43]. Although immobilization techniques are crucial in terms of providing better communication between the enzyme and electrode surface, many biological species undergo very slow heterogeneous electron transfer due to energy barriers associated with their proteic structure, since the active center of redox proteins are usually deeply located in protein matrices [179] and there are a limited distance for the occurrence of this transfer [70]. One pathway to intensify the charge transfer originated by bioelectrochemical reactions is by means of mediated electron transfer [51, 105, 169, 180–183]. Electrochemical mediators are molecules that can easily and reversibly switch between oxidation/reduction states used to facilitate electron transfer processes between redox centers on enzymes and electrodes [51, 105, 169, 180–183]. These mediators could achieve buried active site more readily than the electrode materials, which increases the number of available electrons that are transferred to the circuit [168], resulting in a better performance of the current generated in modified electrodes that can be applied to biosensors [180–182] and BFCs [51, 105, 169, 183]. Electrode preparation with most kinds of redox mediators, such as ferrocene [180, 184], Prussian blue [185, 186], 1,4-benzoquinone [187, 188], hexacyanoferrates (Co, Fe, Ni, Cu), etc., can improve current transport on the electrode surface by several orders of magnitude. Since their discovery in 1951 [189], ferrocene, along with its derivatives, is one of the most popular redox mediators used for amperometric [118, 190], potentiometric [191, 192], and immunosensor [193, 194] development. Several works have reported about the use of redox mediators to enhance charge transport in bioelectrochemical devices [51, 105, 118, 169, 180, 183, 184]. However, the redox mediator should be chosen such that its redox potential is as close as possible to that of the cofactor enzymes [20].

Several approaches have reported the utilization of nanostructured metallic materials, such as noble metal nanoparticles with redox mediators, to intensify the electrochemical response on modified electrodes [126, 151, 194–197]. The synthesis of gold nanoparticles by organic molecules produces hybrids with very interesting electrochemical properties [194, 196, 197]. This involves the utilization of PAMAM dendrimers as gold nanoparticle stabilizing agents for the development of electroactive nanostructured membranes (ENMs) [117, 198, 199]. The system is based on the use of polyelectrolytes with opposite charges for the fabrication of self-assembled multilayer thin films of PSS and PAMAM-AuNPs with a cobalt hexacyanoferrate (CoHCF) mediator electrodeposited around the AuNPs to intensify the electron transfer on the electrode surfaces [117, 198, 199]. In other words, this concept can be extended to many other redox mediators. In addition, this electrochemical approach was successfully applied to a glucose amperometric biosensor, as reproduced in Fig. 6.

Another important approach was described by Katz et al. [200], and utilized the enzyme GOx by the reconstitution of apo-GOx on charge-transporting wires consisting of a relay unit as the charge carrying element, and a flavin adenine dinucleotide cofactor (FAD) as a reconstitution site. In one configuration, pyrroloquinoline quinone (PQQ) was assembled on a gold electrode, and N^6-(2-aminoethyl-flavinadenine dinucleotide) (amino-FAD) was covalently linked to the PQQ [200]. The reconstitution of apo-GOx on the surface-bound FAD site led to a structurally aligned enzyme monolayer immobilized on a gold electrode. This electrode configuration provided an efficient electrical communication between the redox enzyme center and the electrode, leading to an oxygen-insensitive amperometric glucose sensing. Figure 7 shows the preparation of the bioanode, applicable as a recognition element for glucose sensing.

For many years GOx has been used as a powerful enzyme-based bioanode [70, 157, 201, 202], however, its utilization has some issues such as the necessity of the membrane in a biofuel cell once the oxygen usually present in cathode compartment acts like an interference for this enzyme [203]. Thus, in a past few years, biocatalysts such as glucose dehydrogenase (GDh) [60], and alcohol dehydrogenase (ADH) [204] have been used for the development of bioanodes for biofuel cells. In the case of GDh NAD^+-dependent for glucose oxidation, the active sites are unaffected by the presence of molecular oxygen allowing the absence of a membrane to divide the cell in two chambers, which is interesting for implantable biofuel cells studies, for example, since oxygen is present in blood [203]. For ADH, the interest is at the catalysis of the reversible interconversions of alcohols to aldehydes or ketones, which allows the utilization of this enzyme to development new kind of bioanodes [204].

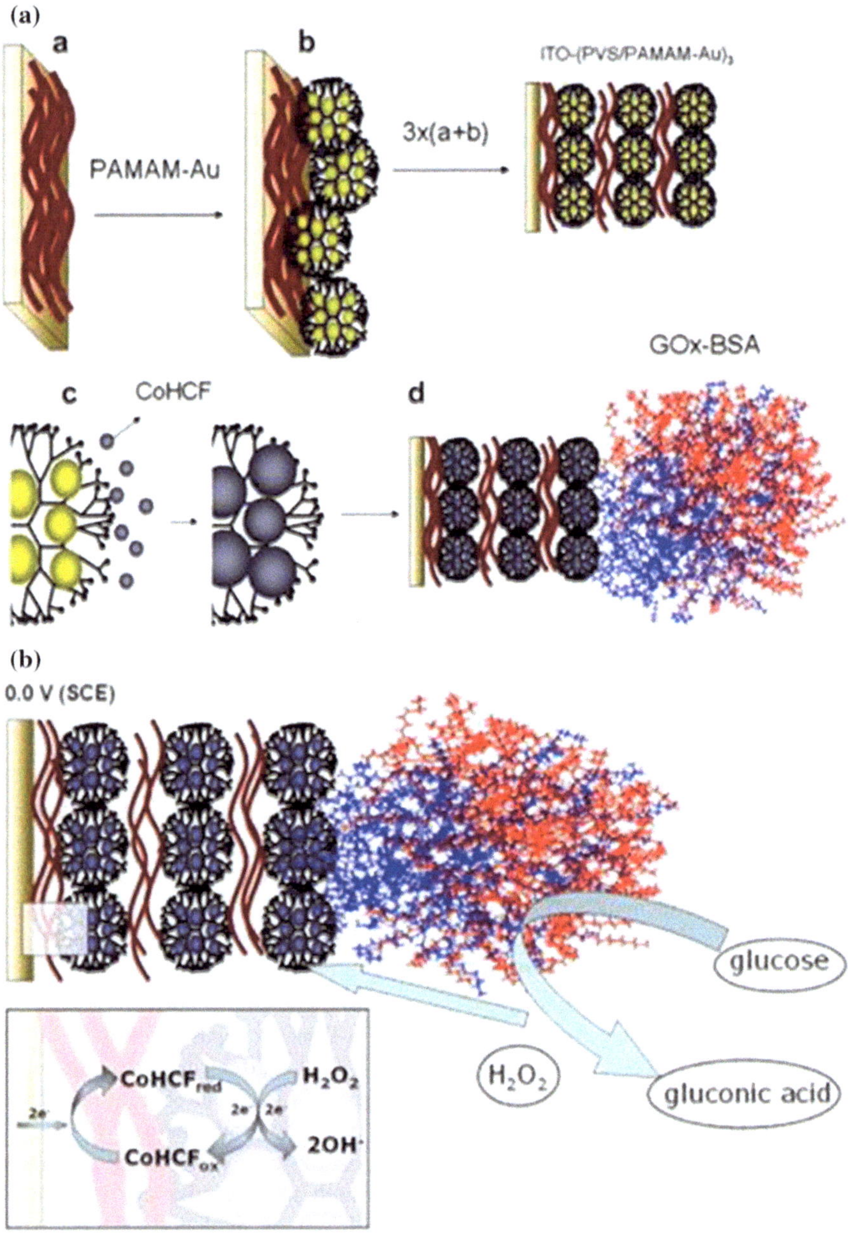

Fig. 6 **a** Schematic fabrication of layer-by-layer (LbL) films comprising PVS and PAMAM-Au. The sequential deposition of LbL multilayers was carried out by immersing the substrates alternately into *a* PVS and *b* PAMAM-Au solutions for 5 min per step. After deposition of 3 layers, an ITO-(PVS/PAMAM-Au)$_3$@CoHCF electrode was prepared by potential cycling (*c*). The enzyme immobilization to produce ITO-(PVS/PAMAM-Au)$_3$@CoHCF-GOx (*d*) was carried out in a solution containing BSA, glutaraldehyde and GOx. **b** Schematic representation of the reaction of glucose at the ITO-(PVS/PAMAM-Au)$_3$@CoHCF-GOx electrode. Reproduced with permission of Elsevier

(a)

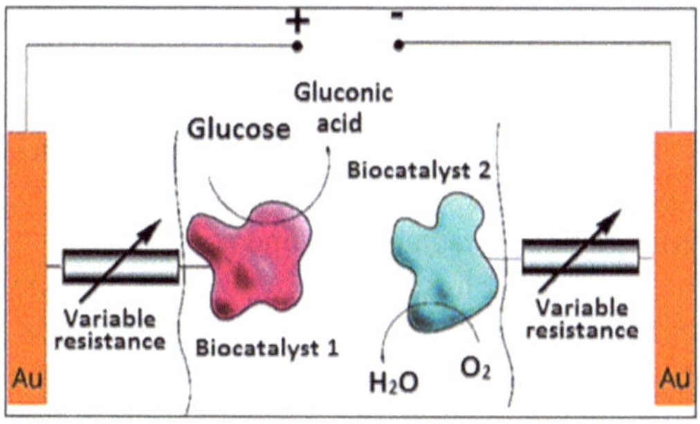

(b)

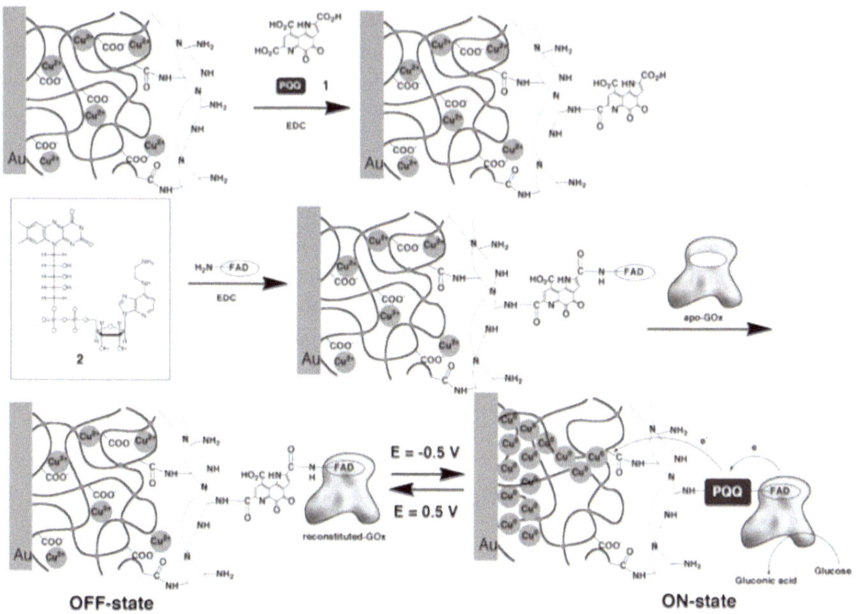

Fig. 7 **a** Schematic representation of the biocatalytic current generation of a BFC. **b** Stepwise covalent binding of PQQ (*1*) and N^6-(2-aminoethyl)-flavin adenine dinucleotide (*2*) to the polymer functionalized electrode, followed by the reconstitution of apo-glucose oxidase and the reversible activation and deactivation of the biocatalytic activity anode by the electrochemical reduction of the Cu^{+}-polymer film and the oxidation of the Cu^{0}-polymer film, respectively. Copyright @ American Chemical Society

6 Microbial-Based Modified Electrodes Applied to BFCs

The present potential application of bacteria and other microorganisms in energy storage originates from living cultures [93, 96, 97, 205, 206]. These living organisms, e.g., *Saccharomyces cerevisiae* or *Escherichia coli*, actuate on biochemical conversion of organic molecules during metabolism [207] and can be used for electrical energy generation. For this purpose, the growth of microorganisms occurs in environment culture generating energy by the oxidation of organic fuels. The utilization of microorganisms in BFCs is also advantageous since it eliminates the isolation of individual enzymes and these microorganisms can be used in different ways for producing electrical energy, such as, they can generate active substances through fermentation or metabolism and it could be used in a microbial bioreactor, which is kept separated from the fuel cell or the process could proceed directly in the anodic chamber of the cell. Another possible configuration involves the utilization of electron transfer mediators, which accept electrons from the biological reaction in the microorganism and transport them to the anode [208].

Anode-based microbial BFCs facilitate oxidation processes of organic fuels by microorganisms, and a cathode, which facilitates the reduction of O_2 with generation of water [93, 95, 99, 101, 155, 209]. In this type of fuel cells, the utilization of the microorganisms has some advantages than the utilization of enzymes, once many biofuels could be used and they can be less susceptible to poisoning and loss of activity under normal operation conditions [5]. In microbial BCFs, the two electrodes are separated by a semi-permeable membrane for the proton exchange that permits the flow of ions through the circuit. However, the major challenge is to establish better communication between electron transport inside the microorganism and the electrode surface [15, 17, 97, 100, 210]; the transfer of electrons across the microorganism membrane leads to a slow flux of ions and, consequently, generates low levels of energy at long times. The problem of current generation was partially resolved by utilization of redox mediators that are able to diffuse along the microbial membrane and transfer electrons to the anode electrode [169, 183]. Typically, a microbial fuel cell utilizing glucose as the organic fuel and a redox mediator operate according to the following reactions.

$$\text{Anodic reaction: } C_6H_{12}O_6 + 6H_2O \rightarrow 6CO_2 + 24e^- + 24H^+ \quad (8)$$

$$\text{Cathodic reaction: } 1/2O_2 + 2H^+ + 2e^- \rightarrow H_2O \quad (9)$$

$$\text{Mediator}_{\text{Ox}} + e^- \rightarrow \text{Mediator}_{\text{Red}} \quad (10)$$

Some interesting strategies have reported the utilization of microbial fuel cells without the use of redox mediators [94]. The microorganism *Rhodoferax ferrireducens* capable of glucose oxidation was reported for the oxidation of glucose [94] in a biological fuel cell generating CO_2 without a redox mediator, using graphite cathode to generate energy. Many different configurations have also been reported according to the fuel or type of catalyst utilized. Fishilevich et al. [211] reported the use of GOx as the redox enzyme displayed on the surface of *S. cerevisiae* (YSD-GOx fuel cells) in the anode compartment with methylene blue as the redox mediator (Fig. 8). The reduction of oxygen was performed by the enzyme laccase at the cathode with 2,2′-azino-bis(3-ethylbenzothiazoline-6-sulfonic acid) (ABTS) as a redox mediator. This configuration demonstrated that GOx immobilized on a microorganism's surface can be used as another important approach in the development of BFCs.

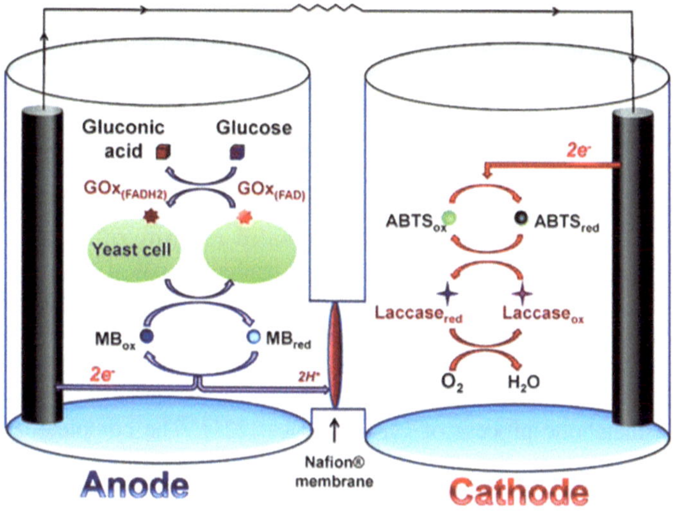

Fig. 8 Principle of operation of the YSD-GOx fuel cells. Copyright @ American Chemical Society

7 Implantable BFCs

Nature in course of evolution provided the most effective catalysts with optimal recognition capability towards biological events, the enzymes. These biological events are part of complex energy-consuming power biochemical reactions, such as single cells in mammals with average power consumption of ~4 pW [212, 213] and also neurons in the brain with a consuming process of ~0.5 nW of power [214]. Owing to the specificity, biomolecules has been investigated in integrated systems opening the opportunity to understand fundamental scientific questions and practical purposes with the so-called bioelectronics [215]. Substantial efforts have been made to efficiently attach biomolecules to the electrode surface, at the same time providing desired functions to biomaterials for practical applications. One of the fundamental requirements for bioelectronics is the effective existence of electronic coupling and communication between biomolecules and the electrode, regards to the preservation of bioactivity [216]. In particular, the continuous progress in the field opens new possibilities for power energy conversion using implantable bioelectronics, which is usually also of very low power generation.

The implant of biodevices faces some issues [217]. Some of faced issues are the miniaturization of the biodevice to be portable, having long lifetime, maintaining the operational efficiency and also the biocompatibility, one of the requisites for a successful implantable system [110]. Concerning the biomedical applications, some difficulties of the implementation of this technology are related to the necessity of biodevices with small size and lightweight, also because sometimes is formed by a non-rechargeable system. For the last, the time-life of the biodevice after the implant is crucial since the surgical procedure is an important aspect to be considered, e.g., for the implant and operational stability of cardiac pacemakers, drug pumps, and neurostimulators. Concerning micro- and nanoelectronic systems, further developments highlights the importance of the synergy between short biological events and small electrode areas, giving rise to low- and ultra-low-power bioelectronics.

The first implanted cardiac pacemaker dates from 1960s [218] by making use of myocardial electrodes with suitable controlled intensity of stimulus. Usually, pacemakers operates over consumption of tenths of μW [219]. The field of research was driven then by the development of many implantable devices and great breakthroughs with improvements in reduced scale electronics have been achieved [220], such as developed in the last decades in glucose/O_2 BFCs. Glucose is abundant in biomass and also is present in human body as an energy source via metabolic process of carbohydrates [221]. The complete oxidation of glucose molecule to carbon dioxide yields 24 electrons (according to Eq. 8). The incomplete or first step oxidation of glucose results in two electrons and formation of gluconolactone molecules, generally the predominant step when enzymes are components in enzymatic BFC anodes. The reduction of molecular O_2 enzymatically is usually a complete process through the formation of four electrons, four protons and H_2O molecules, process taking place in BFC cathodes. These two elements in blood gave rise to a promising idea for energy conversion in vivo [222, 223]. If biologically efficient converted in the final products, glucose/O_2 biofuel cells are able to provide enough power energy for some implantable devices.

Implantable glucose/O_2 fuel cell prototypes date from 70s. In an engineering device operating in vivo, abiotic components for oxidation of glucose/reduction of O_2 on Pt catalysts and porous anode containing catalyst in a disk shape format was reported with power densities of about 2.2 μW cm^{-2} during 30 days of operation. Although important advances in the field was continuously reported since then with the use of different metal catalysts and selective permeable membranes for optimizing architectures, the search for better approaches using biological electrode components owing to the selectivity and specificity is being reported [224]. GOx is capable of oxidation of glucose via two-electron mechanism reducing flavin adenine dinucleotide (FAD^{2+}) to $FADH_2$, which is regenerated with the presence of molecular oxygen in the environment, and is extensively reported as possible anode components in enzymatic BFCs. Aiming to the direct electron transfer between enzyme/electrode [122, 157], the presence of oxygen is somehow a disadvantage concerning miniaturization and efficiency of BFC systems since *cross-over* of fuels and also the presence of chemical inhibitors diminishes the operation of these biodevices. As a way to overcome these operating limitations, some studies reports the use of a semi-permeable membrane [225], once GOx is deactivated by the presence of molecular O_2 in the electrolyte media. One interesting strategy was pioneering reported by using catalase enzyme both with GOx enzyme in BFC anodes to convert H_2O_2 to H_2O and O_2, and preventing the enzyme deactivation [90]. The strategy was also reported in implantable BFC in retroperitoneal space of a rat with the design of a confined graphite biodevice in a dialysis bag [114]. Our research group also reported the obtention of an engineered implantable glucose/O_2 BFC in jugular vein of a rat (*Rattus novergicus*) (Fig. 9a) by surgical implantation of a poly(propylene) catheter containing both, anode and cathode, flexible carbon fiber (FCF) electrodes (Fig. 9b) [203]. Anode FCF electrode was modified with poly(neutral red) redox mediator to improve the communication between GOx enzyme and the electrode surface. For cathode electrode, poly(amido amine) dendrimers generation 4 was employed to synthesize platinum nanoparticles and then used for the oxygen reduction reaction (ORR). In this case, polarization curves presented high current densities over oxidation of glucose/ORR. Further results showed also good performance for the enzymatic BFC in vitro and in vivo with power densities of 200 μW cm^{-2} at 250 mV and 95 μW cm^{-2} at 80 mV,

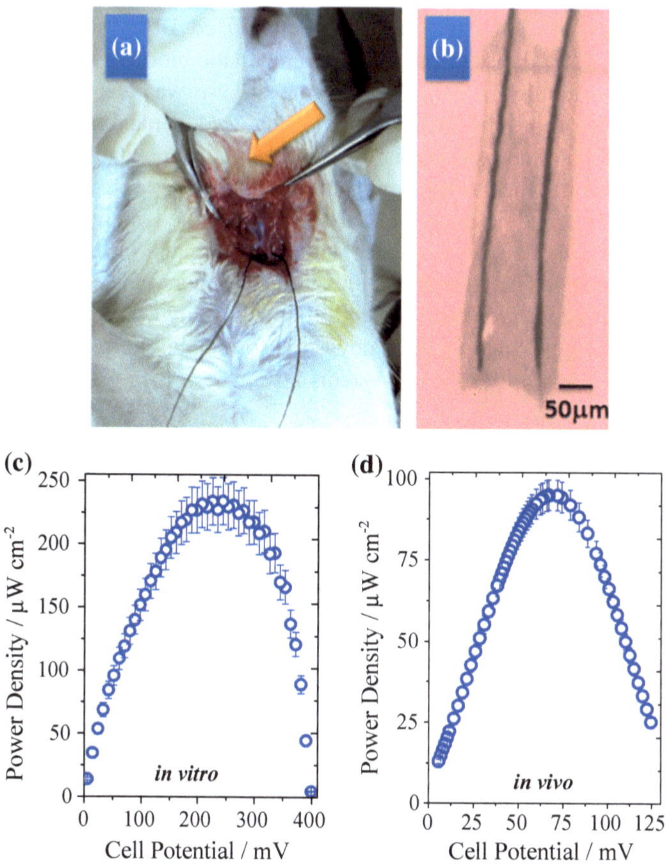

Fig. 9 **a** Optical microscope image of the FCF-based BFC inside if the catheter. **b** The catheter containing bioanode (FCF/NR/GOx) and cathode (FCF–PAMAM–PtNP) was surgically introduced at the ventral surface of the living rat. **c** In vitro micro-BFC and **d** power density versus operating potential for in vivo BFC. Copyright@RSC

respectively. Figure 9c and d shows the power output curves for in vitro and in vivo experiments.

The obtention of enough electric power from body fluids to operate a Light Emitting Diode (LED) was also reported after implant in the abdominal cavity of a rat. The reported biodevices formed a mechanically compressed CNT/enzymatic bioelectrodes affording efficient electrical contact between enzyme and the electrode. The dimensions were 6 mm thickness and 5 mm of diameter, enclosed in a silicone biocompatible tube. For the latter, enzyme GOx plus catalase enzyme and laccase was used as anode and cathode, respectively. The resulting power output density of 193.5 μW cm^{-2} was enough to deliver energy for a miniaturized device [114]. Some other studies also reported implantable BFCs in rat by incision of the bioelectrodes through the skin [226]. For the later, PQQ-dependent glucose dehydrogenase (PQQ-GDh) was immobilized onto a bucky paper, composed of compressed carbon nanotubes, and used as anode. The use of GDh enzyme in fact is being a successful strategy to obtain miniaturized BFCs since the dissolved molecular O_2 is not a limiting factor for the enzymatic biocatalysis [227–230]. The bucky paper containing immobilized enzyme laccase was also used as cathodes platforms. In this case, the implanted BFC showed a maximum power output density up to 70 μW cm^{-2} (external load of 1 kΩ). Implantable enzymatic BFC was also reported in other living organisms such lobster [231], insects [232], snail [111], clam [113], and it has potential to be implanted in human brain [233]. Wireless implantable devices were also subject of research in the last years [234]. Overall, the search for better architectures affording more efficient enzymatic BFC is still subject of intense research owing to the possibility of application in medical (implantable) devices. Such optimal devices also include the possibility to mimic biological behavior in bio-inspired systems.

8 Bio-inspired Catalysts

Proteins are one example of biological component of formidable complexity and functionality and have been a source of inspiration to design synthetic materials [235]. In resume, proteins are arranged in fold chain molecules due to the self-attraction compaction and hydrophobicity [235–237]. The possibility of mimetizing functions of some proteins for viable applications in industry and possibly to reveal scientific questions is nowadays one of the main goals to be reached. In particular, an enzyme serves as inspirational catalysts for new and functional materials [238]. Oxidoreductases are a particular class of enzymes that performs major importance owing to the capability to study catalytic electron transfer reaction from molecule–molecule path or also heterogeneous electron transfer reactions in bioelectronic devices. Such functions are related to the metal center/cluster atoms or also prosthetic groups in an active site surrounded by amino-acid residues, providing accessibility to a specific substrate molecule [239].

Although very effective, some biocatalytic conversions are not available and the demand for the design and development of functional smart materials are desirable [240]. The rapid interest for the use of synthetic materials with functional properties gives rise to the necessity to understand fundamental aspects of atoms/molecules interactions with nanoparticles [241] such as the study of constitutional dynamical systems [242, 243] or supramolecular chemistry. Recently, several self-organized systems have been proposed such by the obtention of layer-by-layer [244] and Langmuir–Blodgett [245, 246] films, organic [247] and biological supramolecular [248] spheres. Focus has also been made on the obtention of photosynthetic and also bio-inspired energy conversion devices using self-organized molecular structures.

The search for novel materials applied to BFC has arised aiming to the obtention of more efficient energy conversion devices [249, 250]. The modern concept of biomimetics allows the possibility to use bio-inspired inorganic materials/organic materials as catalysts for specific chemical reactions, also behaving in a similar fashion to natural bimolecular receptors [249, 250]. Artificial receptors provide new possibilities for bioelectrocatalysis, being obtained by self-organized inorganic–organic molecular chains [250] or by the synthesis of nanostructures [249]. Some of the recent examples are reported by the mimic of enzymes functionality, such as horseradish peroxidases [251] and NADH Peroxidase [252] using carbon dots and graphene oxide, respectively. A biomimetic iron metal center deeply located in an organic matrix was also investigated. Biomimetic Iron (III) nanoparticles stabilized in a poly(diallyldimethylammonium)(PDAC) matrix was also reported as a strategy for reduction of H_2O_2 [249]. In this case, a metastable phase of iron oxide and iron hydroxide nanoparticles (PDAC-FeOOH/Fe_2O_3-NPs) was obtained, showing a quasi-reversible anodic/cathodic electrochemical reaction attributed to Fe^{3+}/Fe^{2+} species, which produces high electrochemical catalytic activity for H_2O_2 from −290 to 390 mV in acidic media (Fig. 10).

The manipulation of metal–organic frameworks is another strategy envisioned for the obtention of highly organized materials onto surfaces with studies in coordination polymers. These organic coordination networks are formed by a connector (metal) and a link (ligant) forming highly porous and flexible materials after self-organization at vacuum–solid or liquid–solid interfaces [253]. For instance, some interesting and specific textbooks and reviews [253, 254] can be founded in the literature given to information about the obtention of metal–organic frameworks. In some cases, the study of biological-inspired electrochemical catalyst systems is shown. For the last, many studies have been focused on the obtention of highly organized metal surfaces for oxygen reduction reaction (ORR) systems [255–259]. Furthermore, molecular switches which regulates the proton-coupled electron transfer in biology and in artificial systems plays an important role such in photosynthesis and ORR reaction mechanism [255, 256] and has been recently subject of intense investigations. As a promising perspective, the control of self-organized organic/metal structures is a versatile tool for obtention of smart and small-scale devices. The future of bio-inspired materials attracts attention,

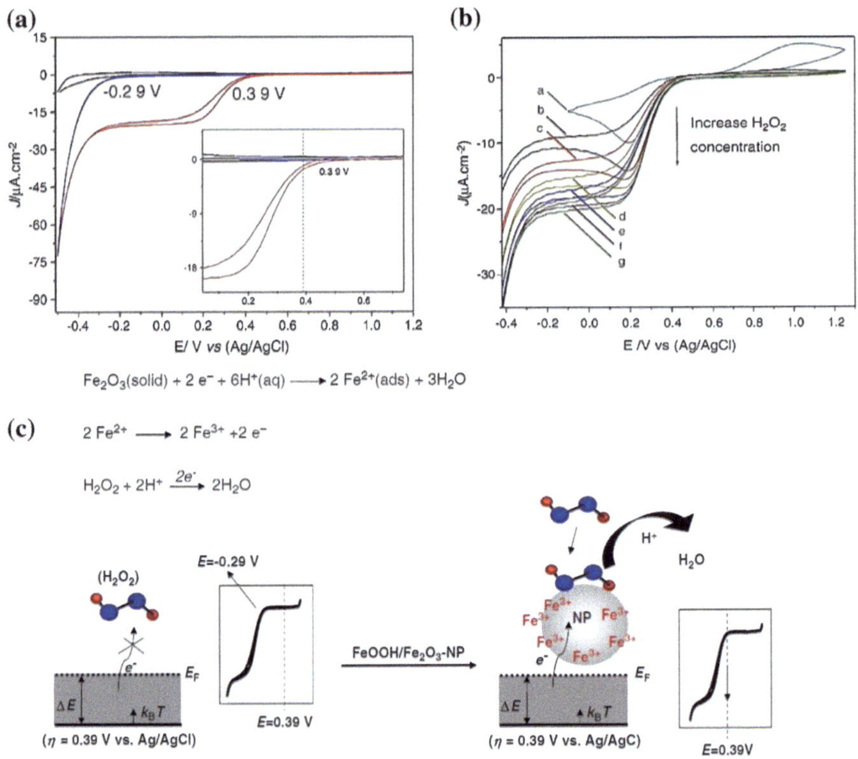

Fig. 10 **a** Cyclic voltammograms of ITO (*black line*) (0.1 mol L^{-1} H_2SO_4 + 11.6-µmol L^{-1} H_2O_2) in the presence (*red line*) and absence (*blue line*) of PDAC–FeOOH/Fe_2O_3–NPs. **b** Solution phase voltammetry of PDAC–Fe_2O_3–FeOOH–NPs at different concentrations of H_2O_2 ($a = 0.0$, $b = 1.16$, $c = 2.32$, $d = 3.48$, $e = 5.80$, $f = 9.12$ and $g = 11.67$ mol L^{-1}). **c** Proposed electrochemical mechanism for H_2O_2 reduction and schematic representation of the electrode process involving PDAC–FeOOH/Fe_2O_3–NPs. Reproduced with permission of Elsevier

highlights for simple synthetic methods that bring the possibility to obtain energy conversion devices operating on-chip.

9 Perspectives on Green Energy Generation and Future Applications

The modern and recent advances in BFC development for practical technological applications are still in an early stage, with barriers to be overcome. However, their potential applications utilizing live organisms and implantable devices for energy generation are unquestionable. Enzymatic BFCs are a promising tool for the development of these new implantable devices; they also possess challenges for developing suitable and more efficient biodevices capable of producing power energy for many applications. An efficient method of enzymatic immobilization and a more suitable modified electrode must be applied. Nanostructured materials applied to modified electrodes are an important strategy for bioanode/biocathode development; metal nanoparticles, carbon nanotubes, and graphene are very interesting materials reported by several authors to provide better performance in the field of biological fuel cell development and a more suitable and biocompatible route for enzyme immobilization. Much attention is concentrated on the improvement of communication between the enzyme's redox center and the electrode surface. The use of redox mediators provides a key for the development of more suitable communication on modified electrodes. In addition, microbial fuel cells offer the possibility of using biological processes of microorganisms like bacteria to produce energy density at long times of operation and using organic molecules as fuel. One important approach has been developed which uses microbial BFCs utilizing glucose as a renewable fuel for power energy generation. On the other hand, biomimetic materials, such as metal-oxide nanoparticles, have been used as an alternative for bioelectrode development.

Some applications for biological fuel cells are implantable biodevices and industrial reactors that use rejects, such as wastewaters, which depend on the concentration of biodegradable organic matter originating from some industries, as fuels for energy production. Thus, the biggest challenge is to develop such systems with a long lifetime of operation, high stability, and with sufficient power density to be used in technological applications. Moreover, the continuing research on biological systems conjugated with nanostructured materials, and studies on biological areas and electrochemical characteristics of biomolecules, need to be improved to a more rapid expansion of this new and clean generation of power energy.

References

1. Hubenova Y, Mitov M (2010) Bioelectrochemistry 78:57–61
2. Giroud F, Gondran C, Gorgy K, Pellissier A, Lenouvel F, Cinquin P, Cosnier S (2011) J Power Sources 196:1329–1332
3. Liu Y, Dong S (2007) Electrochem Commun 9:1423–1427
4. Bullen RA, Arnot TC, Lakeman JB, Walsh FC (2006) Biosens Bioelectron 21:2015–2045
5. Davis F, Higson SPJ (2007) Biosens Bioelectron 22:1224–1235
6. Iost RM, Crespilho FN, Kern K, Balasubramanian K (2016) Nanotechnology 27:29LT01
7. Katz E (2010) Electroanalysis 22:744–756
8. Vielstich W, Yokokawa H, Gasteiger HA (2009) Handbook of fuel cells: fundamentals technology and applications. Wiley, USA
9. Yamamoto O (2000) Electrochim Acta 45:2423–2435
10. Winter M, Brodd RJ (2004) Chem Rev 104:4245–4270
11. Heinzel A, Barragan V (1999) J Power Sources 84:70–74
12. Patil A, Davis J (2011) Coord Chem Rev 255:1970–1980
13. Armstrong FA, Wilson GS (2000) Electrochim Acta 45:2623–2645
14. Galvani L (1972) De Viribus Electricitatis In Motu Musculari. Apud Societatem Typographicam, Bologna

15. Franks AE, Nevin KP (2010) Energies 3:899–919
16. Logan BE, Hamelers B, Rozendal RA, Schrorder U, Keller J, Freguia S, Aelterman P, Verstraete W, Rabaey K (2006) Environ Sci Technol 40:5181–5192
17. Rabaey K, Verstraete W (2005) Trends Biotechnol 23:291–298
18. Wang X, Sjöberg-Eerola P, Eriksson J-E, Bobacka J, Bergelin M (2010) Synth Met 160:1373–1381
19. Zhao T (2009) Micro fuel cells: principles and applications. Academic Press, USA
20. Atanassov P, Apblett C, Banta S, Brozik S, Barton SC, Cooney M, Liaw BY, Mukerjee S, Minteer SD (2007) Interface-Electrochem Soc 16:28–31
21. Cooney MJ, Svoboda V, Lau C, Martin G, Minteer SD (2008) Energy Environ Sci 1: 320–337
22. Hao Yu E, Scott K (2010) Energies 3:23–42
23. Minteer SD, Liaw BY, Cooney MJ (2007) Curr Opin Biotechnol 18:228–234
24. Potter MC (1912) Proc Roy Soc B 84:260
25. Zebda A, Renaud L, Cretin M, Innocent C, Ferrigno R, Tingry S (2010) Sensors and Actuators B: Chemical 149(1):44–50
26. Barton SC, Gallaway J, Atanassov P (2004) Chem Rev 104:4867–4886
27. Karyakin AA, Morozov SV, Karyakina EE, Varfolomeyev SD, Zorin NA, Cosnier S (2002) Electrochem Commun 4:417–420
28. Arechederra RL, Treu BL, Minteer SD (2007) J Power Sources 173:156–161
29. Yahiro A, Lee S, Kimble D (1964) Biochimica et Biophysica Acta (BBA)-Specialized Sect Biophys Subj 88:375–383
30. Nien P-C, Wang J-Y, Chen P-Y, Chen L-C, Ho K-C (2010) Biores Technol 101:5480–5486
31. Courjean O, Gao F, Mano N (2009) Angew Chem Int Ed 48:5897–5899
32. Gao F, Yan Y, Su L, Wang L, Mao L (2007) Electrochem Commun 9:989–996
33. Yan Y, Zheng W, Su L, Mao L (2006) Adv Mater 18:2639–2643
34. Atanassov P, Colon F, Rajendran V (2004) Proc Colloid Chem Div, Am Chem Soc, Wash DC 207
35. Cai C, Chen J (2004) Anal Biochem 332:75–83
36. Tsujimura S, Kano K, Ikeda T (2002) Electrochemistry 70:940–942
37. Pizzariello A, Stred'ansky M, Miertuš S (2002) Bioelectrochemistry 56:99–105
38. Katz E, Willner I, Kotlyar AB (1999) J Electroanal Chem 479:64–68
39. Willner I, Katz E, Patolsky F, Buckmann AF (1998) J Chem Soc, Perkin Trans 2:1817–1822
40. Willner I, Heleg-Shabtai V, Blonder R, Katz E, Tao G, Bückmann AF, Heller A (1996) J Am Chem Soc 118:10321–10322
41. Milton RD, Cai R, Abdellaoui S, Leech D, De Lacey AL, Pita M, Minteer SD (2017) Angew Chem Int Ed 56:2680–2683
42. Bachas LG, Law SA, Gavalas V, Ball JC, Andrews R (2002) Abstracts of papers, 223rd ACS national meeting. Orlando, FL, United States
43. Gregg BA, Heller A (1990) Anal Chem 62:258–263
44. Bullen RA, Arnot T, Lakeman J, Walsh F (2006) Biosens Bioelectron 21:2015–2045
45. Davis F, Higson SPJ (2007) Biosens Bioelectron 22:1224–1235
46. Ticianelli E, Derouin C, Redondo A, Srinivasan S (1988) J Electrochem Soc 135:2209–2214
47. Wang Z, Wang C, Chen K (2001) J Power Sources 94:40–50
48. Topcagic S, Minteer SD (2006) Electrochim Acta 51:2168–2172
49. Liu C, Alwarappan S, Chen Z, Kong X, Li C-Z (2010) Biosens Bioelectron 25:1829–1833
50. Akers NL, Moore CM, Minteer SD (2005) Electrochim Acta 50:2521–2525
51. Barrière F, Ferry Y, Rochefort D, Leech D (2004) Electrochem Commun 6:237–241
52. Chen T, Barton SC, Binyamin G, Gao Z, Zhang Y, Kim H-H, Heller A (2001) J Am Chem Soc 123:8630–8631
53. Coman V, Vaz-Domínguez C, Ludwig R, Harreither W, Haltrich D, De Lacey AL, Ruzgas T, Gorton L, Shleev S (2008) Phys Chem Chem Phys 10:6093–6096
54. Kim H-H, Mano N, Zhang Y, Heller A (2003) J Electrochem Soc 150:A209–A213
55. Yue PL, Lowther K (1986) Chem Eng J 33:B69–B77

56. Siqueira JR Jr, Caseli L, Crespilho FN, Zucolotto V, Oliveira ON Jr (2010) Biosens Bioelectron 25:1254–1263
57. Iost RM, Crespilho FN (2012) Biosens Bioelectron 31:1–10
58. Iost RM, Madurro JM, Brito-Madurro AG, Nantes IL, Caseli L, Crespilho FN (2011) Int J Electrochem Sci 6:2965–2997
59. Crespilho FN, Iost RM, Travain SA, Oliveira ON Jr, Zucolotto V (2009) Biosens Bioelectron 24:3073–3077
60. de Souza JC, Iost RM, Crespilho FN (2016) Biosens Bioelectron 77:860–865
61. Luz RA, Pereira AR, de Souza JC, Sales FC, Crespilho FN (2014) ChemElectroChem 1:1751–1777
62. Küchler A, Yoshimoto M, Luginbühl S, Mavelli F, Walde P (2016) Nat Nanotechnol 11:409–420
63. Schmidt-Dannert C, Lopez-Gallego F (2016) Microb Biotechnol 9:601–609
64. Tello A, Cao R, Marchant MJ, Gomez H (2016) Bioconjug Chem 27:2581–2591
65. Brena B, González-Pombo P, Batista-Viera F (2013) Immobil Enzymes Cells: Third Ed 15–31
66. Sirisha V, Jain A, Jain A (2016) Adv Food Nutr Res 79:179–211
67. Ansari SA, Husain Q (2012) Biotechnol Adv 30:512–523
68. Oliveira ON Jr, Iost RM, Siqueira JR Jr, Crespilho FN, Caseli L (2014) ACS Appl Mater Interfaces 6:14745–14766
69. Migneault I, Dartiguenave C, Bertrand MJ, Waldron KC (2004) Biotechniques 37(790–796):798–802
70. Martins MVA, Pereira A, Luz RAS, Iost RM, Crespilho FN (2014) Phys Chem Chem Phys 17426–17436
71. Olyveira GM, Kim JH, Martins MV, Iost RM, Chaudhari KN, Yu J-S, Crespilho FN (2012) J Nanosci Nanotechnol 12:356–360
72. Caseli L, Crespilho FN, Nobre TM, Zaniquelli MED, Zucolotto V, Oliveira ON (2008) J Colloid Interface Sci 319:100–108
73. Crespilho FN, Emilia Ghica M, Florescu M, Nart FC, Oliveira ON Jr, Brett C (2006) Electrochem Commun 8:1665–1670
74. Caseli L, Oliveira RG, Masui DC, Furriel RP, Leone FA, Maggio B, Zaniquelli MED (2005) Langmuir 21:4090–4095
75. Hianik T, Šnejdárková M, Passechnik VI, Rehák M (1996) Bioelectrochem Bioenerg 41:221–225
76. Cosnier S, Senillou A, Grätzel M, Comte P, Vlachopoulos N, Renault NJ, Martelet C (1999) J Electroanal Chem 469:176–181
77. Naik RR, Tomczak MM, Luckarift HR, Spain JC, Stone MO (2004) Chem Commun 1684–1685
78. Abad JM, Vélez M, Santamaría C, Guisán JM, Matheus PR, Vázquez L, Gazaryan I, Gorton L, Gibson T, Fernández VM (2002) J Am Chem Soc 124:12845–12853
79. Shin KM, Kim SI, So I, Kim SJ (2009) Electrochim Acta 54:3979–3983
80. Kim J, Grate JW (2003) Nano Lett 3:1219–1222
81. Scodeller P, Carballo R, Szamocki R, Levin L, Forchiassin F, Calvo EJ (2010) J Am Chem Soc 132:11132–11140
82. Raitman OA, Katz E, Bückmann AF, Willner I (2002) J Am Chem Soc 124:6487–6496
83. McCreery RL (2008) Chem Rev 108:2646–2687
84. Pimenta MA, Dresselhaus G, Dresselhaus MS, Cancado LG, Jorio A, Saito R (2007) Phys Chem Chem Phys 9:1276–1291
85. Melanitis N, Tetlow PL, Galiotis C (1996) J Mater Sci 31:851–860
86. Nagaoka T, Fukunaga T, Yoshino T, Watanabe I, Nakayama T, Okazaki S (1988) Anal Chem 60:2766–2769
87. Iost RM, Crespilho FN, Zuccaro L, Yu HK, Wodtke AM, Kern K, Balasubramanian K (2014) Chemelectrochem 1:2070–2074
88. Gao F, Viry L, Maugey M, Poulin P, Mano N (2010) Nature communications 1:2

89. Pereira AR, de Souza JCP, Iost RM, Sales FCPF, Crespilho FN (2016) J Electroanal Chem 780:396–406
90. Zebda A, Gondran C, Le Goff A, Holzinger M, Cinquin P, Cosnier S (2011) Nat Commun 2
91. Kwon CH, Lee SH, Choi YB, Lee JA, Kim SH, Kim HH, Spinks GM, Wallace GG, Lima MD, Kozlov ME, Baughman RH (2014) Nat Commun 5
92. Plumere N, Rudiger O, Oughli AA, Williams R, Vivekananthan J, Poller S, Schuhmann W, Lubitz W (2014) Nat Chem 6:822–827
93. Aelterman P, Freguia S, Keller J, Verstraete W, Rabaey K (2008) Appl Microbiol Biotechnol 78:409–418
94. Chaudhuri SK, Lovley DR (2003) Nat Biotechnol 21:1229–1232
95. He Z, Angenent LT (2006) Electroanalysis 18:2009–2015
96. Logan BE (2009) Nat Rev Microbiol 7:375–381
97. Logan BE, Regan JM (2006) Trends Microbiol 14:512–518
98. Montanha EA, Pavinatto FJ, Caseli L, Kaczmarek O, Liebscher J, Huster D, Oliveira ON Jr (2010) Colloids Surf, B 77:161–165
99. Oh S, Min B, Logan BE (2004) Environ Sci Technol 38:4900–4904
100. Sharma V, Kundu PP (2010) Enzyme Microbiol Technol 47:179–188
101. You S, Zhao Q, Zhang J, Liu H, Jiang J, Zhao S (2008) Biosens Bioelectron 23:1157–1160
102. Lee JW, Kjeang E (2010) Biomicrofluidics 4:041301
103. Rahimnejad M, Adhami A, Darvari S, Zirepour A, Oh SE (2015) Alex Eng J 54:745–756
104. Kerres JA (2001) J Membr Sci 185:3–27
105. Coman V, Vaz-Domínguez C, Ludwig R, Harreither W, Haltrich D, Lacey ALD, Ruzgas T, Gorton L, Shleev S (2008) Phys Chem Chem Phys 10:6093–6096
106. Ticianelli EA, Derouin CR, Redondo A, Srinivasan S (1988) J Electrochem Soc 135:2209
107. Du Z, Li H, Gu T (2007) Biotechnol Adv 25:464–482
108. Kim J, Jia H, Wang P (2006) Biotechnol Adv 24:296–308
109. Brett CMA, Brett AMO, Brett A (1996) Electroquímica: princípios, métodos e aplicações, vol 1. Almedina, Coimbra
110. Barton SC, Gallaway J, Atanassov P (2004) Chem Rev 104:4867–4886
111. Halamkova L, Halamek J, Bocharova V, Szczupak A, Alfonta L, Katz E (2012) J Am Chem Soc 134:5040–5043
112. Moore CM, Minteer SD, Martin RS (2005) Lab Chip 5:218–225
113. Szczupak A, Halamek J, Halamkova L, Bocharova V, Alfonta L, Katz E (2012) Energy Environ Sci 5:8891–8895
114. Zebda A, Cosnier S, Alcaraz J-P, Holzinger M, Le Goff A, Gondran C, Boucher F, Giroud F, Gorgy K, Lamraoui H, Cinquin P (2013) Scientific reports 3:1516
115. Kharkats YI, Sokirko AV, Bark FH (1995) Electrochim Acta 40:247–252
116. Gil GC, Chang IS, Kim BH, Kim M, Jang JK, Park HS, Kim HJ (2003) Biosens Bioelectron 18:327–334
117. Crespilho FN, Ghica ME, Zucolotto V, Nart FC, Oliveira ON Jr, Brett C (2007) Electroanalysis 19:805–812
118. Qiu JD, Zhou WM, Guo J, Wang R, Liang RP (2009) Anal Biochem 385:264–269
119. Colvin VL (2003) Nat Biotechnol 21:1166–1170
120. Huczko A (2000) Appl Phys A Mater Sci Process 70:365–376
121. Park MS, Kang YM, Wang GX, Dou SX, Liu HK (2008) Adv Func Mater 18:455–461
122. Holland JT, Lau C, Brozik S, Atanassov P, Banta S (2011) J Am Chem Soc 133:19262–19265
123. Jose MV, Marx S, Murata H, Koepsel RR, Russell AJ (2012) Carbon 50:4010–4020
124. Zhao G-C, Yin Z-Z, Zhang L, Wei X-W (2005) Electrochem Commun 7:256–260
125. Zuo X, He S, Li D, Peng C, Huang Q, Song S, Fan C (2009) Langmuir 26:1936–1939
126. Luz RA, Crespilho FN (2016) RSC Adv 6:62585–62593
127. Moghaddam AB, Ganjali MR, Dinarvand R, Razavi T, Saboury AA, Moosavi-Movahedi AA, Norouzi P (2008) J Electroanal Chem 614:83–92
128. Mohanpuria P, Rana NK, Yadav SK (2008) J Nanopart Res 10:507–517

129. Iijima S (1991) Nature 354:56–58
130. Baughman RH, Zakhidov AA, De Heer WA (2002) Science 297:787
131. Thostenson ET, Ren Z, Chou TW (2001) Compos Sci Technol 61:1899–1912
132. Wang J (2005) Electroanalysis 17:7–14
133. Banks CE, Compton RG (2005) Analyst 130:1232–1239
134. Banks CE, Crossley A, Salter C, Wilkins SJ, Compton RG (2006) Angew Chem Int Ed 45:2533–2537
135. Banks CE, Davies TJ, Wildgoose GG, Compton RG (2005) ChemInform 36
136. Streeter I, Wildgoose GG, Shao L, Compton RG (2008) Sens Actuators B: Chem 133:462–466
137. Wildgoose GG, Banks CE, Leventis HC, Compton RG (2006) Microchim Acta 152:187–214
138. Britto PJ, Santhanam KSV, Ajayan PM (1996) Bioelectrochem Bioenerg 41:121–125
139. Zhang WD, Zhao YD, Chen H, Luo QM (2002) Anal Sci 18:939–941
140. Guiseppi-Elie A, Lei C, Baughman RH (2002) Nanotechnology 13:559
141. Xue H, Sun W, He B, Shen Z (2003) Synth Met 135:831–832
142. Zheng W, Li Q, Su L, Yan Y, Zhang J, Mao L (2006) Electroanalysis 18:587–594
143. Geim AK, Novoselov KS (2007) Nat Mater 6:183–191
144. Hashimoto A, Suenaga K, Gloter A, Urita K, Iijima S (2004) Nature 430:870–873
145. Novoselov KS, Geim AK, Morozov SV, Jiang D, Zhang Y, Dubonos SV, Grigorieva IV, Firsov AA (2004) Science 306:666–669
146. Li D, Müller MB, Gilje S, Kaner RB, Wallace GG (2008) Nat Nanotechnol 3:101–105
147. Stankovich S, Dikin DA, Dommett GHB, Kohlhaas KM, Zimney EJ, Stach EA, Piner RD, Nguyen SBT, Ruoff RS (2006) Nature 442:282–286
148. Pumera M, Ambrosi A, Bonanni A, Chng ELK, Poh HL (2010) TrAC Trends Analytical Chemistry 29:964–965
149. Kang X, Wang J, Wu H, Aksay IA, Liu J, Lin Y (2009) Biosens Bioelectron 25:901–905
150. Shan C, Yang H, Song J, Han D, Ivaska A, Niu L (2009) Anal Chem 81:2378–2382
151. Wu H, Wang J, Kang X, Wang C, Wang D, Liu J, Aksay IA, Lin Y (2009) Talanta 80 (1):403–406
152. Dan Y, Lu Y, Kybert NJ, Luo Z, Johnson ATC (2009) Nano Lett 9:1472–1475
153. Fowler JD, Allen MJ, Tung VC, Yang Y, Kaner RB, Weiller BH (2009) ACS Nano 3:301–306
154. Liu C, Alwarappan S, Chen Z, Kong X, Li CZ (2010) Biosens Bioelectron 25:1829–1833
155. Logan B, Cheng S, Watson V, Estadt G (2007) Environ Sci Technol 41:3341–3346
156. Ramanavicius A, Kausaite A, Ramanaviciene A (2005) Biosens Bioelectron 20:1962–1967
157. Ivnitski D, Branch B, Atanassov P, Apblett C (2006) Electrochem Commun 8:1204–1210
158. Rusling JF, Wang B, S-E. Yun (2008) Electrochemistry of redox enzymes. In: Bioelectrochemistry: fundamentals, experimental techniques and applications. Wiley, USA, pp 39–85
159. Colmati F, Yoshioka SA, Silva V, Varela H, Gonzalez ER (2007) Int J Electrochem Sci 2:195–202
160. Ramanavicius A, Kausaite A, Ramanaviciene A (2008) Biosens Bioelectron 24:761–766
161. Jia H, Zhu G, Vugrinovich B, Kataphinan W, Reneker DH, Wang P (2002) Biotechnol Prog 18:1027–1032
162. Bunte C, Prucker O, König T, Rühe J (2009) Langmuir 26(8):6019–6027
163. Taqieddin E, Amiji M (2004) Biomaterials 25:1937–1945
164. Rege K, Raravikar NR, Kim DY, Schadler LS, Ajayan PM, Dordick JS (2003) Nano Lett 3:829–832
165. Wang P, Sheng Dai SD, Tsao AY, Davison BH (2001) Biotechnol Bioeng 74:249–255
166. de la Garza L, Jeong G, Liddell PA, Sotomura T, Moore TA, Moore AL, Gust D (2003) J Phys Chem B 107:10252–10260
167. Daniel DK, Das Mankidy B, Ambarish K, Manogari R (2009) Int J Hydrogen Energy 34:7555–7560

168. Palmore GTR, Kim H-H (1999) J Electroanal Chem 464:110–117
169. Alferov SV, Tomashevskaya LG, Ponamoreva ON, Bogdanovskaya VA, Reshetilov AN (2006) Russ J Electrochem 42:403–404
170. Inamuddin KM, Kim SI, So I, Kim SJ (2008) Electrochim Acta 54:3979–3983
171. Miyake T, Oike M, Yoshino S, Yatagawa Y, Haneda K, Kaji H, Nishizawa M (2009) Chem Phys Lett 480:123–126
172. Campbell AS, Murata H, Carmali S, Matyjaszewski K, Islam MF, Russell AJ (2016) Biosens Bioelectron 86:446–453
173. Yahiro AT, Lee SM, Kimble DO (1964) Biochimica et Biophysica Acta (BBA)-Specialized Section on Biophysical Subjects 88:375–383
174. Ivanov I, Vidaković-Koch T, Sundmacher K (2010) Energies 3:803–846
175. Bullock C (1995) Sci Prog 78:119–134
176. Kennedy JF, Melo EHM, Jumel K (1990) Chem Eng Prog 86:81–89
177. Tischer W, Kasche V (1999) Trends Biotechnol 17:326–335
178. Tischer W, Wedekind F (1999) Biocatalysis-From Discov Appl 95–126
179. Heller A (1992) J Phys Chem 96:3579–3587
180. Okawa Y, Nagano M, Hirota S, Kobayashi H, Ohno T, Watanabe M (1999) Biosens Bioelectron 14:229–235
181. Smolander M, Livio HL, Räsänen L (1992) Biosens Bioelectron 7:637–643
182. Wang J, Mo JW, Li S, Porter J (2001) Anal Chim Acta 441:183–189
183. Picioreanu C, Katuri KP, van Loosdrecht MCM, Head IM, Scott K (2010) J Appl Electrochem 40:151–162
184. Hodak J, Etchenique R, Calvo EJ, Singhal K, Bartlett PN (1997) Langmuir 13:2708–2716
185. Karyakin AA, Gitelmacher OV, Karyakina EE (1995) Anal Chem 67:2419–2423
186. Ricci F, Palleschi G (2005) Biosens Bioelectron 21:389–407
187. Nakano K, Nakamura K, Iwamoto K, Soh N, Imato T (2009) J Electroanal Chem 628: 113–118
188. Nien PC, Wang JY, Chen PY, Chen LC, Ho KC (2010) Biores Technol 101:5480–5486
189. Kealy TJ, Pauson PL (1951) Nature 168:1039–1040
190. Merchant SA, Tran TO, Meredith MT, Cline TC, Glatzhofer DT, Schmidtke DW (2009) Langmuir 25:7736–7742
191. Ishige Y, Takeda S, Kamahori M (2010) Biosens Bioelectron 26(4):1366–1372
192. Kato R, Sato A, Yoshino D, Hattori T (2011) Anal Sci 27:61–66
193. Kwon SJ, Yang H, Jo K, Kwak J (2008) The Analyst 133:1599–1604
194. Qiu JD, Liang RP, Wang R, Fan LX, Chen YW, Xia XH (2009) Biosens Bioelectron 25:852–857
195. De Cuyper M, Joniau M (1992) Biotechnol Appl Biochem 16:201
196. Mukhopadhyay K, Phadtare S, Vinod VP, Kumar A, Rao M, Chaudhari RV, Sastry M (2003) Langmuir 19:3858–3863
197. Zhang S, Wang N, Yu H, Niu Y, Sun C (2005) Bioelectrochemistry 67:15–22
198. Crespilho FN, Ghica ME, Gouveia-Caridade C, Oliveira ON Jr, Brett C (2008) Talanta 76:922–928
199. Siqueira JR Jr, Crespilho FN, Zucolotto V, Oliveira ON Jr (2007) Electrochem Commun 9:2676–2680
200. Katz E, Filanovsky B, Willner I (1999) New J Chem 23:481–487
201. Fu C, Yang W, Chen X, Evans DG (2009) Electrochem Commun 11:997–1000
202. Wang D, Chen L (2009) Electrochim Acta 54:4316–4320
203. Sales FCPF, Iost RM, Martins MVA, Almeida MC, Crespilho FN (2013) Lab Chip 13:468–474
204. Pereira AR, de Souza JCP, Gonçalves AD, Pagnoncelli KC, Crespilho FN (2017) J Braz Chem Soc 28:1698–1707
205. Lewis K (1966) Microbiol Mol Biol Rev 30:101
206. Park DH, Zeikus JG (2000) Appl Environ Microbiol 66:1292

207. Ishikawa M, Yamamura S, Takamura Y, Sode K, Tamiya E, Tomiyama M (2006) Int J Hydrogen Energy 31:1484–1489
208. Shukla A, Suresh P, Berchmans S, Rajendran A (2004) Curr Sci 87:455–468
209. Cheng S, Liu H, Logan BE (2006) Environ Sci Technol 40:364–369
210. Schröder U, Nießen J, Scholz F (2003) Angew Chem 115:2986–2989
211. Fishilevich S, Amir L, Fridman Y, Aharoni A, Alfonta L (2009) J Am Chem Soc 131:12052–12053
212. Makarieva AM, Gorshkov VG, Li BL, Chown SL, Reich PB, Gavrilov VM (2008) P Natl Acad Sci U S A 105:16994–16999
213. Zhirnov VV, Cavin RK (2015) Microsystems for bioelectronics: scaling and performance limits, William Andrew
214. Sarpeshkar R (2010) Cambridge University Press, Cambridge
215. Willner I, Katz E (2000) Angew Chem Int Edit 39:1180–1218
216. Lau C, Moehlenbrock MJ, Arechederra RL, Falase A, Garcia K, Rincon R, Minteer SD, Banta S, Gupta G, Babanova S, Atanassov P (2015) Int J Hydrogen Energy 40:14661–14666
217. Hannan MA, Mutashar S, Samad SA, Hussain A (2014) Biomed Eng Online 13(1):79
218. Hayes DL, Furman S (2004) J Cardiovasc Electr 15:619–627
219. Cosnier S, Le Goff A, Holzinger M (2014) Electrochem Commun 38:19–23
220. Lee K, Park J, Lee MS, Kim J, Hyun BG, Kang DJ, Na K, Lee CY, Bien F, Park JU (2014) Nano Lett 14:2647–2654
221. Horecker BL (1965) J Chem Educ, 42(5):244
222. Katz E (2015) 6th Ieee international workshop on advances in sensors and interfaces (Iwasi) pp 2–13
223. Katz E (2010) Electroanalysis 22:744–756
224. Cracknell JA, Vincent KA, Armstrong FA (2008) Chem Rev 108:2439–2461
225. Zhang YJ, Chu M, Yang L, Tan YM, Deng WF, Ma M, Su XL, Xie QJ (2014) ACS Appl Mater Interfaces 6:12808–12814
226. Castorena-Gonzalez JA, Foote C, MacVittie K, Halamek J, Halamkova L, Martinez-Lemus LA, Katz E (2013) Electroanalysis 25:1579–1584
227. Yehezkeli O, Tel-Vered R, Reichlin S, Willner I (2011) ACS Nano 5:2385–2391
228. Ivnitski D, Atanassov P, Apblett C (2007) Electroanalysis 19:1562–1568
229. Flexer V, Durand F, Tsujimura S, Mano N (2011) Anal Chem 83:5721–5727
230. Tsujimura S, Murata K, Akatsuka W (2014) J Am Chem Soc 136:14432–14437
231. MacVittie K, Halamek J, Halamkova L, Southcott M, Jemison WD, Lobeld R, Katz E (2013) Energy Environ Sci 6:81–86
232. Rasmussen M, Ritzmann RE, Lee I, Pollack AJ, Scherson D (2012) J Am Chem Soc 134:1458–1460
233. Rapoport BI, Kedzierski JT, Sarpeshkar R (2012) Plos One 7(6)
234. MacVittie K, Conlon T, Katz E (2015) Bioelectrochemistry 106:28–33
235. Banavar JR, Maritan A (2007) Annu Rev Bioph Biom 36:261–280
236. Baker D (2000) Nature 405:39–42
237. Bernal JD (1939) Nature 143:663–667
238. Gutzler R, Stepanow S, Grumelli D, Lingenfelder M, Kern K (2015) Acc Chem Res 48:2132–2139
239. de Silva N, Ha JM, Solovyov A, Nigra MM, Ogino I, Yeh SW, Durkin KA, Katz A (2010) Nat Chem 2:1062–1068
240. Deuss PJ, den Heeten R, Laan W, Kamer PCJ (2011) Chem-Eur J 17:4680–4698
241. Shemetov AA, Nabiev I, Sukhanova A (2012) ACS Nano 6:4585–4602
242. Lehn JM (2007) Chem Soc Rev 36:151–160
243. Alencar WS, Crespilho FN, Martins MVA, Zucolotto V, Oliveira ON, Silva WC (2009) Phys Chem Chem Phys 11:5086–5091
244. Decher G (1997) Science 277:1232
245. Blodgett KB (1934) J Am Chem Soc 56:495
246. Blodgett KB (1935) J Am Chem Soc 57:1007–1022

247. Wang Y, Zeiri O, Raula M, Le Ouay B, Stellacci F, Weinstock IA (2016) Nat Nanotechnol 12(2):170–176
248. Pereira AR, Iost RM, Martins MVA, Yokomizo CH, da Silva WC, Nantes IL, Crespilho FN (2011) Phys Chem Chem Phys 13:12155–12162
249. Martins MVA, Bonfin C, da Silva WC, Crespilho FN (2010) Electrochem Commun 12:1509–1512
250. Sokic-Lazic D, Minteer SD (2008) Biosens Bioelectron 24:939–944
251. Vazquez-Gonzalez M, Liao WC, Cazelles R, Wang S, Yu X, Gutkin V, Willner I (2017) ACS nano 11(3):3247–3253
252. Wang S, Cazelles R, Liao WC, Vazquez-Gonzalez M, Zoabi A, Abu-Reziq R, Willner I (2017) Nano Lett 17:2043–2048
253. Bouju X, Mattioli C, Franc G, Pujol A, Gourdon A (2017) Chem Rev 117(3):1407–1444
254. Barth JV, Costantini G, Kern K (2005) Nature 437:671–679
255. Barile CJ, Tse ECM, Li Y, Sobyra TB, Zimmerman SC, Hosseini A, Gewirth AA (2014) Nat Mater 13:619–623
256. Tse EC, Barile CJ, Kirchschlager NA, Li Y, Gewargis JP, Zimmerman SC, Hosseini A, Gewirth AA (2016) Nat Mater 15:754–759
257. Wang J, Wang K, Wang FB, Xia XH (2014) Nat Commun 5:5285
258. Cao R, Thapa R, H. Kim, X. Xu, M.G. Kim, Q. Li, N. Park, M.L. Liu, J. Cho (2013) Nat Commun 4
259. Grumelli D, Wurster B, Stepanow S, Kern K (2013) Nat Commun 4:2904

Developments in Electrocatalysts for Oxygen Reduction and Ethanol Oxidation

Fabio H.B. Lima, Daniel A. Cantane, F.E.R. Oliveira and Nelson A. Galiote

1 Introduction

1.1 Initial Considerations

Fuel Cells are electrochemical energy conversion devices that transform the chemical energy of a fuel into electrical energy. These devices work as a regular galvanic cell, with the exception that the fuel is supplied externally. Low-temperature fuel cells, such as proton exchange membrane fuel cells (PEMFCs), can be used in a large range of power applications. For the particular case of portable applications, methanol is widely proposed as a possible fuel. The direct oxidation of methanol in fuel cells has been widely investigated, where the usage of carbon-supported Pt-Ru bimetallic nanoparticles have been presented high activity and high faradaic conversion efficiency to CO_2. However, methanol is toxic, possesses large miscibility with water, and it is not a renewable fuel, which brings serious disadvantages for application on commercial devices. On the other hand, ethanol is an attractive fuel for a direct alcohol fuel cell (DAFC), since it can be produced by fermentation of sugar-containing raw materials from agriculture. This is particularly interesting for countries like Brazil, where ethanol is already produced in large scale, and distributed by gas station networks, in order to fuel internal combustion engines of regular cars.

Despite several advances in recent years, the existing fuel cell technology still has two main drawbacks: the inadequate efficiency of energy conversion and the high Pt content of the electrocatalysts. The last one is the main problem associated

F.H.B. Lima (✉) · F.E.R. Oliveira · N.A. Galiote
Institute of Chemistry of Sao Carlos, University of Sao Paulo, Sao Paulo, Brazil
e-mail: fabiohbl@iqsc.usp.br

D.A. Cantane
Battery Laboratory, Itaipu Technological Park, Foz de Iguaçu, Brazil

F.L. de Souza and E. Leite (eds.), *Nanoenergy*,
DOI 10.1007/978-3-319-62800-4_6

with the cathode, where the oxygen reduction, generally, from the air, takes place. In order to overcome these problems, intensity research has been conducted. So, in this text, we will review, briefly, some published works that aimed at developing more active and efficient electrocatalysts for the ethanol electro-oxidation, and oxygen electro-reduction reaction for low-temperature fuel cells, operating in acid and an alkaline electrolyte.

2 Oxygen Reduction Electrocatalysis

The oxygen reduction reaction (ORR) can be represented as follows in acidic and alkaline electrolyte [1]:

A. "Direct" four-electron pathway:

Alkaline Electrolyte:

$$O_2 + 2H_2O + 4e^- \rightarrow 4\,OH^- \quad E_0 = 0.401\,V \tag{1}$$

Acid Electrolyte:

$$O_2 + 4H^+ + 4e^- \rightarrow 2\,H_2O \quad E_0 = 0.229\,V \tag{2}$$

B. "Series" pathway:

Alkaline Electrolyte:

$$O_2 + H_2O + 2e^- \rightarrow HO_2^- + OH \quad E_0 = -0.065\,V \tag{3}$$

Followed by the peroxide electro-reduction:

$$HO_2^- + H_2O + 2\,e^- \rightarrow 3\,OH^- \quad E_0 = 0.867\,V \tag{4}$$

Or by a decomposition reaction:

$$2\,HO_2^- \rightarrow 2\,OH^- + O_2 \tag{5}$$

Acid Electrolyte:

$$O_2 + 2\,H^+ + 2\,e^- \rightarrow H_2O_2 \quad E_0 = 0.67\,V \tag{6}$$

Followed by:

$$H_2O_2 + 2\,H^+ + 2\,e^- \rightarrow 2\,H_2O \quad E_0 = 1.77\,V \tag{7}$$

Or by:

$$2\,H_2O_2 \rightarrow 2\,H_2O + O_2 \tag{8}$$

On Pt-based electrocatalysts, some evidences suggest that the first electron transfer or the O_2 adsorption with the simultaneous electron transfer and the proton addition is the rate-determining step [2, 3].

$$O_2 + Pt \rightarrow Pt\text{ - - - }O_2 \tag{9}$$

Followed by an electrochemical step:

$$Pt\text{ - - - }O_2 + H^+ + e^- \rightarrow \text{products} \tag{10}$$

(the difference to the alkaline electrolyte is that the hydrogen atom for the O–H bond formation comes from the water molecules of the electrolyte, instead of from H_3O^+ species).

From the kinetic point of view, the other steps are less important, since they take place after the rate-determining step. The transfer of the second electron, with the addition of a second proton, generates OH or hydrogen peroxide species. The peroxide can diffuse to the bulk of the solution electrolyte and the reaction terminates with a process of 2 electrons. One additional transfer of 2 electrons and 2 protons involving OH or H_2O_2 completes the reduction via 4 electrons, resulting in H_2O (for the case of acidic media). So, the reaction via 4 electrons, wherein it is of major interest since it involves a higher number of electrons per O_2 molecule, has to deal with the O–O bond breaking and the O–H bond formation, as represented below [4, 5]:

$$O_2 \rightarrow O + O \tag{11}$$

$$O + H \rightarrow OH \tag{12}$$

Thus, electrocatalysts for the ORR have to present electronic structures that result in adsorption forces that strike these two competing steps: while strong adsorption leads a facilitated O–O bond breaking, weak adsorption tends to facilitate the O–H bond formation (hydrogen addition). This produces the so-called "volcano" plot of the activity as a function of the adsorption strength on the catalyst surface [4, 5].

Nørskov and co-authors [6, 7] introduced the *d*-band model. This model correlates the *d*-band density of electronic states of a metal with its ability to form chemisorption bonds, which better rationalize the catalytic activity of the metal catalysts. According to this model, as the *d*-band center shifts up, a distinctive anti-bonding state appears above the Fermi level. The anti-bonding states above the Fermi level are empty, and the bond becomes increasingly stronger as their number increase. Thus, strong bonding occurs if the anti-bonding states are shifted up through the Fermi level (and become empty), and weak bonding occurs if anti-bonding states are shifted down through the Fermi level (and become filled). Since an adsorption step is involved in the mechanism of O_2 reduction, the *d*-band structure of a metal catalyst is considered an important parameter determining the kinetic of a particular reaction. So, this evidences that it can be expected some correlation between the electrocatalyst *d*-band center and its electrocatalytic activity for the ORR.

A particularly difficult problem of the ORR electrocatalysis is the high loss in potential, which is a substantial source of the decline in the efficiency of fuel cells. Another drawback is the high Pt loading in the cathode or the low mass activity (current per mass of active metal) in the regular Pt-based electrocatalysts [8, 9]. Thus, the research in the ORR electrocatalysis aims at developing better electrocatalysts in order to reduce the cathode overpotential and the total mass of Pt.

In a recent work, Lima et al. [5] studied the kinetics of the ORR in alkaline electrolyte in well-ordered single-crystal surfaces of Au(111), Ag(111), Pd(111), Rh(111), Ir(111), and Ru(0001), and on nanoparticles of these metals supported on high-surface-area XC-72 Vulcan carbon. Figure 1a shows the volcano curves obtained for Au(111), Ag(111), Pd(111), Rh(111), Ir(111), and Ru(0001) single-crystals surfaces, and Fig. 1b shows the curves obtained for carbon-supported nanoparticles of these same metals. As can be observed, the nature of the metal has a pronounced effect on the kinetics of the ORR. The results show that Pt has the higher electrocatalytic activity, both as a single-crystal and carbon-supported metal nanoparticle. The limiting current densities (not shown) for Pt, Pd, Rh, and, at lower potentials, for Ir and Ru, have similar values, signifying that the ORR is dominated mainly by a four-electron reduction process in that potential region [2]. With Au, the limiting current for the ORR is half of that for the other metals, which indicates that the O_2 reduction on Au occurs via 2 electrons (the reaction stops at the peroxide stage), in accordance with a previous work [3, 4].

The activity of Pt monolayers, deposited on five different single-crystal surfaces using the Cu UPD technique [10], were investigated in acid and in alkaline electrolyte [5, 11, 12], and the ORR polarization curves are presented in Fig. 2a, b. As can be seen, the Pt monolayer electrocatalysts exhibited support-induced tunable activity by properly selecting the substrate, in which the activity of a Pt monolayer can be enhanced through structural and electronic effects. It can be observed that the most active of all surfaces is Pt_{ML}/Pd(111), and the least active is Pt_{ML}/Ru(0001). The plots of the kinetic current on the platinum monolayers on various substrates at 0.8 V as a function of the calculated *d*-band center, ε_d, generated a volcano-like curve, with Pt_{ML}/Pd(111) showing the maximum activity (Fig. 3a).

The modification of the electronic and chemical properties of Pt(111) surfaces by sub surface 3*d* transition metals was studied using density-functional theory [13]. In each case investigated, the Pt surface *d*-band was broadened and lowered in energy by interactions with the subsurface 3*d* metals, resulting in weaker dissociative adsorption energies of hydrogen and oxygen on these surfaces. The magnitude of the decrease in adsorption energy was largest for the early 3*d* transition metals and smallest for the late 3*d* transition metals.

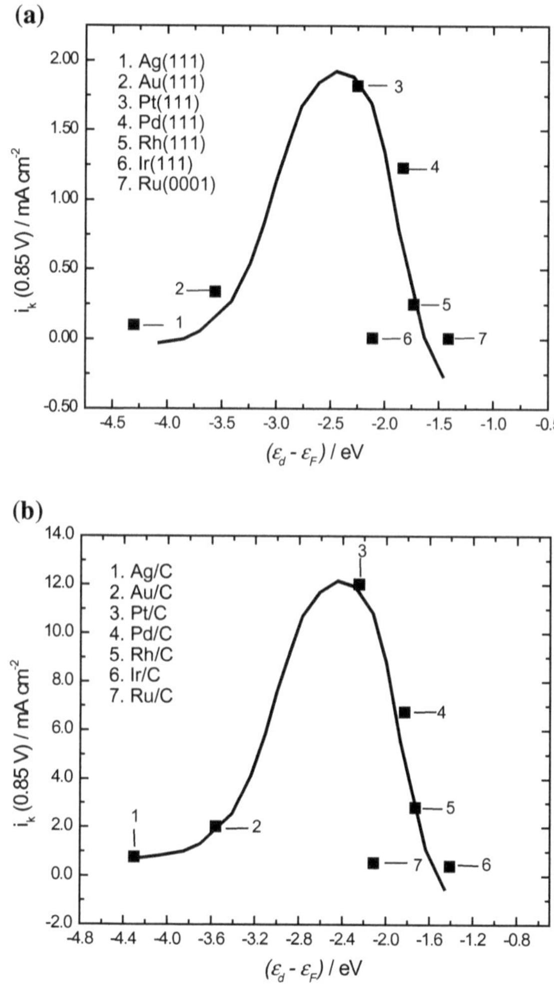

Fig. 1 Kinetic currents (j_k) at 0.80 V for an O_2 reduction on the different metal single-crystals and on carbon-supported metal nanoparticles, in 0.1 mol L^{-1} NaOH solution as functions of the metal *d*-band center (ε_d-ε_F; relative to the Fermi level). Published with the permission of the publisher [5]

The O_2 reduction reaction was carried out on platinum monolayers deposited on and on polycrystalline alloy films of the type Pt_3M (M = Ni, Co, Fe, and Ti) [14]. The Pt_3M alloys were annealed to 1000 K under ultrahigh-vacuum (UHV) conditions; analysis of the low-energy ion-scattering spectra indicated that pure Pt constitutes the first surface layer. The surface enrichment of Pt atoms results from a surface segregation phenomenon, whereby one of the alloy's components (in this case Pt) enriches the surface region [15]. UHV experimental analyses and theory revealed that the strong enrichment of Pt in Pt_3M alloy systems

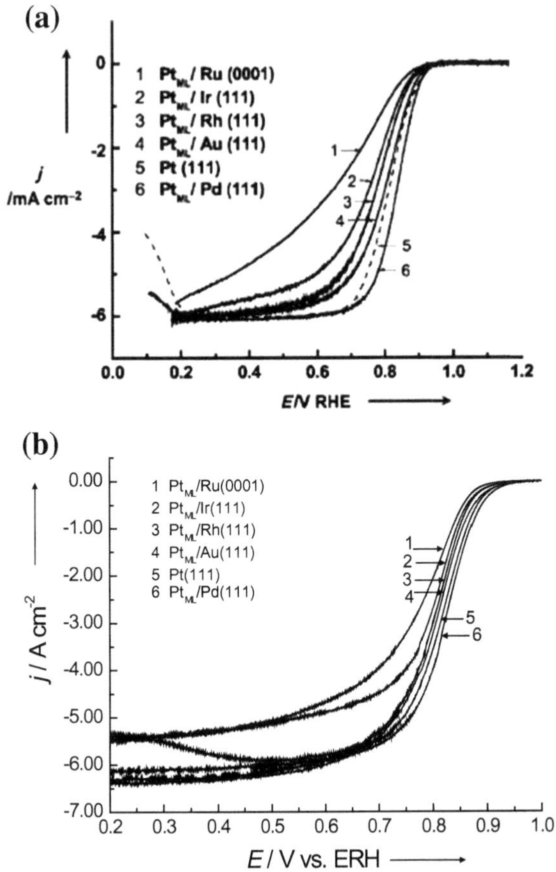

Fig. 2 Polarization curves for an O_2 reduction on Pt monolayers on Ru(0001), Ir(111), Rh(111), Au(111), and Pd(111) in a 0.1 mol L^{-1} $HClO_4$ (**a**) and NaOH 0.1 mol L^{-1} electrolyte (**b**) on a disk electrode at room temperature. For comparison, the curve for Pt(111) in acid media from Ref. [18] is shown. The rotation rate: 1600 rpm; sweep rate: 0.02 V s^{-1} (0.05 V s^{-1} for Pt(111)) (**a**) and 5 mV s^{-1} (**b**); (Published with permission of the publisher [4, 5])

is counterbalanced by the depletion of Pt in the first two or three layers beneath the surface, giving a concentration profile that oscillates around the bulk value [16]. The *d*-band center of these alloy films was measured by synchrotron-based high-resolution photoemission spectroscopy, a methodology described in a previous publication [17]. Figure 4 plots the obtained results of the electrocatalytic activity for the ORR (A) versus the position of the *d*-band center at 0.9 V [14]. It exhibits a classical volcano-shaped dependence, agreeing very well with the activity predicted from DFT calculations (black curve), and with the results for Pt monolayer deposited on different single-crystal substrates, as discussed above.

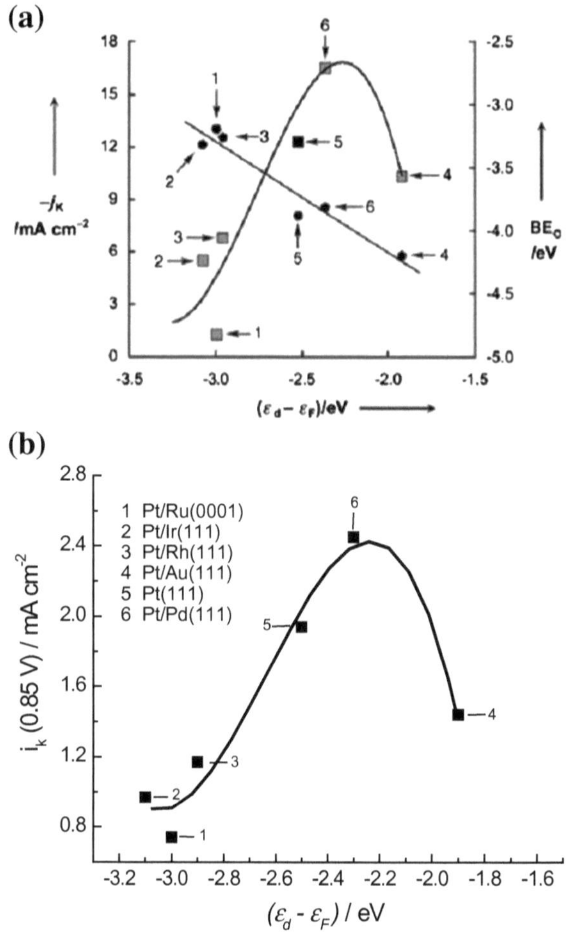

Fig. 3 Kinetic currents (j_K; square symbols) at 0.8 V for an O_2 reduction on platinum monolayers in a 0.1 mol L^{-1} $HClO_4$ (**a**) and on NaOH 0.1 mol L^{-1} solution (**b**) on the different Pt monolayers, as functions of the electrocatalyst *d*-band center. The current data for Pt(111) in (**a**) is obtained from Ref. [18]. Labels: *1* Pt_{ML}/Ru(0001), *2* Pt_{ML}/Ir(111), *3* Pt_{ML}/Rh(111), *4* Pt_{ML}/Au(111), *5* Pt (111), *6* Pt_{ML}/Pd(111) (Published with permission of the publisher [4, 5])

According to the Sabatier's principle, a good catalyst has to strike a balance between two competing influences [19–21]. The competing influences in the ORR mechanism [22–24] seems to be the first charge-transfer step, or O_2 adsorption with simultaneous charge and proton transfer, which is the rate-determining step on pure platinum. The exchange of a second electron with the addition of another proton forms two OH species or hydrogen peroxide (H_2O_2) (in acid media). The latter can escape into the solution phase and terminate the reaction in a two-electron process. An additional exchange of two electrons and two protons in reactions with OH or

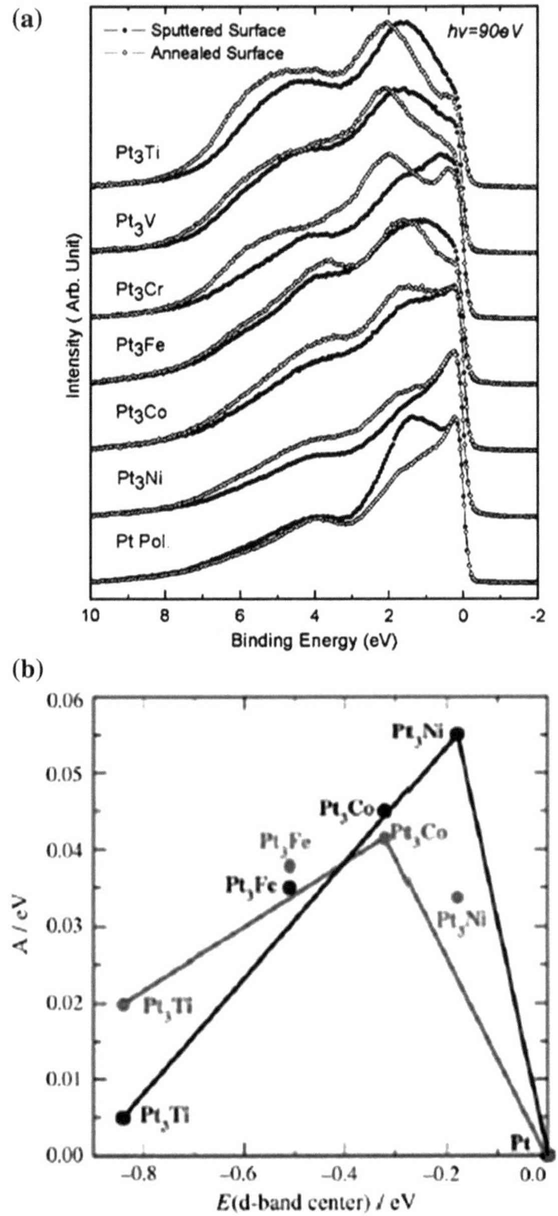

Fig. 4 **a** The background-subtracted VB (valence band) spectra of each Pt_3M polycrystalline alloy versus the pure Pt reference after ion sputtering and thermal annealing, respectively. **b** Activity versus the experimentally measured *d*-band center relative to platinum. The activity predicted from DFT simulations is shown in *black*, and the measured activity is in *gray* (Published with permission of the Publishers [14, 17])

H_2O_2 completes the four-electron reduction of O_2 to H_2O. The O_2 reduction reaction via 4 electrons must involve the breaking of the O–O bond (whether in O_2, HO_2, or H_2O_2) and the formation of an O–H bond. As discussed above, a more reactive surface, such as one characterized by a higher lying ε_d, tends to bind adsorbates more strongly [13], thus enhancing the kinetics of the dissociation reactions. On the other hand, a surface with a lower lying ε_d tends to bind adsorbates more weakly, facilitating the formation of bonds in the reaction adsorbates. So, it is expected that the ε_d of the most active platinum monolayer should have an intermediate value.

Regarding the Geometric Effect, compressive strain tends to down-shift ε_d in energy, whereas tensile strain has the opposite effect, as revealed by DFT studies [6, 7, 13]. The platinum monolayers on Ru(0001), Rh(111), and Ir(111) are compressed compared with Pt(111), whereas Pt_{ML}/Au(111) is stretched by more than 4%. However, the position of the ε_d depends also on the Electronic Effect, in which the magnitude of the ε_d shift depends on the intensity of the electronic interaction between the platinum monolayer and its substrate [25]. This indicates that Pt_{ML}/Ru (0001), Pt_{ML}/Ir(111), and Pt_{ML}/Rh(111) are less active for O_2 reduction than platinum because breaking the O–O bond is more difficult on their surfaces than on Pt(111), while the kinetics of hydrogenation of the oxygen atoms may be hindered on Pt_{ML}/Au(111) due to its stronger binding of oxygen atoms or oxygen-containing fragments and, therefore, it is less active than Pt(111). (In addition, slow rates of O or OH hydrogenation cause an increased in the O or OH surface coverage. This results in a block of the active sites for the O_2 adsorption, dissociation, or hydrogenation). These properties of Pt monolayer on different substrates result in the volcano dependence of kinetic currents as a function of ε_d [4, 5].

Therefore, the results evidence that Pt_{ML}/Pd(111) possesses a better balance between the kinetics of the two opposite steps (Eqs. 11 and 12) compared to that of Pt(111), suggesting that Pt_{ML}/Pd(111) is close to the position of an optimum compromise or an optimum balance. Experimentally, Adzic and co-authors reported that the superior ORR catalytic activity of Pt_{ML}/Pd(111) could be associated with reduced OH coverage, which was observed by in situ XANES (*X-ray Absorption Near Edge Structure*). This result agrees with DFT findings that the binding energy of OH is weaker on Pt_{ML}/Pd(111) than on Pt(111).

In the case of the Pt_3M polycrystalline alloys [14], the activity versus the position of the metal d states relative to the Fermi level also resulted in a Volcano plot. Pt_3Ni and Pt_3Co presented higher activities when compared to Pt_3Ti, Pt_3Fe or even pure Pt. It was also evidenced that the increased activity of these alloys in relation to Pt was due to a reduced Pt-oxygen species interaction, which in turns accelerates hydrogenation step. Thus, in both cases, it was observed that Pt bins oxygen a little too strong. So, in order to conduct the Pt-based catalyst closer to the optimum compromise between the two opposite steps, its reactivity for binding oxygen should be slightly reduced.

2.1 Nanoparticle

Pd seems to be the best substrate for Pt electrocatalyst. Therefore, several works aimed at synthesizing Pt monolayers on carbon-supported Pd nanoparticles. A suitable and efficient method was developed by the Adzic group and involves a galvanic displacement of a UPD Cu monolayer by a Pt monolayer [10]. The formation of this *core-shell* structure/composition can be evidenced and determined by HRTEM (*high-resolution transmission electron microscopy*) of the nanoparticle, and using HAADF-STEM (*high-angle annular dark-field scanning transmission electron microscopy*), as demonstrated recently [26].

Oxygen reduction polarization curves obtained on Pt monolayer deposited on carbon-supported Pd nanoparticles in the alkaline electrolyte are presented in Fig. 5 [27]. The results for the ORR on Pd/C (2 nm) and thermally treated Pd/C (higher particle size ~10 nm) were included for comparison. It can be noted that the ORR activity of the Pt_{ML}/Pd/C is much higher than that of Pd/C, but more importantly, it is higher than that of Pt/C. The Pt content corresponding to Pt_{ML}/Pd/C, calculated using the charge associated with depositing a Cu monolayer at under potentials on Pd/C (after correcting for the double layer charging, revealed that the Pt loading on the *core-shell* nanoparticle was seven times lower than that on Pt/C. This higher mass activity (current for the ORR per mass of Pt) of the Pt_{ML}/Pd/C material is a substantial achievement for practical and/or commercial application of such type of electrocatalysts.

Although this Pt_{ML}/Pd/C material presents a considerable reduction of the Pt mass, it still contains a large amount of Pd. So, an additional reduction of the total noble metal mass is advantageous. Using non-noble metals as a core facilitates the reduction of the noble metal's content while maintaining the full activity of a Pt monolayer by choosing an appropriated *core*. The choice of the metals constituting the *shell* and *core* can be based on considering the segregation properties of the two metals, and their electronic- and strain-inducing effects on the Pt monolayer.

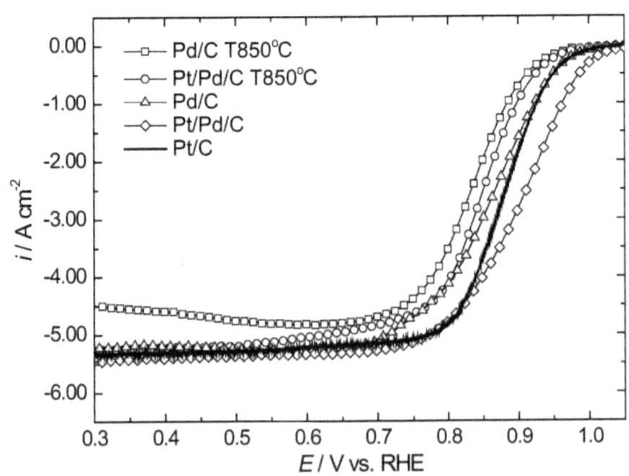

Fig. 5 Comparison of polarization curves for an O_2 reduction on Pt monolayer deposited on Pd/C and on Pd/C *T* 850 °C and on the pure substrates. The electrode geometric area is 0.163 cm^2. Published with permission of the publisher [27]

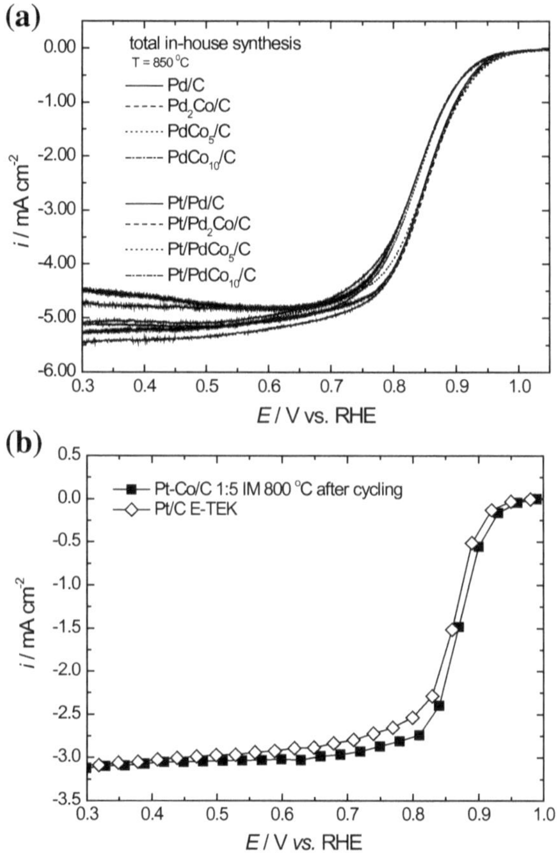

Fig. 6 **a** Comparison of polarization curves for an O_2 reduction on Pt monolayer deposited on the PdCo/C $T = 850$ °C with different atomic ratios and on the pure substrates in 0.1 NaOH electrolyte, and at 0.005 V s^{-1}. **b** Oxygen reduction on Pt–Co/C alloy electrocatalyst obtained after cycling the material in 0.5 mol L^{-1} H_2SO_4 solution, compared to that on Pt/C E-TEK; the electrolyte was 1.0 mol L^{-1} KOH, and the curves were obtained in steady state condition. Published with permission of the publishers [27, 28]

The O_2 reduction on Pt monolayer deposited on carbon-supported PdCo *core-shell* nanoparticles was studied by Lima and Adzic [27]. These carbon-supported PdCo/C *cores* have Pd surface segregated atoms, which was induced by thermal treatment under an H_2 atmosphere at 850 °C. In this work, the Pt monolayer was prepared by the Cu UPD method [10] onto PdCo/C *core-shell* surface. Figure 6 presents the polarization curves for the O_2 reduction reaction on the Pt monolayer deposited on PdCo/C presenting different Pd/Co atomic ratios. Although their half-wave potentials are below those of Pt_{ML}/Pd/C due to their lower surface area (higher particle size due to the thermal treatment), an important and surprising fact is that the half-wave potentials for O_2 reduction remain essentially constant for different Pd/Co atomic ratios (even for the lowest Pd content). This correspondence may reflect surface enrichment by Pd due to the surface segregation of Pd atoms in the substrate particles [6, 7]. The limiting current densities for all electrocatalysts are close to that of Pt/C. So, the results presented in Fig. 6a indicate that the behavior of the ORR is very similar to Pt/C and on the investigated Pt monolayer catalysts [5, 12].

As discussed before, Stamenkovic and co-authors [14] analyzed the activity towards the ORR of several Pt-3d-metal alloys, and the plot of activity versus d-band center showed a "volcano" dependency, with Pt–Co/C situated on the top of this curve. So, it becomes interesting, considering the practical point of view, to synthesize carbon-supported Pt-Co nanoparticles, presenting low Pt content, and with a Pt-enriched surface. This was the focus of a previously published work [28], where Pt-Co/C nanoparticles were synthesized by reducing the metal ions in the H_2 atmosphere, followed by thermal treatment at high temperatures. However, the Pt surface enrichment is difficult to be achieved for this composition (Pd has a higher tendency to segregate on Co), and so, a large amount of the non-noble metal atoms was still present on the surface particle alloys. In acid electrolyte, these Co atoms at the surface may suffer dissolution, resulting in a Pt-enriched surface. However, an alkaline electrolyte, the non-noble metal may form an oxide or hydroxide phase at the catalyst surface, introducing a "screening" effect [29] on neighboring Pt atoms, lowering the number of Pt sites for the oxygen reduction reaction (site blocking). The screening effect may overcome any electronic effect induced by the Co atoms, and this can explain the lower or similar activity of the Pt–Co/C catalysts compared to Pt/C obtained by previous works in alkaline media [9].

In order to overcome this problem, after the thermal treatment of the Pt-Co/C (1:5 atomic ratio) nanoparticles [28], this material was electrochemically cycled in the sulfuric acid electrolyte in order to remove (dissolve) de remaining Co atoms at the nanoparticle surface. Figure 6b shows the polarization curves for the O_2 reduction after cycling the Pt-Co/C material in the H_2SO_4 electrolyte. The ORR curve on the commercial Pt/C E-TEK (20 wt%) was included for comparison. As can be observed, the half-wave potentials of these curves clearly indicate higher activity for the cycled alloy catalyst. The higher activity obtained for the Pt-Co/C catalyst, after the Co dissolution from the particle surface, seems to be a consequence of three aspects: (i) increased surface area; (ii) lower "screening effect"; (iii) modification in the 5*d*-band properties of the surface Pt atoms, mainly caused by lattice mismatch and by a electronic coupling of the Pt in the *shell* and Co atoms in the beneath layer. A lattice mismatch of the Pt atoms on the Pt-rich surface and the Pt–Co alloy core may exist, caused by the Pt–Pt bond contraction due to the presence of the Co atoms, which have atomic radii lower than that of Pt. However, this may conduct to a small compressive strain in the Pt surface layer and, probably, this effect contributes little to the changes in the Pt 5d-band properties. On the other hand, the electronic interaction of the Pt and Co atoms (ligand effect) leads to a significant lowering of the Pt *d*-band center. This effect conducts to a faster electro-reduction of oxygenated intermediates (O–H bond formation) and/or a

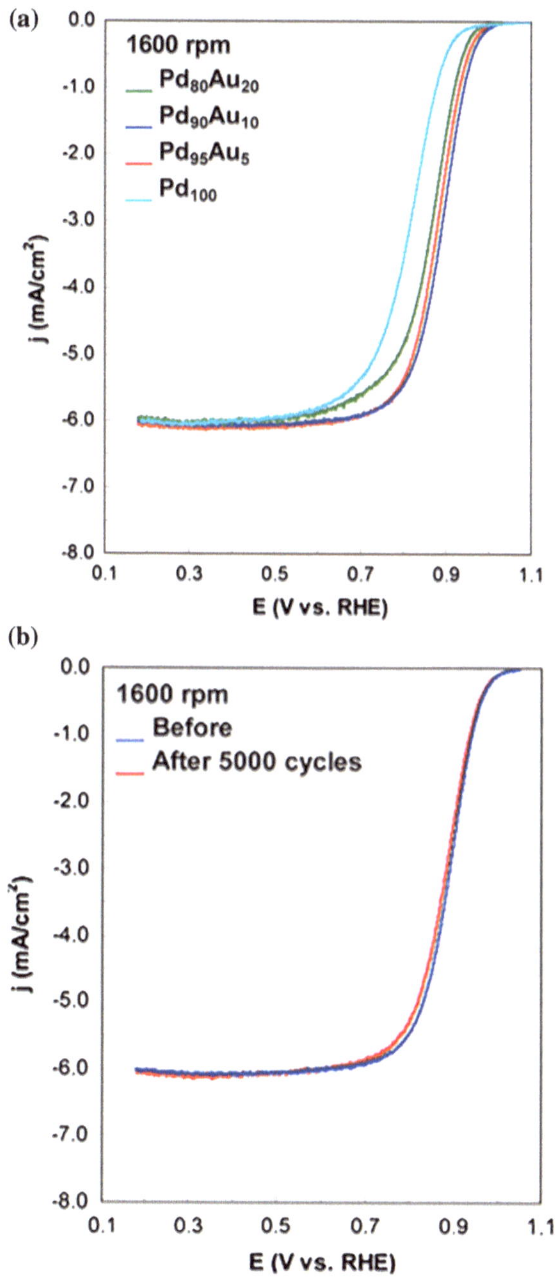

Fig. 7 **a** ORR activities of different catalysts obtained with electrode rotating speed at 1600 rpm in 0.1 M HClO4, showing the best activity with an alloy sublayer of $Pd_{90}Au_{10}$ and the worst without a sublayer; **b** stability of ORR of the catalyst with $Pd_{90}Au_{10}$ alloy sublayer, showing a small deactivation of the catalyst after 5000 cycles. Published with the permission of the publisher [31]

decrease in the Pt-oxides coverage from water activation (lower site blocking effect), in both cases inducing an increase of the ORR kinetics. So, the role of Co atoms is, principally, to promote the increase of the catalyst surface area, after their leaching from the surface, and to lower the Pt *d*-band center, as a result of the electronic interaction of the Pt atoms at the catalyst surface, and the Co atoms in the inner layers.

Regardless considerable recent advances have been achieved in the ORR electrocatalytic activity of nanoparticles, including increased specific activity (A cm^{-2}—total surface area), and increased mass activity (A g^{-1}—active metal—Pt), the stability of these electrocatalysts during the fuel cell operation is of vital importance. One central problem is the substantial loss of the Pt surface area over time in PEMFC during the stop-and-go, and upon holding at constant potentials for an extended period of time. A great step forward was accomplished demonstrating that platinum oxygen reduction fuel cell electrocatalysts can be stabilized against dissolution under potential cycling regimes by modifying Pt nanoparticles with gold clusters [30]. This was ascribed to the decreased oxidation of Pt nanoparticles covered by Au in comparison with the oxidation of pure Pt, as evidenced by in situ XAS (*X-ray Absorption Spectroscopy*).

In the particular case of *core-shell* electrocatalysts, the stability of the core has to be considered. Recently, it was reported a new design of the *core-shell* catalyst consisting of double shells instead of a single one. The two shells were made on a Pd nanoparticle core, with the outermost shell being a Pt_{ML} and the one underneath is an alloy sublayer of $(Pd\text{-}Au)_{ML}$ [31]. This *core-shell* catalyst was obtained by performing Cu UPD twice on a Pd/C catalyst and the Pd:Au atomic composition was varied. The ORR curves obtained on the Pt monolayers deposited on PdAu sublayers with different compositions is presented in Fig. 7a. Analyzing the half-wave potentials, it can be seen the following order of activity: $Pt_{ML}/Pd_{90}Au_{10} > Pt_{ML}/Pd_{95}Au_5 > Pt_{ML}/Pd_{80}Au_{20} > Pt_{ML}/Pd_{100}$.

The presence of Pd–Au alloys sublayer may conduct to lateral strain (compressive and tensile) and radial contraction (compressive strain), which results in an activity improvement for the ORR, and, in that case, the optimal alloy composition was 10 at.% Au. Furthermore, the interaction of the substrate with Pt_{ML} could result in an electronic effect on the ORR activity through charge redistribution. The formation of PtOH is significantly prohibited for the Pt_{ML} on Pd nanoparticles when compared with Pt nanoparticles, as observed by in situ XAS and cyclic voltammetry [12]. Therefore, as hydroxyl formation is considered as a major factor affecting ORR activity because of site blocking effect by OH, the increased activity for the O_2 reduction has an important contribution from the reduced coverage by OH on the Pt sites. So, in this case, the Au in the sublayer of the $Pt_{ML}/PdAu$ catalysts may have contributed to the reduction of surface oxidation, leading to improved ORR activity of the catalysts.

The stability of the catalysts was tested using $Pt_{ML}/Pd_{90}Au_{10}$, which was the best electrocatalyst. Their ORR activities are shown in Fig. 7b before and after 5000 cycles at 1600 rpm. As can be observed, the catalyst has almost no loss in ORR activity, demonstrating that the catalyst is practically stable. Surface oxidation may lead to catalyst dissolution, and so, the stabilization of the catalyst may be due to the reduction of surface oxidation from Au in the sublayer, due to the fact that Au may play an important role in protecting low-coordination sites from being oxidized [32]. Therefore, the Au in the Pd–Au alloy may have improved the catalysts by affecting the lattice contraction (both lateral and radial) and preventing some specific sites of Pt_{ML} from oxidation.

2.2 Non-noble Electrocatalysts for the Oxygen Reduction Reaction

As mentioned in the introduction section, fuel cells appear as a promising alternative as electrochemical energy conversion devices. However, their large-scale commercialization is hindered by the use of costly platinum-based electrocatalysts. Therefore, alternative electrocatalysts based on more abundant elements are mandatory for application in practical fuel cell systems.

Jasinski, in 1964 [33], was the first to report the use non-noble metal-based electrocatalyst for the ORR. He used cobalt phthalocyanines as oxygen reduction electrocatalyst due to its ability of chemisorbed molecular oxygen without active site degradation and also due to its insolubility in basic or acidic medium (mandatory characteristics for fuel cell application). However, in 1970 Jahnke et al. [34], made a breakthrough by employing pyrolyzed phthalocyanines as an electrocatalyst, enhancing both the activity and stability toward the ORR. Additionally, it was observed that these materials did not lose their central $M\text{-}N_4$ macrocycle precursor structures.

This discovering was lately followed by Yeager et al. [35] that used pyrolyzed nitrogen-rich non-macrocycles materials and metals precursors (iron and cobalt nitrates and/or chlorides). The resulting material showed to possess enhanced activity toward the ORR. By this time, authors attributed the ORR activity to $M\text{-}N_4$ structures formed during the synthesis procedure, similar than those structures encountered in the macrocycles materials. However, due to the heterogeneity introduced by the heat treatment, especially for temperatures above 800 °C, there are some concerns about which kind of structure should be related to the ORR activity and stability of these materials. Nowadays, it is widely accepted that there are two types of metal dependent active sites in discussion: (i) metal chelates forming an $M\text{-}N_4$ embedded into a carbon matrix, in which the metal atoms are coordinated with nitrogen atoms inserted into a carbon matrix, represented by metal-N-C and (ii) metal nanoparticles encapsulated by a nitrogen-doped carbon shell structure, represented by Metal@N-C [36]. On behalf of M-N elucidation as active sites, Mayrhofer et al. [37] proposed an investigation of electrocatalysts based on chelated iron species using *on line* inductively coupled plasma mass spectroscopy (ICP-MS) and differential electrochemical mass spectroscopy (DEMS), also coupled with ex situ Mössbauer spectroscopy. The pristine material

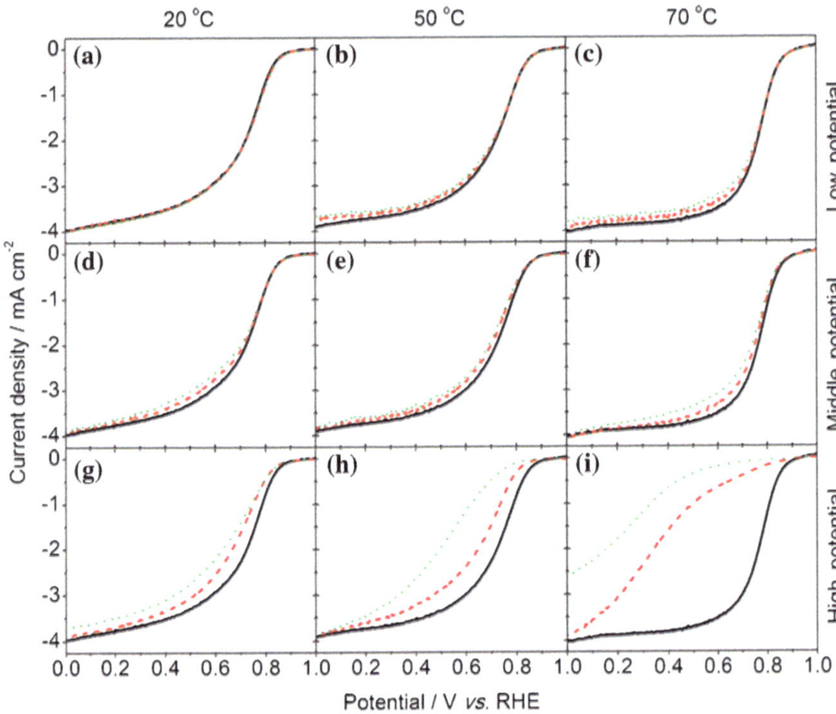

Fig. 8 The decay of the ORR activity following potential cycling. The ORR activity in O_2-saturated 0.1 M $HClO_4$ was measured before (*solid lines*) and after either 2000 potential cycles (*dashed lines*) or 5000 cycles (*dotted lines*). The cycling potential range was 0.1–0.4 V (**a–c**), 0.6–0.9 V (**d–f**), or 1.2–1.5 V (**g–i**). The cycling was performed at 20, 50, or 70 °C (first, second, and third column, respectively). Reproduced with permission from Ref. [37]. Copyright 2015, Wiley Online Library

was identified to contain both referred structures (even after performing an acidic treatment to dissolve unprotected or the excess of metallic species on the electrocatalysts). The pristine material was then submitted to 2000 and 5000 fast voltammetric cycles in the different potential window (below 0.7 V, between 0.7–0.9 V, and above 0.9–1.5 V vs. RHE), and the obtained results are presented in Fig. 8.

When the material was cycled below 0.7 V, there was a continuous iron demetallation, as detected by ICPMS, without leading to differences in the ORR activity (onset potential ≅ 0.90 V vs. RHE in 0.1 M $HClO_4$). In this way, they referred that this iron demetallation is associated with the dissolution of the crystalline metallic iron nanoparticles present on the surface, which are unprotected from the acidic electrolyte by a carbon shell. However, the authors considered them inactive toward ORR, since the ORR onset potential did not change.

In the case of the potential cycling between 0.7 and 0.9 V, there was no demetallation and the material maintained the initial ORR activity. However, when

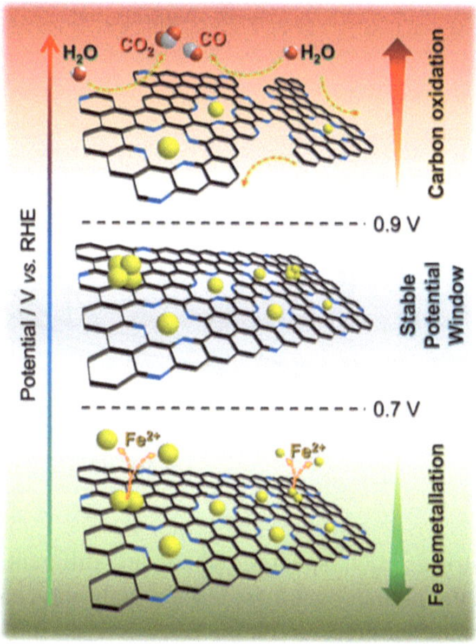

Fig. 9 Scheme showing the demetallation process observed by FeN/C materials during voltammetric cycles and on line ICP-MS measurements. Reproduced with permission from Ref. [37]. Copyright 2015, Wiley Online Library

the electrode experienced potentials higher than 0.9 V, there was an increase in Fe loss (detected by ICP-MS) and ionic signals for $m/z = 44$ and 28, using *on line* DEMS, which are attributed both to CO_2 coming from the electro-oxidation of the carbon support. Using this set of experiments, the authors proposed, based on the scheme shown in Fig. 9 that the stable potential range is found at 0.7–0.9 V, and claimed that the ORR is exclusively associated with the FeN_4 structures.

On the other hand, the encapsulated metallic particles induced by heat treatment of precursors above 900 °C, firstly reported in 1996 by Lalande et al. [38], were indicated to be the responsible for the ORR activity as well as for the stability evidenced during 10 h fuel cell tests at 50 °C. However, according to Miller et al. [39], the pyrolysis of phthalocyanines-derived compounds (FeN_4 dominant structure) thermally treated at temperatures above 900 °C is not enough to lead the total conversion/degradation of FeN_4 structures into N-doped carbon-encapsulated iron nanoparticles. So, in order to credit the higher stability and ORR activity to encapsulated nanoparticles, there is a need to synthesize electrocatalysts with only one type of structure. To address this important issue, Mukerjee et al. [40], claimed to synthesize a FeN_4—free material that is constituted by iron and iron carbides nanoparticles encapsulated by an N-doped carbon shell. The electrocatalyst characterization, carried out by XAS and Mössbauer spectroscopies, showed that most of the material is constituted in encapsulated nanoparticles and a little amount of

Fig. 10 Electrochemical characterization of catalysts. Cyclic voltammograms of ORR on as-prepared, Cl_2-treated and H_2-treated catalysts in 0.1 M $HClO_4$. Reproduced with permission from Ref [41]

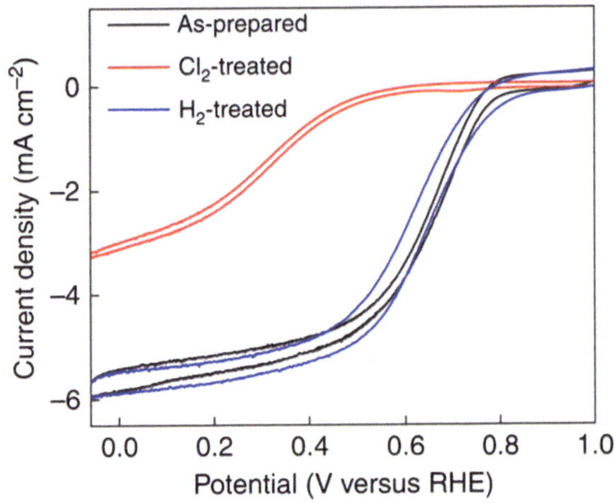

nitrogen-rich iron nitride Fe_zN_x species, which are different from those FeN_xC_y moieties identified in the phthalocyanines-derived materials (commonly represented just as FeN_4). Their material presented high ORR activity in 0.1 M $HClO_4$ with an onset potential at 0.93 V and $E_{1/2}$ = 0.77 V (vs. RHE). Also, they conducted durability tests under fuel cell conditions. After 10,000 potential cycles (the protocol that simulates drive cycle conditions), this material proved to be more durable than the state-of-the-art Pt/C. Besides, the authors concluded that the enhanced activity towards ORR should not be related to those small quantities of nitrides, but to the synergistic relationship between Fe/Fe_zC nanoparticles on the N-doped carbon sheets, allowing the 4-electron mechanism, for which the core of nanoparticle participates as a stabilizer to the peroxide intermediate.

A recent work reported by Varnell et al. [41] had a greater notoriety among others in questions to rule out the ORR activity relationship between the coordinated iron species and encapsulated nanoparticles. In this study, the so-called "as-prepared" material was synthesized from a pyrolysis (900 °C) of a nitrogen-rich polymer and an iron precursor. They subjected the pristine material to a high-temperature (900 °C) treatment in Cl_2 atmosphere and, in a second step, to an H_2 atmosphere (900 °C). They accompanied the evolution of the structure of these materials by XAS and Mössbauer spectroscopy and measured the resulting ORR activity. Surprisingly, the "as-prepared" and the H_2-treated material presented similar ORR activities (onset ≅ 0.90 V vs. RHE in 0.1 M $HClO_4$), and the Cl_2-treated material was less active to ORR (onset 170 mV negatively shifted), as can be seen in Fig. 10.

The effect of the Cl_2 atmosphere treatment at high temperature was to oxidize the iron components (FeN_4 and iron nanoparticles) to $FeCl_3$ and, according to Fig. 11b, when subjected to a reductive treatment in the H_2 atmosphere, all the $FeCl_3$ was transformed to N-doped carbon-encapsulated iron nanoparticles (Figs. 11 and 12).

These electrocatalyst structures were also confirmed by the results of Mössbauer spectroscopy shown in Fig. 12, which evidence the characteristic doublet

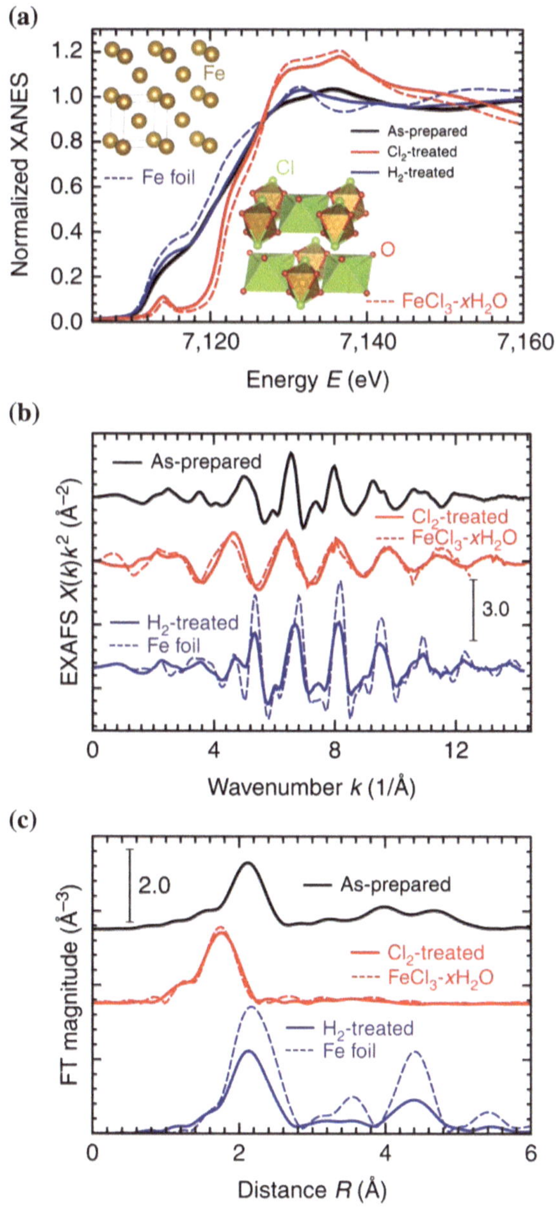

Fig. 11 X-ray absorption spectra for as-prepared, Cl_2-treated and H_2-treated catalysts. XANES (**a**), EXAFS (**b**) and Fourier transforms (*FTs*) of EXAFS (**c**). *Dashed lines* show spectra for reference materials: Fe foil (bcc structure) and hydrated Fe (III) chloride ($FeCl_3xH_2O$). Structures of reference materials are shown in the *insets* of **a**. Reproduced with permission from Ref [41]

associated with FeN_4 structures, along with the signals related to the α-Fe/Fe_3C species for the pristine material. On the other hand, for the H_2-treated sample, only α-Fe/Fe_3C species of the iron nanoparticles were detected. Together, these complementary techniques pointed out that the ORR activity has an intimal correlation

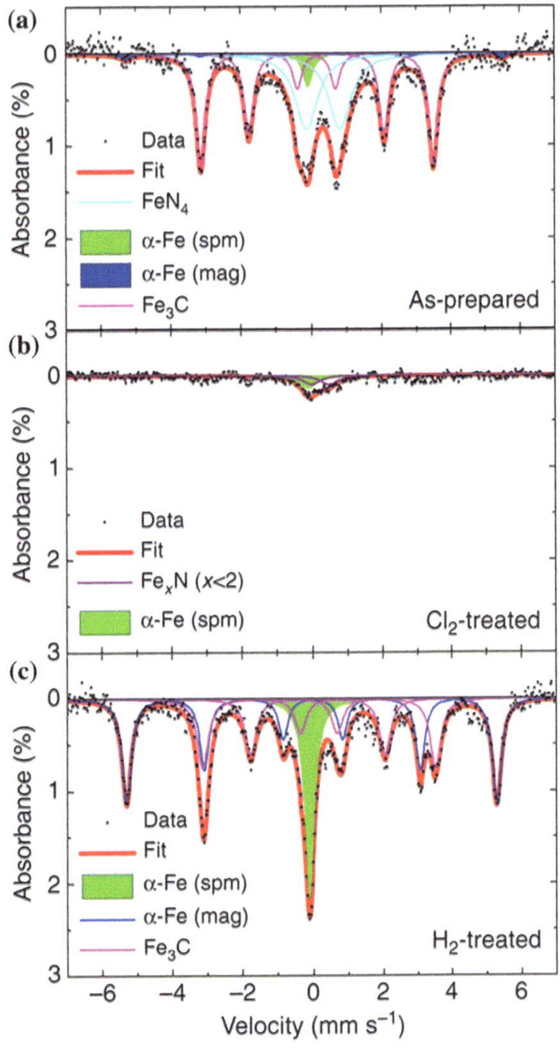

Fig. 12 Mössbauer spectroscopy characterization of catalysts. Spectra and peak fitting of as-prepared **a** Cl_2-treated **b** and H_2-treated **c** catalysts at 300 K. Both magnetically split (*mag*) and superparamagnetic (*spm*) metallic Fe species are observed. Reproduced with permission from Ref. [41]

(or dependence) with structures formed by iron/iron carbide nanoparticles encapsulated by a nitrogen-doped carbon shell (Fig. 12).

In addition, these authors conducted the same synthesis and treatment procedures, but now without adding the iron precursors. The overall ORR activity was significantly lower, indicating that the presence of iron nanoparticles is fundamental to activate the electrocatalyst for the ORR. Recently, this idea was corroborated by the Ohsaka's research group [42], which performed empirical and theoretical experiments appointing the key role of the nitrogen doping level and the thickness of the carbon shell on this kind of electrocatalyst. They showed that by tuning these parameters it is possible to control the oxygen binding energy and, therefore, changing the material's activity for the ORR.

3 Ethanol Electro-oxidatio

Ethanol is interesting fuel for practical application on direct alcohol/air fuel cells. The advantages of the use of ethanol are related to several aspects such as: (i) it is a liquid fuel available at low cost (this is principally the case of Brazil, where the production is already well established); (ii) it is easily handled, transported and stored; (iii) presents lower toxicity, (iv) possesses relatively high theoretical density of energy (8.0 kWh kg^{-1}) when compared the gasoline (10–11 kWh kg^{-1}) [43].

The total ethanol oxidation reaction (EOR), which is the desired reaction to take place at the anode of Direct Ethanol Fuel Cells (DEFCs), is presented in Eq. 13:

$$CH_3CH_2OH + 3H_2O \rightarrow 2CO_2 + 12H^+ + 12\,e^- \quad E_1{}^0 = 0.084\,V \tag{13}$$

And this reaction is coupled to the reaction, that takes place in the cathode:

$$3O_2 + 12H^+ + 12\,e^- \rightarrow 6H_2O \quad E_2{}^0 = 1.229\,V \tag{14}$$

Resulting in a global reaction (16):

$$CH_3CH_2OH + 3O_2 \rightarrow 2CO_2 + 3H_2O \quad E_{eq}{}^0 = E_2{}^0 - E_1{}^0 = 1.145\,V \tag{15}$$

Thus, the total or the complete ethanol oxidation, forming CO_2 and H_2O, involves the transference of 12 electrons (97.3 kJ mol^{-1}) [44] per ethanol molecule, with $E_{eq}{}^0 = 1.145$ V, where $E_{eq}{}^0$ is the equilibrium standard potential (vs. SHE—S*tandard Hydrogen Electrode*). For this system, the theoretical thermodynamic efficiency is 97%. However, due to the loss by incomplete alcohol oxidation, the efficiency drops to around 40% [35].

The ethanol electro-oxidation reaction has a relatively complex mechanism. A good electrocatalyst for achieving the total ethanol oxidation has to facilitate bond breakings (C–H, C–O, and C–C) and to facilitate bond formations (oxygen addition). As most of the investigated electrocatalysts do not satisfy these features, the ethanol electro-oxidation follows parallel reaction pathways, forming, mainly acetaldehyde, acid acetic and little CO_2 [45–48], and some adsorbed reaction intermediates such as CO_{ad} and C_xH_x adsorbed intermediates (hydrocarbon fragments with one or two carbon) [37, 38, 49–51], which will be discussed below.

In order to investigate the reaction mechanism and to determine the reaction products, spectroscopic and spectrometric techniques, such as in situ FTIR (*Fourier Transform Infra red Spectroscopy*) [36, 52, 53], and *online* DEMS (*Differential Electrochemistry Mass Spectrometry*) [54], correlated with electrochemical techniques, such as cyclic voltammetry and chronoamperometry [39], have been currently utilized. In a previously published work, some researchers have shown a high dependence on the crystallographic surface orientation of the electrocatalyst [43, 55], and on the ethanol *bulk* concentration [56–58]. Leung and co-authors [45] investigated the ethanol electro-oxidation pathways on *well*-ordered Pt(111) surface versus disordered Pt(111) (higher density of defects), and on polycrystalline Pt by in situ FTIR. They observed that acetic acid and acetaldehyde were the main produced products. Also, there was a little amount of CO_2 on ordered Pt(111) than disordered Pt(111). In addition, the CO_{ad} degree of coverage was higher on disordered and polycrystalline Pt surface than on the Pt(111) surface. After that, Colmati and co-authors [43] studied the reaction on stepped Pt single crystal electrodes by electrochemical methods and by in situ FTIR. For this purpose, two different series of crystallographic surfaces were investigated: 1—(111) terraces with (100) monoatomic steps and, 2—(111) terraces with (110) monoatomic steps. The electrochemical behaviors of the two investigated crystallographic surfaces were significantly different. As discussed by the authors [43], the surface formed by (111) terraces with (100) monoatomic steps showed no activity alteration in relation to Pt(111) single-crystal. On the other hand, the surface formed by (111) terraces with (110) monoatomic steps presented a twofold effect on the ethanol electro-oxidation: 1—it showed that the step catalyzes the bond break C–C followed by the oxidation of the adsorbed intermediates like CO_{ad} formed below 0.7 V (vs. RHE); and 2—the steps also catalyzes the ethanol oxidation to acetic acid and acetaldehyde at higher potentials (above 0.7 V vs. RHE).

Using in situ FTIR, Camara and Iwasita [50] showed a strong dependence of the reaction product distribution as a function of the ethanol concentration on polycrystalline Pt, and this result is presented in Fig. 13 [50]. It can be noted that at low ethanol concentrations (0.01–0.05 mol L^{-1} of ethanol) the main products were CO_2 and acetic acid. In the case of high ethanol concentration (above 0.2 mol L^{-1} of ethanol), acetaldehyde was the dominant product. In addition, the authors reported that low CO_2 and acetic acid formation, at high ethanol concentration, was probably due to the limited availability of free Pt sites for water adsorption, which is the oxygen species-donor in the oxidation process. This increase in the CO_2 formation

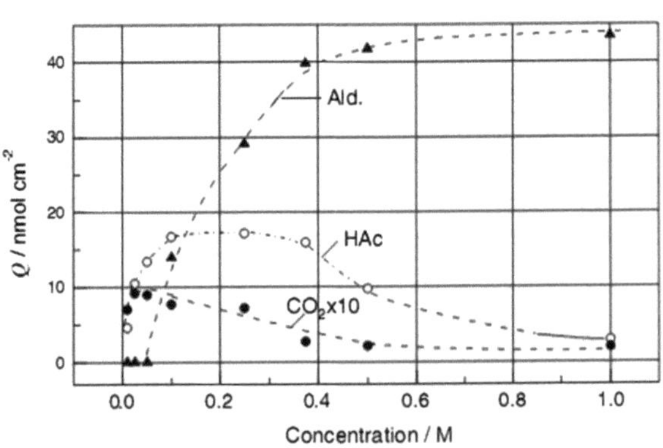

Fig. 13 Amount (Q) of CO_2, acetic acid, and acetaldehyde after 10 min of polarization at 0.5 V for different concentrations of ethanol in 0.1 mol L^{-1} $HClO_4$. CO_2 (*full circle*), acetic acid (*empty circle*) and acetaldehyde (*triangle*), respectively (Published with permission from Elsevier [50])

is explained by transport effect, namely an increasing tendency for re-adsorption and further oxidation of incomplete oxidation products at lower concentrations. So, higher ethanol concentration leads an overall partial oxidation reaction, in which the ethanol electro-oxidation follow parallel pathways forming by-products such as acetaldehyde and acetic acid.

Considering the previous published investigations using *on-line* DEMS, in situ FTIR, and electrochemical techniques, it can be proposed a simplified mechanistic for ethanol oxidation on Pt-based eletrocatalysts [36, 39, 40, 43, 45, 47, 49, 50, 59]:

$$C_2H_5OH + Pt(H_2O) \rightarrow Pt(C_2H_5OH_{ad}) + H_2O \tag{16}$$

$$Pt(C_2H_5OH_{ad}) + Pt \rightarrow Pt(CO_{ad}) + Pt(CH_{x-ad}) + xH^+ + xe^- \tag{17}$$

$$Pt(C_2H_5OH_{ad}) \rightarrow Pt(CH_3CHO) + 2H^+ + 2e^- \tag{18}$$

$$Pt(H_2O) \rightarrow Pt(OH) + H^+ + e^- \tag{19}$$

$$Pt(CH_3CHO) + Pt(OH) \rightarrow CH_3COOH + H^+ + e^- + 2Pt \tag{20}$$

$$Pt(CO_{ad}), Pt(CH_x) + Pt(OH) \rightarrow CO_2 + Pt + xH^+ + xe^- \tag{21}$$

where, Pt-$C_2H_5OH_{ad}$, Pt-CO, and Pt-$CH_{x\text{-}ad}$ are adsorbed intermediates on Pt surface as ethanol, carbon monoxide, and hydrocarbon fragments, respectively. These steps (16–21) do not necessarily represent the sequential ethanol electro-oxidation.

In order to obtain an efficient electrochemical conversion of the ethanol molecules to CO_2, experimental variables such as temperature [60], and the nature of the nanostructured metal electrocatalyst [61] has been studied. The ethanol electro-oxidation on carbon-supported Pt-based electrocatalysts was studied by Sun and co-authors [52]. They have determined, quantitatively, the amount of CO_2 produced over a wide range of reaction temperatures using by *online* DEMS with a dual thin layer flow cell [52]. The electrochemical measurements showed that the faradaic current and CO_2 formation rate, performed at 3 bar overpressure, revealed significant effects of the temperature, and of the ethanol concentration on the reaction efficiency. Elevated values of electrochemical conversion efficiency to CO_2 were obtained for high temperature, low concentration and low electrode overpotentials (nearby 90% current efficiency for CO_2 formation at 100 °C, 0.01 mol L^{-1}, 0.48 V vs. RHE). On the other hand, at high ethanol concentrations (0.1 mol L^{-1}), high potentials or low temperatures, the current efficiency for CO_2 dropped to values of a few percent.

As discussed by the authors [52], at room temperature, CO_2 formation in the positive-going scan is essentially due to oxidation of CO_{ad} preformed at lower potentials in the preceding negative-going scan, and in the negative-going scan, the CO_2 formation is inhibited. With increasing temperature, the preformed CO_{ad} can be oxidized at increasingly lower potential, allowing subsequent bulk oxidation to CO_2 via C–C bond breaking and subsequent oxidation of the resulting $C_{1,ad}$ species in the positive-going scan, and this contributes increasingly more to the CO_2 formation charge in the CO_2 ion current peak. In the negative-going scan, C–C bond breaking can occur at potentials where subsequent oxidation of the resulting fragments is still possible, and the latter reaction extends to lower potentials, where C–C bond breaking is increasingly active until the surface is saturated with adsorbed $C_{1,ad}$ fragments again.

Aiming at increasing the Pt electrocatalytic activity and the rate of CO_2 production, Lima et al. [62] and Camara et al. [63] studied the EOR by in situ FTIR on Pt-Ru bimetallic electrocatalysts. The results showed an increase in the overall reaction kinetics when compared to that of pure Pt. This fact was associated with the formation of Ru-OH species (O-donor) at low potentials, conducting the Pt-CO/Ru-OH and Pt-CH_x/Ru-OH coupling. Nevertheless, the efficiency to remove the adsorbed intermediates (probably Pt-CO and Pt-CH_x) was low, resulting in the formation of acetaldehyde and acetic acid as main reaction products.

On the other hand, electrocatalysts composed by the combination between Pt and Rh have shown the higher production of CO_2 when compared to that for other Pt-based bimetallic materials [37, 38, 53–55, 64]. The results showed that the addition of rhodium atoms produces a significant decrease in the acetaldehyde yield, compared to pure platinum electrodes, even at ambient temperature (25 °C). This behavior was attributed to the higher activity of the Rh atoms in relation to Pt for the dissociative ethanol adsorption, resulting in a C–C bond breaking [65]. Although few is known about of the rhodium's role for the ethanol electro-oxidation, one explanation for the decreased acetaldehyde amount could be related with the existence of more available sites for the C–C bond breaking. However, as in the case of Pt, CO_2 is detected only in the positive-going scan. This also means that the CO_2 is produced only due to the oxidation of the reaction adsorbed intermediate species. Furthermore, this change in the reaction route, or in the CO_2/acetaldehyde ratio, is still very low and does not conduct to an efficient oxidation of *bulk* ethanol to CO_2, as required for application of fuel cells operating at ambient temperature.

An efficient electrocatalyst for the total electro-oxidation to CO_2 and H_2O must present a balance between two competitive steps: bond breaking and bond formation (fast stripping of the adsorbed species) [19]. So, the challenge is to develop nanostructured electrocatalysts that allow controlling those effects modifying the electronic structure or the chemical reactivity of the active metals. In this direction, the *core-shell* structure has shown to be a promising kind of electrocatalyst. This is because of two important characteristics of such structure: (1) they allow electronic "modulation" of the atoms at metal surface, through two major effects: (i) *Electronic* [66]—interaction between the electronic bands of the composing metals, and (ii) *Geometric* [6]—due to the presence of defects and lattice difference in the lattice parameters of the metals in the *shell* and in the *core* of the nanoparticle. So, this structure allows the change of the adsorption strength of the reacting molecules [58, 67], thus changing the catalytic activity; (2) they present a relatively higher mass activity (faradaic current per unit of mass of the active metal, A g^{-1}) due to the presence of the active noble metal only on the particle surface. Therefore, if the particle *core* is a non-noble metal, this kind of electrocatalyst becomes attractive for commercial application due to its reduced cost.

Notably, this type of nanostructure is recent in the electrocatalysis area, and it possesses a particularly strong electronic modification of the active metal when compared to that of bimetallic alloy materials [59, 68]. These nanostructures materials would provide a change in the behavior of the ethanol electro-oxidation reaction in relation to the commonly employed nanoparticles such as Pt-Sn [69], Pt-Ru [54, 55] and Pt-Rh [37, 56] alloys.

Kowal and co-authors [38] studied the ethanol electro-oxidation on Pt-Rh/SnO_2/C nanoparticles. The ternary Pt-Rh/SnO_2/C electrocatalyst was synthesized by the *cation-adsorption-reduction-galvanic-displacement method*, resulting in a *core-shell* structure [38]. The electrochemical measurements showed that the current density obtained for the ternary electrocatalyst was more than two orders of magnitude larger than that of commercial Pt/C. In addition, in situ FTIR experiments showed that the material presents a high activity for the C–C bond breaking. The authors suggested that, in this material, the C–C bond is broken directly, without going through the acetaldehyde step. In addition, Adzic and co-authors [53] have investigated the ethanol electro-oxidation on Pt-Rh-SnO_2/C material with different atomic ratios. The results showed lower reaction onset potential and higher faradaic current for the composition Pt:Rh:Sn = 3:1:4. These ternary-electrocatalysts presented, effectively, a higher rate for CO_2, in relation to the pure Pt/C.

In a recent work [70], Lima and co-authors have investigated the ethanol oxidation products on *core-shell* nanoparticles formed by a carbon-supported Au *core*, and with a *shell* composed of the combination of Pt, Rh, and Ru atoms. The results for the faradaic currents, the ion currents signal for acetaldehyde (m/z = 29), and for CO_2 (m/z = 22) for the different electrocatalysts are presented in Fig. 14. The curves show similar onset potentials for all investigated nanoparticles. At high potential values, Pt/C shows higher current densities, which evidences superior overall reaction rate at high overpotential. Although Pt/C has presented higher current densities, it shows the larger formation of acetaldehyde. Interestingly, for the Rh/Pt/Au/C material, the curves evidenced some increase of the CO_2 production and a reduction of the acetaldehyde formation. These results are in agreement with some other investigations using Pt–Rh alloys [54, 56], where the presence of Rh in the electrocatalyst surface induces higher CO_2/acetaldehyde ratio when compared to materials composed of pure Pt or by bimetallic Pt–Ru.

Therefore, it is important to mention that these results showed that *core–shell* nanoparticles can be developed containing only a fractional amount of Pt, having higher mass-specific activity than Pt/C. Also, it was demonstrated that the overall reaction rate could be hampered by increasing the Pt *shell* thickness (not shown here), and the electrocatalyst efficiency could be improved by adjusting the particle *shell* composition. However, the activity and efficiency of this type of electrocatalyst have also to be tested in other reaction conditions of ethanol concentration and reaction temperature.

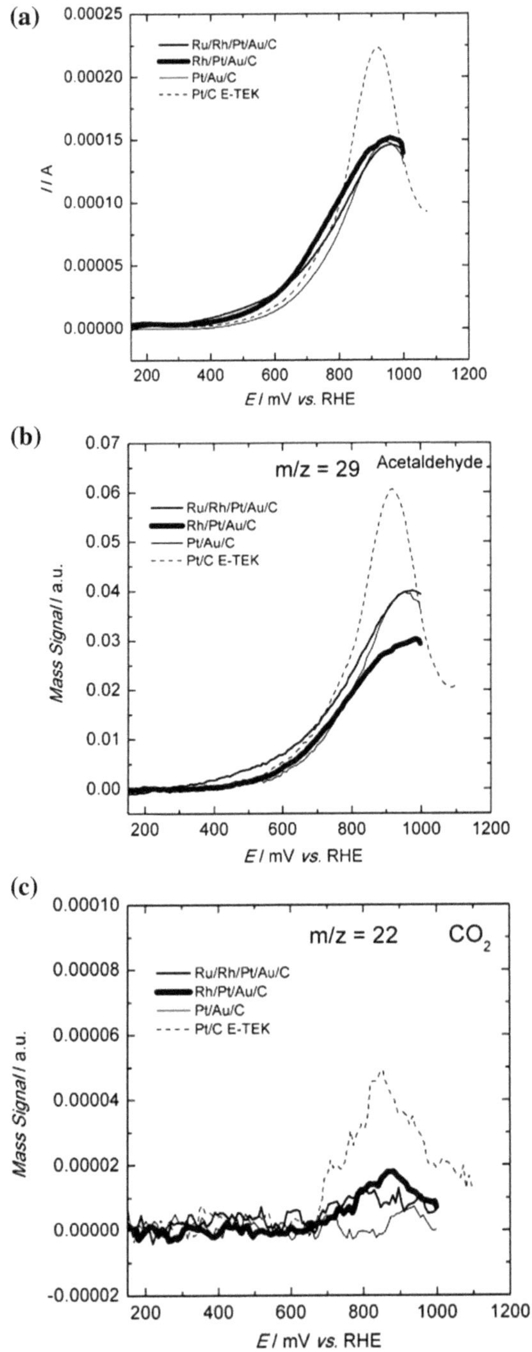

Fig. 14 DEMS experiments for the ethanol electro-oxidation reaction (positive scan) catalyzed by the different electrocatalysts: **a** faradaic currents; **b** ion currents for the acetaldehyde formation (*m/z* = 29, CHO+); and **c** ionic currents for CO_2 formation (*m/z* = 22, CO_2^{2+}), in 0.5 mol L^{-1} C_2H_5OH/0.5 mol L^{-1} $HClO_4$ solution. Scan rate of 10 mV s^{-1}. Published with the permission of the publisher [62]

4 Conclusions

4.1 *Oxygen Reduction on Noble Metal Electrocatalyst*

The results presented in this text showed that the activity of different electrocatalysts correlates well with the *d*-band center of the metals. The O_2 reduction activity of these electrode surfaces showed a volcano-type dependency on the *d*-band center of the metal catalyst. *Core-shell* type electrocatalysts seems to by promising due to the reduced amount of the total Pt loading, and the possibility of a significant alteration of the Pt electronic structure, which can enhance the reactivity of Pt atoms. The studies of long-term stability showed that the addition of Au, forming a PdAu sublayer showed increased catalyst stability during potential cycling, which is interesting for practical fuel cell systems.

4.2 *Oxygen Reduction on Non-noble Metal Electrocatalyst*

The aforementioned works show that important steps were recently achieved to advance in the understanding of the parameters that govern the activity of this class of non-noble metal electrocatalysts for the ORR. The results show that both FeN_4 moieties and N-doped carbon-encapsulated metal nanoparticles (possibly only for those with a very thin layer of the carbon shell) are active for the ORR. Nevertheless, long-term durability tests in real fuel cell systems are still missing for both structures in order to determine the most stable material under ORR conditions. These findings will allow new perspectives to the design and synthesis of non-precious metal electrocatalysts for the commercial applications of fuel cell systems.

4.3 *Ethanol Oxidation*

The experimental results showed in the review indicated that, on Pt, the overall reaction rate and the rate for CO_2 formation increase significantly with increasing reaction temperature, and with decreasing ethanol concentration. The increase with temperature seems to be related to the decrease in the activation energy. The increase for CO_2 formation as the ethanol concentration is decreased where it would be associated with an increasing tendency for re-adsorption and further oxidation of incomplete oxidation products at lower concentrations. However, in order to achieve practicable current densities for direct ethanol fuel cells, the ethanol concentration should be increased. In this case, the experimental results indicated a drastic decrease in the reaction efficiency, with only a few percent of CO_2 production. The results also showed that the addition of Rh atoms to the Pt-based electrocatalysts, and the increase of the temperature, are the possible routes to be investigated in a near future, since both resulted in a higher CO_2 formation, which increases the fuel cell energy conversion efficiency.

Acknowledgements N.A. Galiote, F.E.E. Oliveira, D.A. Cantane, and F.H.B. Lima acknowledge financial support from Fundação de Amparo à Pesquisa do Estado de São Paulo, FAPESP, and Conselho Nacional de Desenvolvimento Científico e Tecnológico, CNPq, Brazil.

References

1. Yeager E (1984) Electrocatalysts for O_2 reduction. Electrochim Acta 29(11):1527–1537
2. Kinoshita K (1992) Electrochemical oxygen technology. Wiley-Interscience, New York, p 431
3. Adzic R (1998) Recent advances in the kinetics of oxygen reduction. In: Lipkowski J, Ross PN (eds) Electrocatalysis, vol 197. Wiley-VCH, New York
4. Zhang J, Vukmirovic MB, Xu Y, Mavrikakis M, Adzic RR (2005) Controlling the catalytic activity of platinum-monolayer electrocatalysts for oxygen reduction with different substrates. Angew Chem Int Ed 44:2132–2135
5. Lima F, Zhang J, Shao M, Sasaki K, Vukmirovic M, Ticianelli E, Adzic R (2007) Catalytic activity-d-band center correlation for the O_2 reduction reaction on platinum in alkaline solutions. J Phys Chem C 111(1):404–410
6. Hammer B, Nørskov JK (2000) Theoretical surface science and catalysis–calculations and concepts. Adv Catal 45:71–129
7. Greeley J, Nørskov JK, Mavrikakis M (2002) Electronic structure and catalysis on metal surfaces. Annu Rev Phys Chem 53(1):319–348
8. Mukerjee S, Srinivasan S, Soriaga MP, McBreen J (1995) Effect of preparation conditions of Pt alloys on their electronic, structural, and electrocatalytic activities for oxygen reduction-XRD, XAS, and electrochemical studies. J Phys Chem 99(13):4577–4589
9. Lima FHB, Ticianelli EA (2004) Oxygen electrocatalysis on ultra-thin porous coating rotating ring/disk platinum and platinum-cobalt electrodes in alkaline media. Electrochim Acta 49 (24):4091–4099
10. Brankovic S, Wang J, Adzic R (2001) Pt submonolayers on Ru nanoparticles: a novel low Pt loading, high CO tolerance fuel cell electrocatalyst. Electrochem Solid-State Lett 4:A217
11. Adzic RR, Zhang J, Sasaki K, Vukmirovic MB, Shao M, Wang J, Nilekar AU, Mavrikakis M, Valerio J, Uribe F (2007) Platinum monolayer fuel cell electrocatalysts. Top Catal 46(3):249–262
12. Zhang J, Mo Y, Vukmirovic M, Klie R, Sasaki K, Adzic R (2004) Platinum monolayer electrocatalysts for O_2 reduction: Pt monolayer on Pd (111) and on carbon-supported Pd nanoparticles. J Phys Chem B 108(30):10955–10964
13. Kitchin JR, Nørskov JK, Barteau MA, Chen J (2004) Role of strain and ligand effects in the modification of the electronic and chemical properties of bimetallic surfaces. Phys Rev Lett 93 (15):156801
14. Stamenkovic V, Mun BS, Mayrhofer KJJ, Ross PN, Markovic NM, Rossmeisl J, Greeley J, Nørskov JK (2006) Changing the activity of electrocatalysts for oxygen reduction by tuning the surface electronic structure. Angew Chem 118(18):2963–2967
15. Ruban A, Skriver HL, Nørskov JK (1999) Surface segregation energies in transition-metal alloys. Phys Rev B 59(24):15990
16. Bardi U, Atrei A, Zanazzi E, Rovida G, Ross P (1990) Study of the reconstructed (001) surface of the Pt80Co20 alloy. Vacuum 41(1–3):437–440

17. Mun BS, Watanabe M, Rossi M, Stamenkovic V, Markovic NM, Ross PN Jr (2005) A study of electronic structures of Pt_3M (M = Ti,V,Cr,Fe,Co,Ni) polycrystalline alloys with valence-band photoemission spectroscopy. J Chem Phys 123:204717
18. Markovic N, Gasteiger H, Grgur B, Ross P (1999) Oxygen reduction reaction on Pt (111): effects of bromide. J Electroanal Chem 467(1–2):157–163
19. Somorjai GA, Li Y (2010) Introduction to surface chemistry and catalysis. Wiley, New York
20. Chorkendorff I, Niemantsverdriet JW, Wiley J (2003) Concepts of modern catalysis and kinetics, vol 138. Wiley Online Library, Weinheim
21. Nørskov JK, Bligaard T, Logadottir A, Bahn S, Hansen LB, Bollinger M, Bengaard H, Hammer B, Sljivancanin Z, Mavrikakis M (2002) Universality in heterogeneous catalysis. J Catal 209(2):275–278
22. Clouser S, Huang J, Yeager E (1993) Temperature dependence of the Tafel slope for oxygen reduction on platinum in concentrated phosphoric acid. J Appl Electrochem 23(6):597–605
23. Yeager E, Gervasio D, Razaq M, Razaq A, Tryk D (1993) Dioxygen reduction in various acid electrolytes. Serb J Chem Soc 57:819
24. Sidik RA, Anderson AB (2002) Density functional theory study of O_2 electroreduction when bonded to a Pt dual site. J Electroanal Chem 528(1–2):69–76
25. Sasaki K, Wang J, Balasubramanian M, McBreen J, Uribe F, Adzic R (2004) Ultra-low platinum content fuel cell anode electrocatalyst with a long-term performance stability. Electrochim Acta 49(22–23):3873–3877
26. Gong K, Su D, Adzic RR (2010) Platinum-monolayer shell on AuNi0.5Fe nanoparticle core electrocatalyst with high activity and stability for the oxygen reduction reaction. J Am Chem Soc 845–910
27. Lima FHB, Zhang J, Shao M, Sasaki K, Vukmirovic M, Ticianelli E, Adzic R (2008) Pt monolayer electrocatalysts for O_2 reduction: PdCo/C substrate-induced activity in alkaline media. J Solid State Electrochem 12(4):399–407
28. Lima F, De Castro J, Santos L, Ticianelli E (2009) Electrocatalysis of oxygen reduction on carbon-supported Pt-Co nanoparticles with low Pt content. J Power Sources 190(2):293–300
29. Obradovic M, Grgur B, Vracar LM (2003) Adsorption of oxygen containing species and their effect on oxygen reduction on Pt3Co electrode. J Electroanal Chem 548:69–78
30. Zhang J, Sasaki K, Sutter E, Adzic R (2007) Stabilization of platinum oxygen-reduction electrocatalysts using gold clusters. Science 315(5809):220
31. Xing Y, Cai Y, Vukmirovic MB, Zhou WP, Karan H, Wang JX, Adzic RR (2010) Enhancing oxygen reduction reaction activity via Pd–Au alloy sublayer mediation of Pt monolayer electrocatalysts. J Phys Chem Lett 1:3238–3242
32. Wang C, van der Vliet D, More KL, Zaluzec NJ, Peng S, Sun S, Daimon H, Wang G, Greeley J, Pearson J Multimetallic Au/FePt3 nanoparticles as highly durable electrocatalyst. Nano Lett 4–107. doi:10.1021/nl102369k
33. Jasinski R (1964) A new fuel cell cathode catalyst. Nature 201:1212
34. Jahnke H, Schönborn M, Zimmermann G (1976) Organic dyestuffs as catalysts for fuel cells. In: Boschke FL (ed) 61 topics in current chemistry—physical and chemical applications of dyestuffs. Springer, Berlin, pp 133–188
35. Gupta S, Tryk D, Bae I, Aldred W, Yeager E (1989) Heat-treated polyacrylonitrile-based catalysts for oxygen electroreduction. J Appl Electrochem 19:19–27
36. Wu G, Santandreu A, Kellogg W, Gupta S, Ogoke O, Zhang H, Wang H-L, Dai L (2016) Carbon nanocomposite catalysts for oxygen reduction and evolution reactions: from nitrogen doping to transitionmetal addition. Nano Energy 29:83–110
37. Choi CH, Baldizzone C, Grote J-P, Schuppert AK, Jaouen F, Mayrhofer KJJ (2015) Stability of Fe-N-C catalysts in acidic medium studied by operando spectroscopy. Angew Chem Int Ed 54:12753–12757
38. Lalande G, Faubert G, Cote R, Guay D, Dodelet JP, Weng LT, Bertrand P (1996) Catalytic activity and stability of heat-treated iron phthalocyanines for the electroreduction of oxygen in polymer electrolyte fuel cells. J Power Sources 61:227

39. Miller HA, Bellini M, Oberhauser W, Deng X, Chen H, He Q, Passaponti M, Innocenti M, Yang R, Sun F, Jiang Z, Vizza F (2016) Heat treated carbon supported iron(ii) phthalocyanine oxygen reduction catalysts: elucidation of the structure-activity relationship using X-ray absorption spectroscopy. Phys Chem Chem Phys 18:33142–33151
40. Strickland K, Miner E, Jia Q, Tylus U, Ramaswamy N, Liang W, Sougrati M-T, Jaouen F, Mukerjee S (2015) Highly active oxygen reduction non-platinum group metal electrocatalyst without direct metalnitrogen coordination. Nat Commun 6:7343
41. Varnell JA, Tse ECM, Schulz CE, Fister TT, Haasch RT, Timoshenko J, Frenkel AI, Gewirth AA (2016) Identification of carbon-encapsulated iron nanoparticles as active species in non-precious metal oxygen reduction catalysts. Nat Commun 7:12582
42. Noh SH, Seo MH, Kang J, Okajima T, Han B, Ohsaka T (2016) Towards a comprehensive understanding of FeCo coated with N-doped carbon as a stable bi-functional catalyst in acidic media. NPG Asia Mater 8:e312
43. Lamy C, Lima A, LeRhun V, Delime F, Coutanceau C, Leger JM (2002) Recent advances in the development of direct alcohol fuel cells (DAFC). J Power Sources 105(2):283–296
44. Iwasita T, Nart F (1997) In situ infrared spectroscopy at electrochemical interfaces. Prog Surf Sci 55(4):271–340
45. De Souza J, Queiroz S, Bergamaski K, Gonzalez E, Nart F (2002) Electro-oxidation of ethanol on Pt, Rh, and PtRh electrodes. A study using DEMS and in-situ FTIR techniques. J Phys Chem B 106(38):9825–9830
46. Kowal A, Li M, Shao M, Sasaki K, Vukmirovic M, Zhang J, Marinkovic N, Liu P, Frenkel A, Adzic R (2009) Ternary Pt/Rh/SnO_2 electrocatalysts for oxidizing ethanol to CO_2. Nat Mater 8(4):325–330
47. Iwasita T, Pastor E (1994) A DEMS and FTIR spectroscopic investigation of adsorbed ethanol on polycrystalline platinum. Electrochim Acta 39(4):531–537
48. Xia X, Liess HD, Iwasita T (1997) Early stages in the oxidation of ethanol at low index single crystal platinum electrodes. J Electroanal Chem 437(1–2):233–240
49. Souza JPI, Queiroz SL, Nart FC (2000) The use of mass spectrometry in electrochemical measurements-the DEMS technique. Quím Nova 23(3):384–391
50. Jiang L, Colmenares L, Jusys Z, Sun G, Behm R (2007) Ethanol electro-oxidation on novel carbon supported Pt/SnOx/C catalysts with varied Pt: Sn ratio. Electrochim Acta 53(2):377–389
51. Colmati F, Tremiliosi-Filho G, Gonzalez ER, Berná A, Herrero E, Feliu JM (2009) The role of the steps in the cleavage of the C–C bond during ethanol oxidation on platinum electrodes. Phys Chem Chem Phys 11(40):9114–9123
52. Gao P, Chang SC, Zhou Z (1989) Electro-oxidation pathways of simple alcohols at platinum in pure nonaqueous and concentrated aqueous environments as studied by real-time FTIR spectroscopy. J Electroanal Chem 272(1–2):161–178
53. Leung LWH, Chang SC, Weaver MJ (1989) Real-time FTIR spectroscopy as an electrochemical mechanistic probe: Electro-oxidation of ethanol and related species on well-defined Pt (111) surfaces. J Electroanal Chem 266(2):317–336
54. Bruckenstein S, Gadde RR (1971) Use of a porous electrode for in situ mass spectrometric determination of volatile electrode reaction products. J Am Chem Soc 93:793–794
55. Colmati F, Tremiliosi-Filho G, Gonzalez ER, Berná A, Herrero E, Feliu JM (2008) Surface structure effects on the electrochemical oxidation of ethanol on platinum single crystal electrodes. Faraday Discuss 140:379–397
56. Cantane DA, Gonzalez E (2009) Mechanistic aspects of ethanol electro-oxidation in unsupported platinum nanoparticles. ECS Trans 25:1161–1168
57. Giz M, Camara G (2009) The ethanol electro-oxidation reaction at Pt (1 1 1): the effect of ethanol concentration. J Electroanal Chem 625(2):117–122
58. Camara G, Iwasita T (2005) Parallel pathways of ethanol oxidation: the effect of ethanol concentration. J Electroanal Chem 578(2):315–321
59. Lai SCS, Koper MTM (2008) Electro-oxidation of ethanol and acetaldehyde on platinum single-crystal electrodes. Faraday Discuss 140:399–416

60. Sun S, Halseid MC, Heinen M, Jusys Z, Behm R (2009) Ethanol electro-oxidation on a carbon-supported Pt catalyst at elevated temperature and pressure: a high-temperature/high-pressure DEMS study. J Power Sources 190(1):2–13
61. Adzic R, Li M, Kowal A, Sasaki K, Marinkovic N, Su D, Korach E, Liu P (2010) Ethanol oxidation on the ternary Pt–Rh–SnO_2/C electrocatalysts with varied Pt: Rh: Sn ratios. Electrochim Acta 55 (BNL–93871-2010-JA)
62. Lima F, Gonzalez E (2008) Ethanol electro-oxidation on carbon-supported Pt-Ru, Pt-Rh and Pt-Ru-Rh nanoparticles. Electrochim Acta 53(6):2963–2971
63. Camara G, De Lima R, Iwasita T (2005) The influence of PtRu atomic composition on the yields of ethanol oxidation: a study by in situ FTIR spectroscopy. J Electroanal Chem 585 (1):128–131
64. Lima F, Profeti D, Lizcano-Valbuena W, Ticianelli E, Gonzalez E (2008) Carbon-dispersed Pt-Rh nanoparticles for ethanol electro-oxidation. Effect of the crystallite size and of temperature. J Electroanal Chem 617(2):121–129
65. Houtman C, Barteau M (1991) Divergent pathways of acetaldehyde and ethanol decarbonylation on the Rh (111) surface. J Catal 130(2):528–546
66. Sasaki K, Wang J, Balasubramanian M, McBreen J, Uribe F, Adzic R (2004) Ultra-low platinum content fuel cell anode electrocatalyst with a long-term performance stability. Electrochim Acta 49(22–23):3873–3877
67. Kristian N, Wang X (2008) Ptshell-Aucore/C electrocatalyst with a controlled shell thickness and improved Pt utilization for fuel cell reactions. Electrochem Commun 10(1):12–15
68. Shao M, Sasaki K, Marinkovic NS, Zhang L, Adzic RR (2007) Synthesis and characterization of platinum monolayer oxygen-reduction electrocatalysts with Co-Pd core-shell nanoparticle supports. Electrochem Commun 9(12):2848–2853
69. Colmati F, Antolini E, Gonzalez E (2008) Effect of thermal treatment on phase composition and ethanol oxidation activity of a carbon supported Pt50Sn50 alloy catalyst. J Solid State Electrochem 12(5):591–599
70. Lima FHB, Profeti D, Chatenet M, Riello D, Ticianelli E, Gonzalez E (2010) Electro-oxidation of ethanol on Rh/Pt and Ru/Rh/Pt sub-monolayers deposited on Au/C nanoparticles. Electrocatalysis 1(1):72–82

Nanocomposites from V_2O_5 and Lithium-Ion Batteries

Fritz Huguenin, Ana Rita Martins and Roberto Manuel Torresi

1 Introduction

The properties of metallic lithium, such as its equivalent weight (M = 6.94 g mol^{-1}), specific gravity (ρ = 0.53 g cm^{-1}), and electropositivity ($\approx$−3.04 V vs. SHE) have motivated the development of lithium-ion batteries, which present high energy densities [1, 2]. Moreover, the discovery of inorganic materials that allow for the chemically reversible insertion/extraction of lithium ions into their structures, denominated intercalation materials, has also contributed to this development.

In these batteries, the chemical energy is converted into electrical energy via the discharge step, followed by the conversion of electrical energy into chemical energy via the charge step. The anode (negative electrode) is the source of lithium ions, which migrate and diffuse through the electrolyte toward the cathode (positive electrode) during the discharge step (Fig. 1).

The first generation of lithium batteries emerged in the 70s, together with the primary (nonrechargeable) batteries. The positive electrodes usually consisted of particulate transition metal oxides (TMO) or chalcogeneous materials of millimetric dimension, meanwhile the negative electrodes were composed of metallic lithium.

F. Huguenin (✉) · A.R. Martins
Departamento de Química, Faculdade de Filosofia,
Ciências e Letras de Ribeirão Preto, Universidade de São Paulo,
Ribeirão Preto, (SP) 14040-901, Brazil
e-mail: fritz@ffclrp.usp.br

R.M. Torresi
Instituto de Química, Universidade de São Paulo,
São Paulo, SP CP 26077, 05513-970, Brazil

F.L. de Souza and E. Leite (eds.), *Nanoenergy*,
DOI 10.1007/978-3-319-62800-4_7

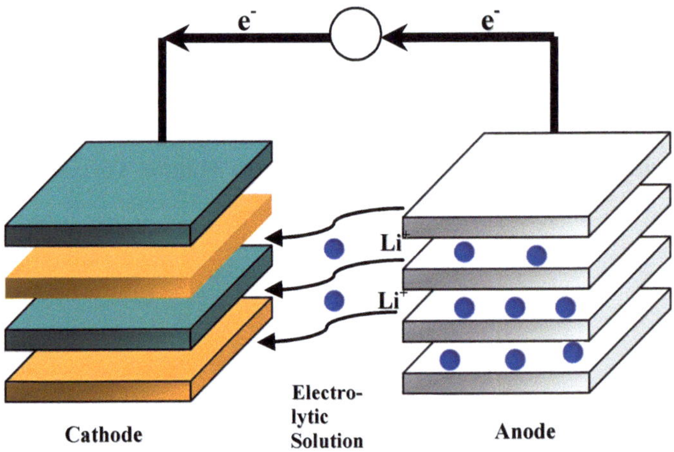

Fig. 1 Schematic representation of lithium-ion batteries during the discharge step

The half-reactions involved in these electrodes can be represented by the following reactions:

$$M + x\mathrm{Li}^+ + x\mathrm{e}^- \leftrightarrows \mathrm{Li}_x M \quad \text{(positive electrode)} \tag{1}$$

$$\mathrm{Li}^+ + \mathrm{e}^- \leftrightarrows \mathrm{Li}^0 \quad \text{(negative electrode)}, \tag{2}$$

where M is a host matrix for lithium-ion intercalation/deintercalation during the discharge/charge step. Electrons are generated due to the oxidation reaction at the negative electrode, flowing through the external circuit and bringing forth work. They are injected into the conduction band of the positive electrode, which must provide an electrical potential far greater than that of the negative electrode. An outer power source is used between the electrodes for the charge step, so that the reaction is reversed.

According to equation

$$\Delta G = -nF\Delta E, \tag{3}$$

where ΔG is the reaction free energy, n is the number of electrons involved per mol of consumed reactants, F is the Faraday constant, and ΔE is the potential difference between the positive (cathode) and negative (anode) electrodes of the battery and is related to the chemical potential difference between the electrodes. As this difference becomes larger and a material with lower weight or volume is employed, the supplied energy density should be greater. Moreover, these devices are associated with the global reaction rate, especially when they are used in applications requiring high power densities and current, as in the case of electric vehicles (EV). Therefore, the intercalation electrodes should offer high ionic conductivity and electronics to minimize their internal resistance. Another factor limiting the performance of lithium-ion batteries is the inner lithium-ion diffusion coefficient in solid matrices, which controls the insertion/extraction rates [3]. Depending on the material selected for the cathode, power sources with higher energy density can be obtained [4–12].

In the beginning, secondary lithium batteries were comprised of TiS_2 cathodes and metallic lithium anodes, which gave an acceptable performance, i.e., high chemical reversibility that allowed cycling for over 1100 cycles [13]. However, the cell curtailment resulting from formation of dendritic lithium could result in hazardous explosions. From 1977 to 1979, button cells composed of LiAl anodes and TiS_2 cathodes were commercialized by Exxon as part of watches and other tiny devices [10, 14]. The safety of the cells was improved through the use of a LiAl anode. However, alloy electrodes held out only a restricted number of cycles resulting in extreme volume changes during the process. In the meanwhile, intercalation materials were becoming significantly advanced.

Due to the safety concerns surrounding the use of metallic lithium, researchers have striven to find optional approaches including modifications within the electrolyte and the negative electrode. A first attempt was to substitute metallic lithium for a second insertion material. Then, the lithium-ion technology was demonstrated at the late 1980s and early 1990s. The presence of lithium in its ionic rather than its metallic state solved the dendritic issue surrounding the use of lithium and, at least in principle, lithium-ion cells are intrinsically safer than metallic lithium cells. To compensate the increased potential in the negative electrode, high-potential insertion compounds can be selected to compose the positive electrode, and the layered-type transition-metal disulfides are shifted to the three-dimensional or layered-type transition-metal oxides.

The first lithium-ion batteries were commercialized by Sony® in 1991, after the discovery of the highly reversible and low-voltage Li intercalation/deintercalation process in carbonaceous material. In these batteries, lithium ions were stored into graphite anodes (Li_xC_6) that were then intercalated into $LiCoO_2$ electrodes during the discharge step [15, 16] with specific energy density around 180 W h Kg^{-1} under a potential discharge of approximately 3.8 V. High current densities (10 mA cm^{-2}) can be achieved in this case, thanks to the diffusivity of lithium ions in the $LiCoO_2$ electrodes (diffusion coefficient values close to 5×10^{-9} cm^2 s^{-1}) [17]. Despite the significant increase in energy density per volume unit ($\approx$400 W h l^{-1}) and mass ($\approx$170 mA h g^{-1}), new materials have been investigated since they were made commercially available because of the scarcity, cost, and toxicity of cobalt.

To be used as electrodes in batteries, these materials must have some basic features. For instance, they must be easy to oxidize and reduce, and they must have high cyclability, high voltage, high capacity of charge storage per mass and volume units, high ionic and electronic conductivity. Moreover, they must be inexpensive and exhibit low toxicity. Advances in solid-state chemistry and innovations in the development of new materials have improved battery performance, which is intrinsically related to the properties of the materials comprising the positive and negative electrodes. Within this context, nanostructured materials have effectively contributed to the performance of lithium-ion batteries because of their small size, which enhances the insertion/extraction rates and thus increases their charge transfer capacity.

2 Vanadium Pentoxide

Among various TMOs studied as cathodes for use in lithium-ion batteries, V_2O_5 must be mentioned. This oxide displays suitable structural and electrochemical properties for Li^+ insertion/extraction [2, 18–22]. As can be observed below, the V_2O_5 xerogel and its nanocomposites are the focus of the present review, although it is inevitable to mention crystalline V_2O_5, which was one of the earliest studied oxides and has been exploited for 30 years [23].

The layered structure of crystalline V_2O_5 with weak vanadium–oxygen bonds between its layers is already known as well as its capacity to react by an intercalation mechanism. The oxide basic unit is an octahedral VO_6 which shares the edges to form a double chain and can be distorted to a square pyramid with a short vanadyl bond and a very long V–O bond. It is observed an alternation among the apexes of the VO_5 pyramids in an up-up-down-down sequence, with the third rows being unoccupied as a consequence of the corner sharing of the double chains. Lithium insertion within the structure of the layered V_2O_5 is rather complex. Initial lithium intercalation into the structure results in the formation of the α-phase ($x < 0.01$) and then the ε-phase ($0.35 < x < 0.7$), in which the layers are more folded. The δ-phase is obtained at $x = 1$, where one layer changes rapidly out of two layers. Nevertheless, significant structural changes can occur if more than one lithium ion per V_2O_5 is discharged, giving rise to the γ-phase and enabling cycles in the $0 \leq x \leq 2$ range. Figure 2 shows that the VO_5 square pyramids in the α-, ε-, and δ-phases, which compose V_2O_5, are set in rows with apexes in an up-up-down arrangement. The structural behavior comes to up-down-up-down in the highly folded γ-phase and, as long as more lithium is discharged, a rock-salt architecture is formed and the ω-$Li_3V_2O_5$ phase arises. This latter phase has a tetragonal structure and becomes a rock-salt structure with the formula $Li_{0.6}V_{0.4}O$ [10]. In addition, the cycling capacity of the initial vanadium oxide phase is quickly lost, making it uninteresting for practical applications.

As a result of the intercalation of 0.5 Li per vanadium, the specific capacity around 147 mA h g^{-1} is comparable to that of $LiCoO_2$, which is also able to cycle 0.5 Li with no irreversible phase changes. However, $LiCoO_2$ demonstrates much better cycling stability and performance at high current densities as compared to V_2O_5. Furthermore, a profound discharge results in significantly lower amount of cycled lithium, a consequence of kinetic limitations due to a rise in current density [24].

In the gel form, vanadium oxides present improved reactivity. The V_2O_5 xerogel can be produced from inorganic or organometallic precursors and remains stable for long periods when placed under adequate conditions. It has already been reported

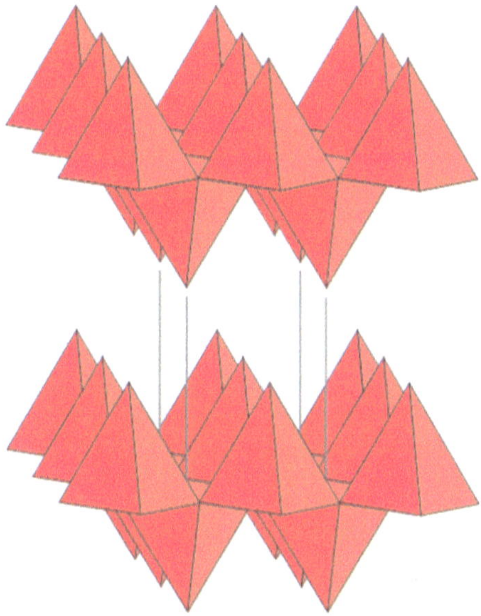

Fig. 2 Structure of crystalline V_2O_5. Adapted with permission from Whittingham [10]. Copyright 2011 American Chemical Society

that vanadium alkoxides $VO(OR)_3$ (R = Bu, t-Am) are good precursors that react through hydrolysis and condensation processes [19]. Also, V_2O_5 gels can be prepared outright from the oxide. Crystalline V_2O_5 reacts vigorously with hydrogen peroxide, resulting in a red gelatinous product [25]. V_2O_5 gels can also be produced by spilling the molten oxide heated at 800 °C into water [26]. Moreover, amorphous V_2O_5 can be hydrated, resulting in the formation of V_2O_5 gels [27]. Alternatively, the amorphous oxide can be produced by splat cooling of the molten oxide and hydration, which has been described as the thickening of a polymeric network in a solvent. Pulverization of the amorphous oxide powder with water results in the formation of gels. Further addition of water yields sols. V_2O_5 films prepared by vapor deposition or even vanadium oxide spheres dissolved in water display a similar behavior, resulting in gels or colloidal solutions [28, 29]. As will be subsequently detailed, vanadium pentoxide gels can be prepared by a most reliable technique based on the hydrolysis and condensation of vanadates or vanadium alkoxides.

When a metavanadate aqueous solution is passed through a proton exchange resin, a pale yellow solution of decavanadic acid ($[H_2V_{10}O_{28}]^{4-}$) is obtained. It is demonstrated in ^{51}VNMR spectra of V^{5+} solutions that barely two V^{5+} species, $[H_2V_{10}O_{28}]^{4-}$ and $[VO_2]^+$, are observed before gelation occurs, in solution round pH $\approx$ 2. Some neutral species form as an intermediate while the decavanadate anion dissociates into the $[VO_2]^+$ cation and one of these neutral species could be written as $[VO(OH)_3(OH_2)_2]^0$. This precursor presents octahedral geometry with the H_2O molecules coordinated to vanadium on the axial and equatorial plane, on opposite direction to vanadyl group (V=O) and hydroxyl group (OH), respectively. The subsequent condensation process occurs only on the equatorial plane due to the

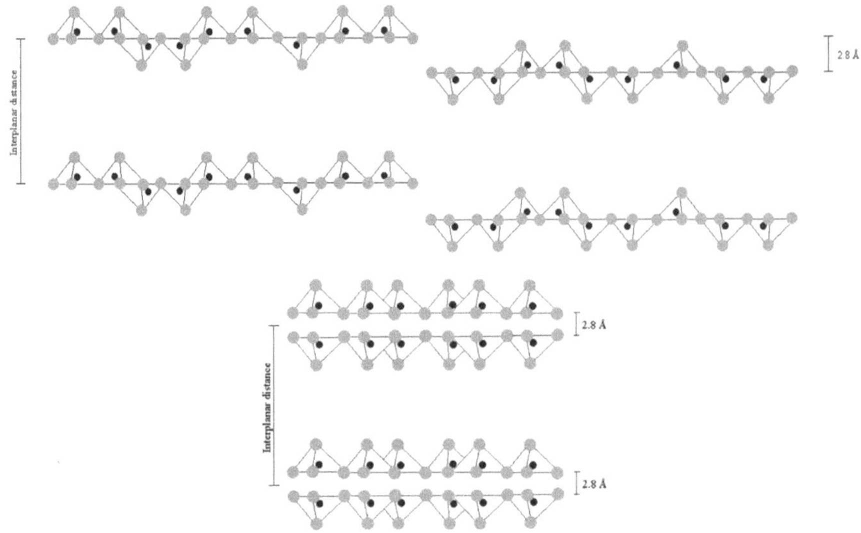

Fig. 3 Schematic representation of V_2O_5 xerogel: single (*upper*) and bilayer (*down*) structural models. Adapted with permission from Giorgetti et al. [37]. Copyright 2011 American Chemical Society

absence of V–OH groups on the axial direction, creating two-dimensional V_2O_5 bilayers perpendicular to the substrate and soaked into amorphous regions [30–32]. The condensation reactions (Eqs. 4–6) and the V_2O_5 xerogel structure are shown below (Fig. 3). In reason of some H_2O molecules coordinate to vanadium atom on axial position and other establishes hydrogen bonds with V=O groups, the distance between the oxide planes varies with the intercalated H_2O amount, approximately estimated in 11.55 Å to $V_2O_5{\cdot}1.8H_2O$ [30, 33].

$$\mathrm{VOH + VOH \rightarrow VOV + H_2O} \quad \text{(Oxolation)} \tag{4}$$

$$\mathrm{VOH + V(ROH) \rightarrow VO(H)V + ROH} \quad \text{(Olation)} \tag{5}$$

$$\mathrm{VOH + V(H_2O) \rightarrow VO(H)V + H_2O} \quad \text{(Olation)} \tag{6}$$

Metal alkoxides, $VO(OR)_3$ (R = Et, n-Bu, n-Pr, i-Pr), are versatile precursors for the synthesis of sol–gel metal oxides under aqueous conditions [33]. In the presence of excessive water, these alkoxides undergo a hydrolysis process (Eq. 7) and an expansion of metal coordination (V) by nucleophilic addition of two water molecules. Afterwards, polymeric chains are obtained via condensation process.

$$\mathrm{VO(OR)_3 + 3H_2O \rightarrow VO(OH)_3 + 3ROH} \quad \text{(Hydrolysis)} \tag{7}$$

Electron microscopy shows that the gel is constituted of flat ribbons with dimensions around 10^3, 10^3, and 10 Å, length x width x thickness, respectively [34,

35]. The behavior of the structure bears close similarity with that of the (*ab*) plane of crystalline V_2O_5 and its arrangement, i.e., pyramidal entities of orthorhombic V_2O_5 with 3.6 Å periodicity along the *b*-axis [19]. In the stacking direction (*c*-axis), a 2.8 Å amplitude is induced in the ribbon corrugation between the fibrils. The stacking of the ribbons is illustrated in Fig. 3 and corresponds to the structure that emerges when the gel is deposited onto a substrate. The ribbon has a thickness of 8.75 Å. Livage has proposed a model using a single layer. However, Oka et al. have put forward a different model entailing a bilayer for characterization of the ribbon structure in the xerogel [36]. This model can be seen in Fig. 3, where the V_2O_5 sheets face each other at 2.8 Å distance.

The electrical properties of the $V_2O_5{\bullet}nH_2O$ xerogels have been extensively researched due to the conductive nature of these materials. However, the conductivity of these materials varies significantly, approximately from 10^{-7} to $10^{0}\ \Omega^{-1}\ cm^{-1}$, depending on many parameters such as the preparation method, employed precursors, gel aging time, and thickness of the sample [38–43]. This conductivity has a mixed character involving electronic and ionic transport that are, respectively, proportional to the water concentration and the vanadium oxidation state [44]. The electrons skipping between the V^{4+} and V^{5+} sites are responsible for the electronic conductivity [45]. However, strong electron–phonon coupling decreases the mobility of charge carriers, with values in the order of 10^{-5}–$10^{-6}\ cm^2\ V^{-1}\ s^{-1}$ [33]. In the presence of excess water, acidic dissociation of V–OH groups occurs at the oxide/water interface, generating a significant amount of charged H_3O^+ species (pH $\approx$ 2) [30], as shown in Eq. (8).

$$V{-}OH + H_2O \Leftrightarrow V{-}O^- + H_3O^+ \tag{8}$$

Analysis of the electrochemical properties of V_2O_5 suggests that this oxide is an excellent candidate for use as cathode in lithium-ion batteries, with a redox potential greater than 3 V (vs. Li/Li^+), specific energy density close to 600 W h Kg^{-1}, and specific capacity around 250 A h Kg^{-1} [46, 47]. Lithium insertion/extraction taking place during the electrochemical process can be described by the following equation:

$$V_2O_5 + xLi^+ + xe^- \leftrightarrow Li_xV_2O_5, \tag{9}$$

where Li^+ compensates the electronic charge involved in the steps of vanadium reduction and oxidation to form vanadium bronze [48]. Four electrons and four Li^+ are involved in this process for each V_2O_5 species in the case of xerogel materials [47]. However, owing to some limitations such as slow diffusion of the intercalating ions into the structure and low electronic conductivity of V_2O_5, only reduction of V^{+5} sites to V^{+4} is usually observed (Fig. 4).

Other studies [20, 46, 49] have reported that the diffusion coefficient (D) values can vary between 10^{-9} and $10^{-17}\ cm^2\ s^{-1}$, depending on the amount of inserted lithium ions. Thus, considerable effort has been made in attempt to raise the rate of lithium-ion diffusion and electronic conductivity of these cathodes. For example,

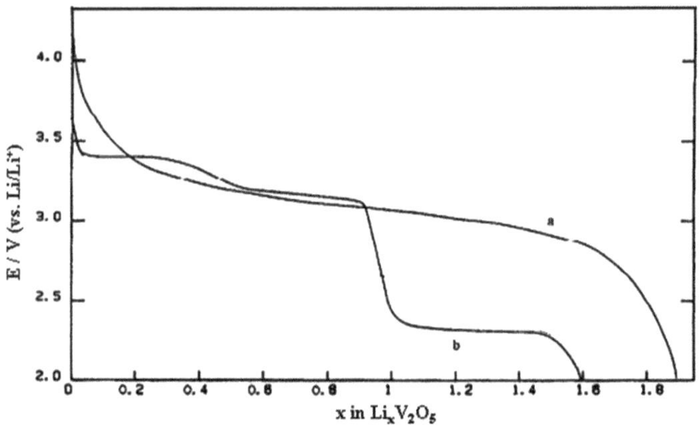

Fig. 4 Potential versus stoichiometric ratio between the amount of lithium ion and V_2O_5 (x in $Li_xV_2O_5$) for **a** V_2O_5 xerogel and **b** orthorhombic V_2O_5. Adapted from Ref. [46]

Parente et al. [50] have described a substantial increase in the lithium-ion diffusion rate by synthesizing V_2O_5 xerogel on synthetic nickel fibers. Dong et al. [51] have used the *sticky carbon* method to obtain higher ionic pore accessibility. Other synthetic routes are being employed in order to obtain new structures and morphologies. For instance, V_2O_5 thin films obtained by spin coating allow for insertion of 3–4 equivalents of Li^+ ions [47, 52–54]. This optimization has been attributed to the amorphous nature and low anisotropy of these materials. In many cases, powdered carbon black is utilized as a conductive agent, to connect metal oxide particles. Passerini et al. [55] have added carbon black in the early stages of V_2O_5 gelation, thereby ensuring higher charge capacity, possibly related to loss of long-range structural ordination. However, carbon black is electrochemically inactive, limiting the ability of the cathode to store charge. Le et al. [53, 56] have held a supercritical drying process for V_2O_5 gels, which produced an amorphous material with very high porosity, thus reducing the diffusional path length. As a consequence, the overall reaction was no longer limited by diffusional processes. These aerogels thus have extremely high charge capacity (560 A h Kg^{-1}) and energy density (1700 W h Kg^{-1}) [22, 54, 56].

Vanadium pentoxide may undergo some structural changes in the presence of organic solvents. One example is the intercalation of propylene carbonate (PC) in interplanar region by replacing the water molecules and resulting in larger distance between the 001 layers [57]. In situ X-ray diffraction (XRD) analysis of the material subjected to Li^+ intercalation revealed reversible structural changes [46], with layer stacking being destroyed during lithium-ion insertion and formed during the detachment process. However, there is a significant decrease in the specific capacity when the electrode is subjected to several charge/discharge cycles. This fact is attributed to mechanical stress in the structure and/or dimensional changes caused by expansion/contraction of the layers due to insertion/extraction of solvent and ions.

3 Nanomaterials

Several research groups from different areas have shown the importance of miniaturization. For instance, it allows for the fast operating speed involving objects designed on a nanometric scale. In fact, the time required for a given system to perform a specific function can be minimized by displacement of electrons, ions, and molecules in a diminished space [58]. Moreover, physical forces that are negligible in the macroscopic world can become decisive in nanometric materials, allowing molecular self-ordering [59]. In addition, the conductivity parallel to the interfaces of a composite can significantly rise when the thickness of the layers' components reaches very low values, leading to an overlap of the charge spaces [60]. Thus, the design of nanoscale materials may contribute to the optimization and emergence of new properties, enabling the use of these systems for various applications in the fields of optics, electronics, catalysis, biotechnology, and solid-state electrochemistry [61–65].

The development of nanostructured electrodes has led to the better performance of lithium-ion batteries, since they ensure higher current and energy densities. In nanostructured systems, the distances which lithium ions must cover are dramatically decreased, thus reducing the free diffusional path, so that lithium ions are able to enter and exit the structure more rapidly. Considering that the time taken for ion intercalation is proportional to the square size of the particle, the current values can significantly increase with the use of nanostructured electrodes.

The intercalation properties can be significantly improved by new developments in the synthesis and characterization of vanadium oxide nanostructures such as nanorolls, nanobelts, nanowires, and ordered nanorods arrays [66]. The galvanostatic discharge performance of V_2O_5 nanorods (with 115 mm diameter and 2 mm length) has been compared to that of a thin film electrode with the same geometric area and equal V_2O_5 mass [67]. The Li^+ storage capacity at low discharge rates (C/20) was equivalent for both the nanostructured and thin film electrodes. Nonetheless, at high discharge rate, a higher capacity was demonstrated for the nanostructured electrode as compared to the thin film one. More specifically, the capacity was three and four times higher for the nanostructured electrode at a rate of 200 C and above 500 C, respectively, as compared to the thin film electrode.

Spahr et al. have synthesized and analyzed V_2O_5 nanotubes, which represent a new structure for cathode materials for use in lithium-ion batteries with smaller diffusion lengths [68]. An initial high capacity of 300 mA h g^{-1} was demonstrated for the V_2O_5 nanotube arrays, almost twice as high as the initial capacity of 140 mA h g^{-1} observed for the V_2O_5 film. The large surface area and short diffusion distance of the nanotube array were responsible for this improvement, but in the second and third cycles the capacity of the nanostructured material decayed to 200 and 180 mA h g^{-1}, respectively. In further cycles, the decrease was less pronounced and a stable capacity of 160 mA h g^{-1} was reached after the sixth cycle, which still was 30% higher than that of the V_2O_5 film. The nanotube array presented a more drastic decay than the film during cycling possibly as a result to the structural flexibility and fragility [69].

The discharge capacities and cyclic performance of platelet and fibrillar-structured V_2O_5 films prepared by solution methods have been compared to those of the conventional, simple structured film. The behavior of these structured V_2O_5 films varied depending on the orientation: the platelet film which is comprised of V_2O_5 particles with sizes of 20–30 nm presented a random orientation while the fibrillar film displayed randomly oriented fibers, most of which protrude from the substrate surface. Comparing the initial discharge value of the plain V_2O_5 film (260 mA h g^{-1}) with the initial discharged capacities of platelet and fibrillar-structured V_2O_5 films (1240 and 720 mA h g^{-1}, respectively), it is clear that the latter capacities were far higher than the former. As a result of the reduced distance of Li^+ diffusion, which avoids the situation in which lithium ions are concentrated in the electrode, and of the poor V_2O_5 interlayer cross-linking, of which promotes intercalation of larger quantities of lithium ions, high discharge capacity values were achieved. However, these nanostructured V_2O_5 films suffered degradation during the electrochemical tests [70].

Bearing in mind, the importance of interfaces for electrochemistry, researchers are developing new systems based on nanocomposites, which are characterized by the nanometric particle size and/or distance between their constituents [71, 72]. Some physical–chemical characteristics of composite materials, which are unique due to the interactions established between the components' phases, can be optimized with the aid of nanoscience. Actually, the design of these materials can maximize short-range interactions between the components, generating synergistic effects in the properties of the system. Thus, another way to optimize the electrochemical properties of V_2O_5 is to intercalate polymers into its lamellar space.

Some one- and bidimensional matrices, consisting of layers, channels, or tunnels, or which are completely amorphous, represent a class of materials made of flexible pores whose size can be adapted to other dimensions of the intercalating species. Given these characteristics, the scientific community has designed and investigated new materials obtained by inclusion of organic polymers into inorganic matrices, to form composites with an intimate contact with their constituents on a scale between the classical molecular and microscopic scales [73–81]. There are several methods for the preparation of these nanohybrids, including: polymerization of monomeric molecules in the inorganic matrix through chemical, thermal, or photo-induced treatment in situ; intercalation of the previously formed polymer; use of self-assembly technique; and intercalated redox polymerization, where the monomers are oxidized by the host matrix, which in turn is reduced and thus promotes formation of a polymer within its structure [82].

One of the most often investigated inorganic matrices is the V_2O_5 xerogel. The interaction between its lamellae is weak enough to allow for the merge of chemical species with different sizes [19, 33]. Moreover, its facile synthesis by the sol–gel methodology allows for their preparation at low temperatures, which has encouraged several research groups to develop and analyze nanocomposites based on V_2O_5 and organic polymers [2, 19, 20, 83]. One reason for the development of these systems is the optimization of energy storage properties, considering the potential use of this oxide as cathode for lithium-ion batteries. With a view to maximizing the energy density and specific capacity, electronic conducting polymers (ECPs) can be interspersed in the xerogel thereby offering an alternative conductive path; facilitating the access of electrons to the vanadium ions, and uniting electrically isolated particles. Besides this role, ECPs still participate in the redox process, directly contributing to increasing the battery capacity, as opposed to the carbon powder normally employed in the cathodes. Furthermore, the presence of ECPs can modify the morphology, porosity, and electrostatic interaction between the lithium ions and the oxygen atoms of the host matrix, thus maximizing ion mobility and consequently optimizing the performance of the electrochemical device [84–91].

Kanatzidis et al. [92, 93] were the first researchers to synthesize and characterize nanocomposites consisting of V_2O_5 and ECPs, which contributed to a better understanding of the polymerization processes of organic molecules interspersed in a TMO matrix. Polymerization of aniline intercalated in the V_2O_5 xerogel resulted in interlayer distance expansion from 11.55 to 13.94 Å, due to the replacement of the H_2O molecules with the aniline polymer (PANI) [94]. Equation (10) shows the overall reaction process, where it is assumed that aniline oxidation triggers a polymerization process accompanied by reduction of the V^{+5} ions. However, molecular oxygen also participates in the reaction, providing an increase in the length of the polymeric chain and partial reoxidation of the vanadium ions.

$$x(C_6H_5NH_2) + V_2O_5 + (y/2)O_2 \rightarrow (C_6H_4NH)_xV_2O_5^{(2x-2y)-} + yH_2O + (2x-2y)H^+ \tag{10}$$

The aniline intercalation/polymerization process was investigated by insertion of the $C_6H_5NH_3I$ salt into the inorganic matrix under inert atmosphere [95], thus enabling oxidation of the iodide ions by V_2O_5 and preventing polymerization of the anilinium ion monomers ($C_6H_5NH_3^+$) in the interlayer. Subsequent exposure of the material to ambient air resulted in PANI formation in the inorganic host matrix which was unexpected, since anilinium ion monomers are not oxidized to polyaniline in the presence of atmospheric oxygen. This suggests that V_2O_5 acted as a catalyst in the polymerization stage, which in turn was associated with electron transfer to molecular oxygen.

The insertion of a conductive path parallel to the V_2O_5 chains caused significant changes in electronic transport, since the charge transport in these materials is commanded by the spatial charge effects at the interface located between the conductive phase (PANI) chains and the oxide [96]. Thanks to the hydrogen bonds formed between the oxygen atom of V_2O_5 and the hydrogen atoms of PANI (NH···OV), chains are held together due to a rise in the interfacial area, thereby leading to an even greater effect on electronic conductivity [95]. The intimate contact between the components can be observed from the Electron Paramagnetic Resonance spectra, indicating proximity (<10 Å) between the PANI *polaron* and the V^{4+} paramagnetic centers [95]. Electrostatic interactions between these polymers have been clearly observed in self-assembled hybrids, for which Raman and UV-Vis spectra enable visualization of electronic transfer from PANI to the oxide during stacking of the monolayers, while the level balance of Fermi was maintained [97, 98]. Furthermore, X-ray photoelectron spectra have suggested short-range interactions between PANI and V_2O_5 through binding between the nitrogen and oxygen atoms (Fig. 5) [96].

Depending on the amount of polymer, morphology, and chain length of the polymer, the electronic conductivity of these nanocomposites can be elevated by about 4 degrees of magnitude as compared to that of the V_2O_5 xerogel. Conductivity in the order of 0.1 S cm^{-1} at room temperature can be achieved after an aging period of 18 months [95]. However, despite the high conductivity value, other preparation methods can be used, in order to ensure faster growth of the

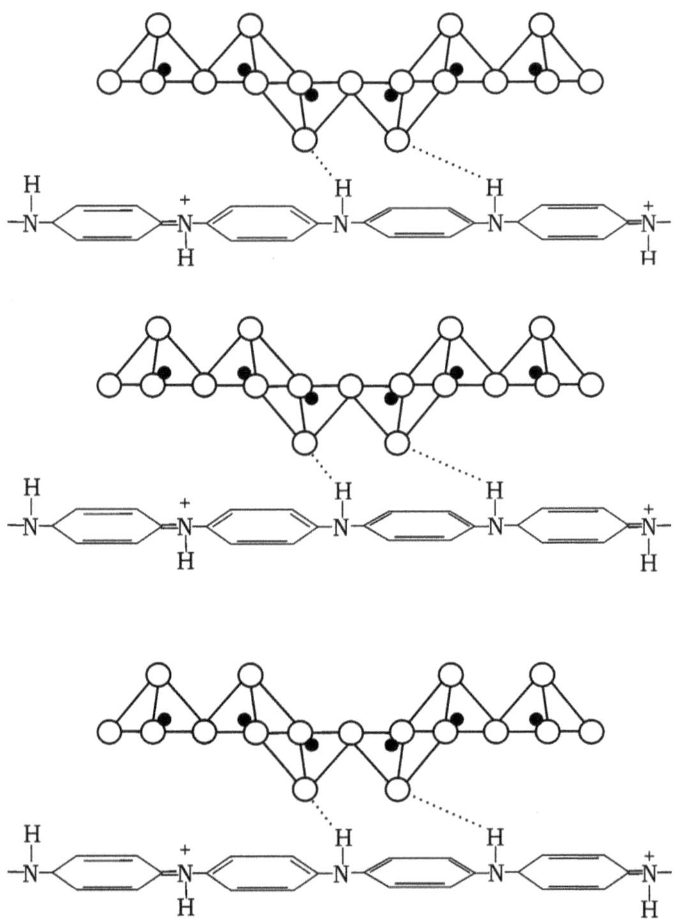

Fig. 5 Schematic representation of PANI/V_2O_5 nanocomposite

macromolecules and consequent achievement of obtaining high electronic conductivity in shorter aging periods. One example is PANI intercalation into an array of mesoporous V_2O_5/Mg in atmospheric air [99], which has also given a nanocomposite with an electrical conductivity (10^{-2} S cm^{-1}) of around 4 orders of magnitude higher than that of xerogel after 2 months of aging, approximately. This relatively short time is attributed to the larger size of the pores, which facilitated diffusion of the polymeric chains into the, and allowed for faster growth of macromolecules.

The first studies involving lithium-ion intercalation into these nanocomposites were carried out by Nazar et al. [100], who demonstrated the potential use of these materials as cathodes for lithium-ion batteries. After an oxidative treatment based on O_2, seeking reoxidation of V^{4+} centers formed during the synthesis stage, a specific capacity of 165 A h kg^{-1} was obtained for the nanocomposite $[PANI]_{0.4}V_2O_5$ in 0.5 M $LiClO_4$/PC. This result evidenced that the charging capacity of the nanohybrid is greater than the sum of the capacities of its components, corresponding to a gain of approximately 40% as compared to the V_2O_5 xerogel. This optimization of electric charge storage properties is assigned to a larger ionic transport rate, since the presence of the ECP weakens the interaction between the V_2O_5 oxygen atoms and lithium ions, while increasing their mobility [100, 101]. In a study on the oxidative treatment of PANI/V_2O_5 nanohybrids based on O_2 at different temperatures [102], the influence of PANI, molecular oxygen, and heat treatment capacity was verified. The V_2O_5 xerogel exhibited a specific capacity (124 A h kg^{-1}) lower than that of $[PANI]_{0.6}V_2O_5$ (161 A h kg^{-1}), which was maximized when the nanocomposite was exposed to O_2 at 30 and 80 °C, with values of 220 and 302 A h kg^{-1}, respectively.

Although the reaction mechanism was not investigated in these nanocomposites, determination of the chemical species present in the hybrid materials consisting of PANI and V_2O_5 crystalline particles following the charge/discharge steps provided good insight into the effect of the interaction between PANI and the oxide on the charge compensation process. On the basis of elemental analysis of the composites (V_2O_5 comprised 85% of the total mass), quantity of anions was six times lower (0.07 mol of Cl^- and ClO_4^- per mole of aniline) than that usually seen in a PANI film (0.4 mol of Cl^- per mole of aniline) [103]. These experimental results were attributed to the presence of negative charges on the V_2O_5 particles, which acted as counteranions for the ECP in the oxidized state. However, these assignments are not conclusive, since participation of the PANI redox process in the presence of V_2O_5 has not been confirmed.

Huguenin et al. [104] have employed another method for the preparation of nanocomposites from V_2O_5 xerogel and PANI, namely in situ oxidative polymerization of aniline in the V_2O_5 dispersion. Electrochemical impedance was measured by these authors and a value of 2×10^{-11} cm^2 s^{-1} for the lithium-ion diffusion coefficient was found and compared to the value of 3×10^{-12} cm^2 s^{-1} for V_2O_5. The V_2O_5/PANI nanocomposite exhibited higher lithium-ion diffusion coefficient and larger electronic conductivity as compared to the V_2O_5 xerogel, a result of the

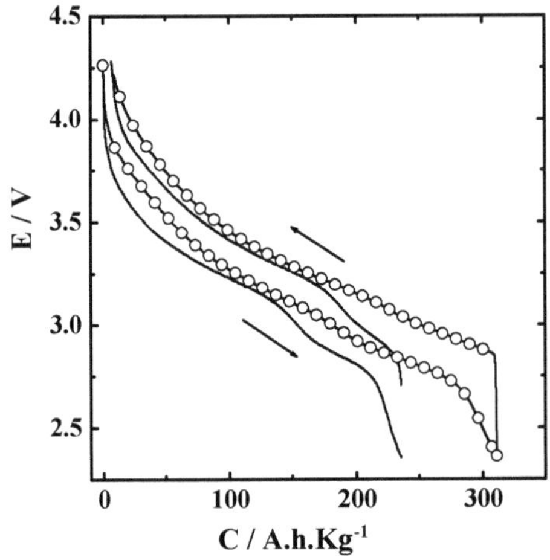

Fig. 6 **a** Chronopotentiometric curves for V_2O_5 (*solid line*) and $[PANI]_{0.3}V_2O_5$ (*line with circle*) in 0.5 M $LiClO_4$, j = 10 μA cm^{-2}. *Sample mass* 12 μg. Adapted from Ref. [104]

increased charge capacity (313 A h Kg^{-1} for $[PANI]_{0.3}V_2O_5$ and 234 A h Kg^{-1} for V_2O_5), as presented in Fig. 6. These researchers also conducted measurements using an electrochemical quartz crystal microbalance (EQCM), and they proposed that a parallel solvent (propylene carbonate) transport occurred during the charge compensation process in V_2O_5. In contrast with the expected result, knowing that PANI charge compensation occurs by anion transport [105], it has been observed that the charge compensation proceeds with lithium-ion and solvent transport in the V_2O_5/PANI nanocomposite in an average of 5% on a molar basis (i.e., 0.05 mol of propylene carbonate per mole of Li^+). This can be related to mingling of the PANI and V_2O_5 chains at molecular level which enables PANI to be oxidatively doped. Nevertheless, only anionic vanadium oxide chains are present as anions during formation of the nanocomposite, so it is possible that the electroneutrality condition for the p-doped PANI is achieved by charge compensation from the firm anionic sites on the vanadium oxide lattice when the nanocomposite microstructure is initially established. In accordance with the observed results, charge compensation for PANI should occur via lithium-ion transport, since the anionic sites on the vanadium oxide lattice should have no mobility.

In some cases, the ionic transport during redox exchange can be significantly improved by manipulation of the ionic transport, so that charge compensation is predominantly achieved by lithium-ion movement. This happens when PANI is employed as a cathode material in lithium-ion batteries. This improvement in the properties of PANI results in reduction of the demand for excess supporting electrolyte and increased in energy density for the cathode. Moreover, it is known that solvent transport, and therefore volumetric changes within the thin film systems, biases the film life cycle. For instance, in the case that $LiClO_4$/PC was utilized as supporting electrolyte, V_2O_5/PANI nanocomposites demonstrated higher electrochemical stability as well as greater charge capacity than the V_2O_5 xerogel after several charge/discharge cycles [104, 106].

Raman spectroscopy measurements carried out in situ on the poly(*N*-propane sulfonic acid aniline) PSPAN/V_2O_5 nanohybrid have attested to the oxidation and reduction of the PSPAN component during its redox cycling, although the electrochemical profile of the PSPAN in the presence of the V_2O_5 component remains unknown [107]. In contrast to the V_2O_5/PANI nanocomposite, only lithium ion participated in the charge compensation process in the PSPAN/V_2O_5 nanocomposite, hence avoiding anion and solvent contribution and accounting for the lower specific capacity as compared to that of the V_2O_5/PANI nanocomposite. In situ EQCM and cyclic voltammetry measurements have revealed the major participation of lithium ions. Taking into consideration the mass conservation principles and the electroneutrality during the redox process, determination of the electron, cation, anion, and solvent flux was feasible. Thus, once there was no significant modification in the composition in the presence of small volume of the electrolyte, which served as a path for lithium-ion transport, these cathodes can be utilized in high-energy secondary batteries. Moreover, the minimum insertion/extraction of solvent minimized expansion/contraction of the structure, resulting in enhanced electrode cycling [107, 108]. Nevertheless, the lithium-ion diffusion coefficient values are virtually the same for PSPAN/V_2O_5 and V_2O_5 [107].

The reasons for subtle difference among the diffusion coefficients of the PSPAN/V_2O_5, V_2O_5/PANI, and V_2O_5 host matrices were elucidated by ultraviolet–visible (UV-Vis), Fourier transform infrared (FTIR), and nuclear magnetic resonance (NMR) experiments. Several previously published works have proposed that the Li^+–O=V is more shielded interaction in the PANI/V_2O_5 material, which would explain higher lithium-ion diffusion rate observed for the PANI/V_2O_5 nanocomposite as compared to the V_2O_5 and the PSPAN/V_2O_5 nanocomposites.

The band-gap energies related to the electronic transition from O 2p to V 3d was detected in the UV-Vis absorption spectra of both the oxidized and reduced states ($E = 0.4$ V and $E = -1.0$ V, respectively) of the V_2O_5 xerogel and the $[PANI]_{0.3}V_2O_5$ nanocomposite. This corresponds to a shift from 2.40 to 2.75 eV for the V_2O_5 xerogel and from 2.15 to 2.20 eV for the $[PANI]_{0.3}V_2O_5$ nanocomposite. The shifting of band-gap energy is lower for the $[PANI]_{0.3}V_2O_5$ nanocomposite as compared to the V_2O_5 xerogel because PANI induces a partial shield between the lithium ions and the oxygen atoms of V_2O_5, thus resulting in a tiny reduction in the O 2p band as compared to the V 3d band [104, 109]. Huguenin et al. have carried out a characterization of the V_2O_5 xerogel and V_2O_5/PANI by Fourier transform infrared (FTIR) analyses, which elucidated another aspect surrounding the molecular level interaction. Figure 7 brings the FTIR spectra for the V_2O_5 xerogel, $Li_{1.3}V_2O_5$ xerogel, $[PANI]_{0.3}V_2O_5$, and $Li_{1.3}[PANI]_{0.3}V_2O_5$ [110]. The doublet wavenumbers at 1022 and 1011 cm^{-1} correspond to the V=O vibrations for the V_2O_5, and it shifts to about 978 cm^{-1} for $Li_{1.3}V_2O_5$. Moreover, when $[PANI]_{0.3}V_2O_5$ is reduced, the doublet shifts from 1016–1006 to 995 cm^{-1}. The shift is smaller for $Li_{1.3}[PANI]_{0.3}V_2O_5$ as compared to $Li_{1.3}V_2O_5$ because of a partial blinding

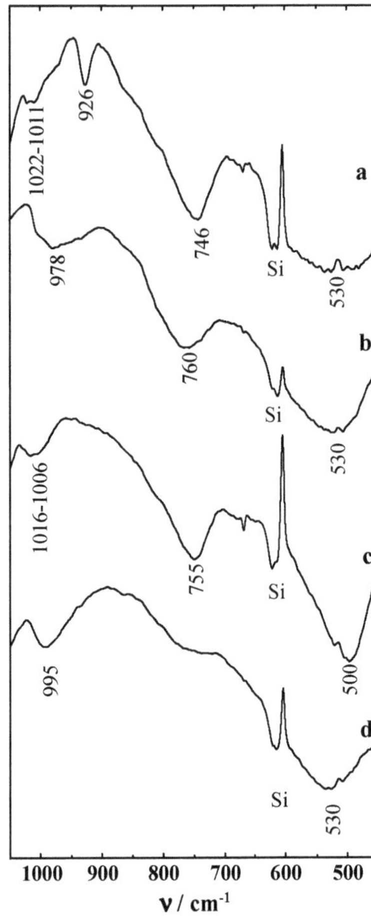

Fig. 7 FTIR spectra of **a** oxidized and **b** reduced V_2O_5, and **c** oxidized and **d** reduced $[PANI]_{0.3}V_2O_5$. Adapted from Ref. [110]

prompted by PANI in the Li^+–O=V interactions, which is due to an innermost contact between the organic and inorganic components. These data corroborated with literature reports describing that the NH–O=V hydrogen bonds and the electrostatic interactions between PANI and V_2O_5 are responsible for hindering the interactions between the lithium ions and the vanadyl groups [95, 97, 98].

Solid-state nuclear magnetic resonance (NMR), particularly magic angle spinning (MAS), better elucidated the Li^+–O interaction for the $Li_xV_2O_5$, $Li_x[PANI]_{0.3}V_2O_5$, and $Li_x[PSPAN]_{0.3}V_2O_5$ nanocomposites at the molecular level [111]. The ^7Li MAS NMR spectra displayed a wide peak for $Li_xV_2O_5$ as a consequence of the presence of lithium ions in bulk sites. This peak shifted upfield as more lithium ions were intercalated, indicating that the oxygen atoms shielded the lithium ions from a result of a more difficult access to the magnetic field inner bulk sites [111, 112]. In contrast, for the $Li_x[PANI]_{0.3}V_2O_5$ nanocomposite the wide peak was shifted downfield, a consequence of less effective shielding from the magnetic field due to the smaller amount of lithium. These results confirmed the partial blinding effect of PANI on the interactions between the lithium ions and the oxygen atoms of V_2O_5. The ^{7}Li MAS NMR spectra for the poly(*N*-propane sulfonic acid) (PSPAN)/V_2O_5 nanocomposite was also analyzed, so that the influence of oxygen atoms on shielding of lithium ions from the magnetic field could be understood [111]. As the amount of intercalated lithium ions increased the wide peak due to $Li_x[PSPAN]_{0.3}V_2O_5$ slightly shifted upfield. This change occurred in the same direction as that observed for the $Li_xV_2O_5$. It is noteworthy that the difference between PSPAN and PANI lies on the presence of the sulfonic propane group in the former, so the intercalated lithium ions can interact with the oxygen atoms, which allows for a partial blinding effect on the interaction between the lithium ions and the oxygen atoms of V_2O_5. Finally, the difference between the shifts detected for the nanostructures consisting of PSPAN, PANI, and V_2O_5 as well as the difference in the lithium diffusion coefficient values were analyzed (Fig. 8).

Ppy/V_2O_5 nanocomposites are also good candidates for use as cathodes in batteries of high energy density. However, depending on the preparation method, these materials may exhibit high coulombic irreversibility [114]. One way to facilitate extraction of ions from the host matrix and enhance the charge capacity is to employ a supercritical drying process based on CO_2 and organic solvents, which raises the porosity of the nanocomposite [115]. Depending on the utilized conditions, it is possible to reversibly merge up to 3.5 mol of Li^+ to a mole of material in these hybrid aerogels, which represents a 40% increase as compared to V_2O_5 aerogels. Another approach that can be used to optimize the electrochemical performance is to employ thin films, which give rise to high specific capacity and coulombic reversibility under application of a high current/mass ratio [116].

Distribution of the ECP in the inorganic matrix interferes with the structural, electrical, and electrochemical properties of these electrodes which can be controlled depending on the preparation method [117]. When the V_2O_5 gel was exposed to Ppy dispersion, microcomposites were formed. The short-range

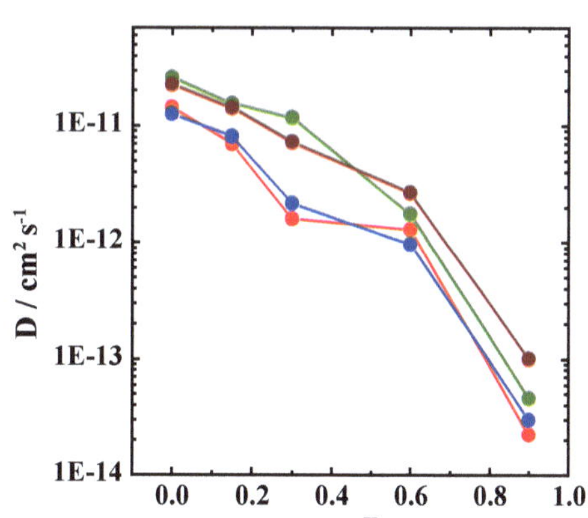

Fig. 8 Logarithm of lithium-ion diffusion coefficient as a function of x for (*red circle*) $Li_xV_2O_5$ xerogel, (*green circle*) $Li_x[PANI]_{0.3}V_2O_5$, (*blue circle*) $Li_x[PSPAN]_{0.3}V_2O_5$, and (*brown circle*) $Li_x[Ppy]_{0.3}V_2O_5$. Adapted with permission from Huguenin and Torresi [113]. Copyright 2011 American Chemical Society

interactions between components of these microcomposites were restricted, thereby minimizing the synergistic effects on the energy storage properties. When pyrrole was polymerized in the presence of the V_2O_5 xerogel, nanohybrid materials were formed, and the amount of Ppy necessary to produce percolation of a conductive path was much smaller than in the case of the microcomposite. However, there were Ppy aggregates in the films, which diminished the homogeneity of the conductive phase in the inorganic matrix. When the organic and inorganic precursors were polymerized simultaneously, homogeneous nanocomposites with higher specific capacities were generated, attesting to the influence of intimate contact between the ECP and V_2O_5 on these hybrid materials. Effects between the components caused by the interfaces have also been observed by Demets et al., who analyzed modified electrodes deposited on V_2O_5, Ppy/V_2O_5 nanocomposites, and V_2O_5 deposited on Ppy [118]. These authors commented on the different electrical and electrochemical behaviors of these materials and showed the importance of the nature, composition, and architecture of the film regarding the materials' properties.

A two-component nanocomposite constituted of conducting polymers (polyaniline, polypyrrole, and polythiophene) and vanadium oxide has been prepared by a mechanochemical method developed by Posudievsky et al. [119]. The nanocomposite based on polypyrrole exhibited the greatest capacity. The structure stability of this nanocomposite was demonstrated by cycling at a current of $\approx 0.2\ A\ g^{-1}$, and its capacity was retrieved at lower current cycling. As a result of the higher lithium-ion diffusion rate and the default of film passivation on the surface of the nanoparticles, lithium-ion intercalation in the nanocomposites was more effective than its intercalation into the V_2O_5 xerogel. The stacking of the conducting polymer and the inorganic compound in the nanocomposite material can be related to the stability during extended charge/discharge cycles, higher discharge capacity, and enhanced diffusion coefficient of the lithium ions.

The morphology of the host matrix also interfered in lithium-ion intercalation. In fact, during the copolymerization process, the Ppy chains inhibited the growth of V_2O_5 chains, resulting in a low planning degree and larger pore size (160 Å) as compared to the xerogel (80 Å), which facilitated ion intercalation [102]. Addition of surfactants (nonylphenol EON2) to the V_2O_5 xerogel also elicited an increase in structural disorder, resulting in elevated charge capacity and allowing for the merge of 2.7 mol of Li^+ per mole of V_2O_5 in the nanocomposite against 1.6 mol of Li^+ in the xerogel [49]. The increase in the flatness of the nanocomposites consisting of poly(3,4-ethylenedioxythiophene) (PEDOT)/V_2O_5, which is due to larger interlayer spacing, was considered one of the reasons for higher specific capacity as well as elevated electronic conductivity [120, 121].

Depending on the medium, the high oxidation potential of V_2O_5 can cause undesired reactions with the electrolyte, culminating in the passivation of the electrode surface. Because Ppy has a lower oxidation potential, the polymer can cover the metal oxide particles and thus prevent degradation reactions from taking place in the electrolyte solution. As noted by Kuwabata et al., coating of crystalline V_2O_5 particles with Ppy assured a high cyclability of the composite, and avoided the possible reactions between V_2O_5 and poly(methyl methacrylate) (PMMA) [122].

The reverse micelle method has been applied for preparation of V_2O_5 and V_2O_5/PANI nanofibers. The cycling behavior of the V_2O_5/PANI nanofibers was greater as compared to the V_2O_5. Moreover, there was higher capacity impairment for V_2O_5 nanofibers during cycling. It was found that V_2O_5/PANI nanofibers with a 30% PANI molar ratio exhibited a stable capacity of about 300 mA h g^{-1}. The morphology of V_2O_5 and V_2O_5/PANI nanofibers during charge/discharge cycles was examined by scanning electron microscopy (SEM), and it was found that the morphology of the V_2O_5 nanofibers changed after 10 charge/discharge cycles, while morphology of the V_2O_5/PANI nanofibers was preserved after the electrochemical cycling. Thus, the presence of the polymer within the nanohybrid material may have stabilized the capacity via homogeneous distribution during cycling [123].

Another method for obtaining nanofibers is to react the V_2O_5/PANI nanocomposite and hexadecylamine after the hydrothermal treatment, which usually gives V_2O_5/PANI nanofibers with dimensions varying from 1 to 10 μm length and 15–300 nm width. Electrochemical measurements evidenced a specific capacity of about 150 Ahk/g during the 10 initial cycles. Moreover, electrochemical impedance spectroscopy estimated the apparent diffusion coefficient in about 1×10^{-12} cm^2/s^{-1}. In contrast, replacing hexadecylamine with sodium cations by submitting the nanofibers through a reflux treatment led to an apparent diffusion coefficient that was at least three orders magnitude higher as compared to the vanadium oxide template [124].

Searching for a better performance, ECPs and other redox polymers have been used in the preparation of new nanocomposites based on V_2O_5. An example is the use of functionalized polymers such as poly(aniline-*co*-*N*-(4-sulfophenylacetic) aniline (PAPS), whose solubility in H_2O renders homogeneous nanocomposites with better defined stoichiometry as compared to PANI/V_2O_5. However, despite the increase in conductivity promoted by this polymer, the electrochemical stability was lower as compared to the xerogel. Indeed, this electrode became electroinactive after 30 voltammetry cycles in 0.1 M $LiClO_4$/acetonitrile [125]. The synthesis and characterization of other nanocomposites consisting of poly(3-decylpyrrole) (P3DP) or poly(hexadecylpyrrole) (P3HDP) and V_2O_5 has been described by Huguenin et al. [126]. Compared to the Ppy/V_2O_5 nanocomposite, the P3DP/V_2O_5 and P3HDP/V_2O_5 nanocomposites demonstrated greater electrochemical stability. Possibly, this was a result of the side chain of the P3DP and P3HDP polymers, which guaranteed higher specific capacity for the P3DP/V_2O_5 and P3HDP/V_2O_5 nanocomposites (115 Ah/Kg and 106 Ah/Kg, respectively), as compared to the Ppy/V_2O_5 nanocomposite (50 Ah/Kg), after 50 electrochemical cycles. Wang G et al. have described intercalation of poly(*N*-[5-(8-hydroxyquinoline)methyl]aniline) into the V_2O_5 xerogel (PNQA/V_2O_5), a hybrid that was synthesized by the in situ intercalative polymerization method [127]. The authors found that aging in air atmosphere facilitated PNQA chain growth between the V_2O_5 lamellae, resulting in increased electrical conductivity: 1.9×10^{-4} S cm^{-1} for the PNQA/V_2O_5 hybrid versus 4.2×10^{-6} S cm^{-1} for the V_2O_5 xerogel.

The slow redox coupling of thiol–disulfide makes poly[2,5-dimercapto (1,3,4-thiadiazole)] (PDTT) unfeasible for application in batteries [128].

Meanwhile, hybrid systems formed from the polymerization of the dimer 2,5-dimercapto1,3,4-thiadiazole (diDMcT) inserted into the V_2O_5 structure exhibited increased charge in relation to the oxide [129], which was attributed to the reduction and oxidation of the organic polymer PDTT, for which the reductive cleavage of S–S bond was facilitated in the inorganic matrix. Nevertheless, after several voltammetry cycles the charge capacity resemble that of V_2O_5, while molecules of dimercapto-1,3,4-thiadiazole (DMcT) produced during reduction of the oligomers diffused out of the host matrix. Considering the good performance of PANI/V_2O_5 nanocomposites, Park et al. intercalated PDMcT and PANI into the inorganic matrix [130]. Although the specific capacity of this ternary material (220 A h kg^{-1}) was lightly higher than that of the xerogel (205 A h kg^{-1}), the binary materials are still considered a better alternative as energy source.

Poly(ethylene oxide) (PEO) is another interesting polymer because of its ion conductivity, a result of its alkali-metal ions complexes, and the PEO/V_2O_5 nanocomposite has been demonstrated to lead to interesting lithium-ion intercalation and display relevant photochemical and electrical properties [131–134]. Investigations into electrochromic processes and electrical conductivity have shown enhanced charge capacity and reversibility due to lithium-ion insertion/extraction. The inserted (Q_c) and extracted (Q_a) charge densities of PEO/V_2O_5 film for the 10th cycle as measured in the study of Jin et al. [135] were 18.1 and 17.9 mC cm^{-2}, respectively, which are about two orders of magnitude higher than those of the V_2O_5 xerogel film (9.6 and 9.8 mC cm^{-2}, respectively). Reversibility was calculated by the ratio between the inserted and extracted charge densities (Q_a/Q_c), and revealed a greater reversibility value (98.9%) achieved for PEO/V_2O_5 as compared to the V_2O_5 xerogel (98%).

The layer-by-layer technique has already been described in the literature and can be applied to the manufacture of these nanocomposites [136, 137]. This method can result in the spontaneous formation of nanoarchitectures, thanks to the ionic attraction between oppositely charged materials. Furthermore, some physical–chemical properties of the film can be tailored by controlling film thickness, uniformity, and composition. Additionally, an improvement of the ionic mobility and the diffusion rate can be achieved by generation of an electrostatic shield between the chemical species involved in the diffusion bounce. The highest negative charge sites bounce into the host matrices as a result of innermost contact between the polymer chains and the metal oxide [113]. Although the interaction of vanadyl group and the lithium ion led to a diminished ionic mobility and energy dissipation during the electroinsertion process, the layer-by-layer technique allowed for organic polymer adsorption onto strategic sites of the V_2O_5, promoting shielding of the ion–dipole interaction [138, 139].

The layer-by-layer (LbL) technique can improve the charge storage capacity, giving rise to new electrochromic effects and even control of the ionic conductivity. Therefore, several works on the manipulation of V_2O_5/PANI nanocomposites at molecular level using this technique have been developed. The amount of PANI participating in the redox processes can be controlled by shifting the film structure at the molecular level, so that the chromogenic properties of PANI on the nanometric scale can be obtained, while the intrinsic electrochemical properties of the V_2O_5 xerogel are retained. By controlling the absorption of V_2O_5/PANI LbL films onto a cast PANI film, the properties of the nanoarchitectures are achieved. A study involving immersion of the PANI-V_2O_5/PANI system into a $LiClO_4$/ propylene carbonate (PC) solution has been conducted by varying the immersion time. The EQCM technique enabled monitoring of the mass gain/loss (Δm) as a function of the charge as well as control of the insertion/extraction species. Thus, the charge compensation process as a consequence of the shift from anionic to cationic contributions resulted in mass reduction and enhancement of the specific capacity of secondary lithium batteries.

In addition, Huguenin et al. [139] have studied nanoarchitectures of V_2O_5/ chitosan/PEO (V_2O_5/blend) obtained by addition of 1% chitosan to the PEO dispersion. It was found that chitosan was only effective regarding assistance in the manufacturing of the films. Indeed, the V_2O_5/blend performed better in comparison to V_2O_5/chitosan (8.03 and 3.41 mC cm^{-2}, respectively), which was attributed to a larger number of electrochemically active sites and an increase in the diffusion rate of lithium ions within the host matrix, as determined from electrochemical and spectroelectrochemical measurements [140]. The enhanced diffusion process arose from the complexation of lithium ions with the oxygen atoms of PEO, guaranteeing higher charge capacity during the charge/discharge steps.

4 Conclusions

When positive electrode is employed in lithium-ion batteries, the V_2O_5 xerogel poses some limitations, which made it uninteresting for workable applications. Nanocomposites constituted of conducting polymers and V_2O_5 have been demonstrated to be an alternative to the V_2O_5 xerogel for electrochemical energy storage. The preparation and investigation of novel manufacture nanoarchitectured systems from these components can overcome these restrictions, improving some features that are essential for lithium-ion batteries, such as stability, charge transport, energy density, specific capacity, and control of solvent insertion. Furthermore, studies on lithium-ion insertion can significantly contribute to further applications such as electrochromic devices and sensors.

Acknowledgements Financial support from FAPESP, CNPq, and Capes is gratefully acknowledged.

References

1. Cheng F, Tao Z, Liang J, Chen J (2008) Template-Directed Materials for Rechargeable Lithium-Ion Batteries. Chem Mater 20:667
2. Tarascon J-M, Armand M (2001) Issues and challenges facing rechargeable lithium batteries. Nature 414:359

3. Bruce PG, Scrosati B, Tarascon J-M (2008) Nanomaterials for rechargeable lithium batteries. Angew Chem Int Ed 47:2930
4. Wang Y, Cao G (2008) Developments in nanostructured cathode materials for high-performance lithium-ion batteries. Adv Mater 20:2251
5. Liu C, Li F, Ma LP, Cheng H-M (2010) Advanced materials for Energy Storage. Adv Mater 22:E28
6. Li H, Wang Z, Xeng L, Huang X (2009) Research on advanced materials for Li-ion batteries. Adv Mater 21:4593
7. Guo Y-G, Hu J-S, Wan L-J (2008) Nanostructured materials for electrochemical energy conversion and storage devices. Adv Mater 20:2878
8. Balaya P, Bhattacharyya AJ, Jamnik J, Zhukovkii YF, Kotomin EA, Maier J (2006) Nano-ionics in the context of lithium batteries. J Power Sources 159:171
9. Armand M, Tarascon J-M (2008) Building better batteries. Nature 451:652
10. Whittingham MS (2004) Lithium batteries and cathode materials. Chem Rev 104:4271
11. Tarascon J-M (2010) Key challenges in future Li-battery research. Philos Trans R Soc A Math Phys Eng Sci 368:3227
12. Brodd RJ, Bullock KR, Leising RA, Middaugh RL, Miller JR, Takeuchi E (2004) Batteries, 1977 to 2002. J Electrochem Soc 151:K1
13. Whittingham MS (1976) Electrical energy storage and intercalation chemistry. Science 192:1126
14. Rao BML, Francis RW, Christopher HA (1977) Lithium-aluminum electrode. J Electrochem Soc 124:1490
15. Nagaura T, Tozawa K (1990) Lithium ion rechargeable battery. Prog Batteries Solar Cells 9:209
16. Ozawa K (1994) Lithium-ion rechargeable batteries with $LiCoO_2$ and carbon electrodes: the $LiCoO_2$/C system. Solid State Ionics 69:212
17. Goodenough JB, Mizuchima K (1981) Electrochemical cell with new fast ion conductors US Patent 4,302,518
18. Ohzuku T (1995) In: Pistoia G (ed) Lithium batteries—new materials, developments and perspectives, vol 5, 2a edn. Elsevier, Amsterdam, p 239
19. Livage J (1991) Vanadium pentoxide gels. Chem Mater 3:578
20. Pereira-Ramos JP, Baffier N, Pistoia G (1995) In: Pistoia G (ed) Lithium batteries—new materials, developments and perspectives, vol 5, 2a edn. Elsevier, Amsterdam, p 281
21. Manthiram A, Kim J (1998) Low temperature synthesis of insertion oxides for lithium batteries. Chem Mater 10:2895
22. Owens BB, Smyrl WH, Xu JJ (1999) R&D on lithium batteries in the USA: high-energy electrode materials. J Power Sources 81:150
23. Whittingham MS (1976) Role of ternary phases in cathode reactions. J Electrochem Soc 123:315
24. Chernova NA, Roppolo M, Dillon AC, Whittingham MS (2009) Layered vanadium and molybdenum oxides: batteries and electrochromics. J Mater Chem 19:2526
25. Ostermann W (1922) Wiss Ind Hamburg 1:17
26. Müller E, Chem Z (1911) Ind Kolloide 8:302
27. Gharbi N, R'Kha C, Ballutaud D, Michaud M, Livage J, Audiere JP, Shiffmacher G (1981) A new vanadium pentoxide amorphous phase. J Non-Cryst Solids 46:247
28. Sanchez C, Livage J, Audière JP, Madi A (1984) Influence of the quenching rate on the properties of amorphous V_2O_5 thin-films. J Non-Cryst Solids 65:285
29. Kittaka S, Sasaki S, Ogawa N, Uchida N (1988) Hydrous phase in the crystalline vanadium-oxide spheres. J Solis State Chem 76:40
30. Livage J (1999) Optical and electrical properties of vanadium oxides synthesized from alkoxides. Coord Chem Rev 190:391
31. Legendre JJ, Livage J (1983) Vanadium pentoxide gels. 1. Structural study by electron-diffraction. J Colloid Interface Sci 94:75

32. Pelletier O, Davidson P, Bourgaux C, Coulon C, Regnault S, Livage J (2000) A detailed study of the synthesis of aqueous vanadium pentoxide nematic gels. Langmuir 16:5295
33. Livage J, Henry M, Sanchez C (1988) Sol-gel chemistry of transition-metal oxides. Prog Solid State Chem 18:259
34. Legendre JJ, Livage J (1983) Structures of vanadium pentoxide gels. 1. Structural study by electron-diffraction. J Coll Int Sci 94:75
35. Legendre JJ, Livage J (1983) Structures of vanadium pentoxide gels. 2. Structural study by x-ray-diffraction. J Coll Bit Sci 94:84
36. Yao T, Oka Y, Yamamoto N (1992) Layered structures of vanadium pentoxide gels. Mater Res Bull 116:279
37. Giorgetti M, Passerini S, Smyrl WH (2000) Evidence of bylayer structure in V_2O_5 xerogel. Inorg Chem 39:1514
38. Bullot J, Gallais O, Gauthier M, Livage J (1980) Semiconducting properties of amorphous V_2O_5 layers deposited from gels. Appl Phys Lett 36:986
39. Bullot J, Cordier P, Gallais O, Gauthier M, Livage J (1981) Experimental-determination of the disorder energy in amorphous V_2O_5 Layers deposited from gels. Phys Status Solidi 68:357
40. Sanchez C, Morineau R, Livage J (1983) Electrical-conductivity of amorphous V_2O_5. Phys Status Solidi 76:661
41. Sanchez C, Babonneau F, Morineau R, Livage J (1983) Semiconducting properties of V_2O_5 gels. Phil Mag B 47:279
42. Livage J (1996) Sol-gel chemistry and electrochemical properties of vanadium oxide gels. Solid States Ionics 86:935
43. Anaissi FJ, Demets GJ-F, Alvarez EB, Politi MJ, Toma HE (2001) Long-term aging of vanadium(V) oxide xerogel precursor solutions: structural and electrochemical implications. Electrochim Acta 47:441
44. Bardoux P, Morineau R, Livage J (1988) Protonic conductivity in hydrates. Solid State Ionics 27:221
45. Bullot J, Cordier P, Gallais O, Gauthier M (1984) Thin-layers deposited from V_2O_5 gels. 2. An optical-absorption study. J Non-Cryst Solids 68:135
46. Baddour R, Pereira-Ramos JP, Messina R, Perichon J (1991) A thermodynamic, structural and kinetic-study of the electrochemical lithium intercalation into the xerogel V_2O_5 1.6 H_2O in a propylene carbonate solution. J Electroanal Chem 314:81
47. Tipton AL, Passerini S, Owens BB, Smyrl WH (1996) Performance of lithium/V_2O_5 xerogel coin cells. J Electrochem Soc 143:3473
48. Park H-K, Smyrl WH, Ward MD (1995) V_2O_5 serogel films as intercalation hosts for lithium I. Insertion stoichiometry, site concentration, and specific energy. J Electrochem Soc 142:1068
49. Mège S, Levieux Y, Ansart F, Savariault JM, Rousset A (2000) Electrochemical properties of a new V_2O_5 xerogel. J Appl Electrochem 30:657
50. Parent MJ, Passerini S, Owens BB, Smyrl WH (1999) Composites of V_2O_5 aerogel and nickel fiber as high rate intercalation electrodes. J Electrochem Soc 146:1346
51. Dong W, Rolison DR, Dunn B (2000) Electrochemical properties of high surface area vanadium oxide aerogels. Electrochem Solid State Lett 3:457
52. Park HK, Smyrl WH (1994) V_2O_5 xerogel films as intercalation hosts for lithium. J Electrochem Soc 141:L25
53. Le DB, Passerini S, Tipton AL, Owens BB, Smyrl WH (1995) Aerogels and xerogels of V_2O_5 as intercalation hosts. J Electrochem Soc 142:L102
54. Owens BB, Passerini S, Smyrl WH (1999) Lithium ion insertion in porous metal oxides. Electrochim Acta 45:215
55. Passerini S, Smyrl WH, Berrettoni M, Tossici R, Rosolen M, Marassi R, Decker F (1996) XAS and electrochemical characterization of lithium intercalated V_2O_5 xerogels. Solid State Ionics 90:5

56. Le DB, Passerini S, Guo J, Ressler J, Owens BB, Smyrl WH (1996) High rate electrodes of V_2O_5 aerogel. J Electrochem Soc 143:2099
57. Lemordant D, Bouhaouss A, Aldebert P, Baffier N (1986) Intercalation of organic-solvents in the lamellar structure of V_2O_5 xerogels. J Chim Phys 83:105
58. Hupp J (2001) Emerging Nanoscience and Functional Artificial Nanoarchitectures. Interface 10:21
59. Lu Y, Lang Y, Sellinger A, Lu M, Huang J, Fan H, Haddad R, Lopez G, Burns AR, Sasaki DY, Shelnutt J, Brinker CJ (2001) Self-assembly of mesoscopically ordered chromatic polydiacetylene/silica nanocomposites. Nature 410:913
60. Chadwick AV (2000) Nanotechnology—Solid progress in ion conduction. Nature 408:925
61. Nazar LF, Goward G, Leroux F, Duncan M, Huang H, Kerr T, Gaubicher J (2001) Nanostructured materials for energy storage. Int J Inorg Mater 3:191
62. Bourgeat-Lami E (2002) Organic-inorganic nanostructured colloids. J Nanosci Nanotech 2:1
63. Maia DJ, De Paoli MA, Alves OL, Zarbin AJG, das Neves S (2000) Conductive polymer synthesis in solid host matrices. Quim Nova 23:204
64. Gimenez IF, Alves OL (1999) Formation of a novel polypyrrole porous phosphate glass ceramic nanocomposite. J Brazil Chem Soc 10:167
65. Kryszewski M (2000) Nanointercalates—novel class of materials with promissing properties. Synth Met 109:47
66. Wang Y, Cao G (2006) Synthesis and enhanced intercalation properties of nanostructured vanadium oxides. Chem Mater 18:2787
67. Patrissi CJ, Martin CR (1999) Sol-gel-based template synthesis and Li-insertion rate performance of nanostructured vanadium pentoxide. J Electrochem Soc 146:3176
68. Spahr ME, Bitterli PS, Nesper R, Haas O, Novák P (1999) Vanadium oxide nanotubes. A new nanostructured redox-active material for the electrochemical insertion of lithium. J Electrochem Soc 145:2780
69. Wang Y, Takahashi K, Shang H, Cao G (1995) Synthesis and electrochemical properties of vanadium pentoxide nanotube arrays. J Phys Chem B 109:3085
70. Lee K, Wang Y, Cao G (2005) Dependence of electrochemical properties of vanadium oxide films on their nano- and microstructures. J Phys Chem B 109:16700
71. Gangopadhyay R, De A (2000) Conducting polymer nanocomposites: a brief overview. Chem Mater 12:608
72. Kerr TA, Wu H, Nazar LF (1996) Concurrent polymerization and insertion of aniline in molybdenum trioxide: formation and properties of a [poly(aniline)]0.24MoO_3 nanocomposite. Chem Mater 8:2005
73. Nazar LF, Wu H, Power WP(1993) Synthesis and properties of a new $(PEO)_x[Na(H_2O)]_{0.25}MoO_3$ nanocomposite. J Mater Chem 5:1985
74. Kanatzidis MG, Marcy HO, McCarthy WJ, Kannewurf CR, Marks TJ (1989) Insitu intercalative polymerization chemistry of feocl—generation and properties of novel, highly conductive inorganic-organic polymer microlaminates. Solid States Ionics 32:594
75. Maia DJ, Das Neves S, Alves OL, De Paoli MA (1999) Photoelectrochemical conversion by SnP-C/Fe/PAni: an integrated chemical system. Synth Met 102:1153
76. Maia DJ, Alves OL, De Paoli M-A (1997) Growth of linear polyaniline chains in a layered tin(IV) phosphonate host. Synth Met 90:37
77. Wu C-G, DeGroot DC, Marcy HO, Schindler JL, Kannewurf CR, Bakas T, Papaefthymiou V, Hirpo W, Yesinowski JP, Liu Y-J, Kanatzidis MG (1995) Reaction of aniline with feocl—formation and ordering of conducting polyaniline in a crystalline layered host. J Am Chem Soc 117:9229
78. Liu Y-J, Kanatzidis MG (1995) Postintercalative polymerization of aniline and its derivatives in layered metal phosphates. Chem Mater 7:1525
79. Vaia RA, Vasudevan S, Krawiec W, Scanlon LG, Giannelis EP (1995) New polymer electrolyte nanocomposites: melt intercalation of poly (ethylene oxide) in mica-type silicates. Adv Mater 7:154

80. Oriakhi CO, Lerner MM (1996) Rapid and quantitative displacement of poly(ethylene oxide) from $MnPS_3$ and other layered hosts. Chem Mater 8:2016
81. Kerr TA, Leroux F, Nazar LF (1998) Surfactant-mediated incorporation of poly (p-phenylene) into MoO_3. Chem Mater 10:2588
82. Schöllhorn R (1996) Intercalation systems as nanostructured functional materials. Chem Mater 8:1747
83. Lev O, Wu Z, Bharathi S, Glezer V, Modestov A, Gun J, Rabinovich L, Sampath S (1997) Sol-gel materials in electrochemistry. Chem Mater 9:2354
84. Harreld J, Wong HP, Dave BC, Dunn B, Nazar LF (1998) Synthesis and properties of polypyrrole vanadium oxide hybrid aerogels. J Non-Cryst Solids 225:319
85. Lira-Cantú M, Gómez-Romero P (1999) Synthesis and characterization of intercalate phases in the organic-inorganic polyaniline/V_2O_5 system. J Solid State Chem 147:601
86. Oliveira HP, Graeff CFO, Brunello CA, Guerra EM (2000) Electrochromic and conductivity properties: a comparative study between melanin-like/V_2O_5 center dot nH(2)O and polyaniline/V_2O_5 center dot nH(2)O hybrid materials. J Non-Cryst Solids 273:193
87. Posudievsky OY, Biskulova SA, Pokhodenko VD (2004) New hybrid guest-host nanocomposites based on polyaniline, poly(ethylene oxide) and V_2O_5. J Mater Chem 14:1419
88. Guerra EM, Brunello CA, Graeff CFO, Oliveira HP (2002) Synthesis, characterization, and conductivity studies of poly-o-methoxyaniline intercalated into V_2O_5 xerogel. J Solid State Chem 168:134
89. Demets GJ-F, Toma HE (2003) Strong electric fields promote oriented intercalative polymerization of pyrrole inside the lamellar matrices of vanadium pentoxide. Electrochem Commun 5:73
90. Yatabe T, Matsubayashi G (1996) Intercalation of 2-, 4-sulfanylpyridine, 2,2'- and 4,4'-dithiobispyridine into VOPO4, and gel-V_2O_5 interlayer spaces. J Mater Chem 6:1849
91. Wang J, Gonsalves KE (1999) A combinatorial approach for the synthesis and characterization of polymer/vanadium oxide nanocomposites. J Comb Chem 1:216
92. Kanatzidis MG, Wu C-G, Marcy HO, Kannewurf CR (1989) Conductive polymer/oxide bronze nanocomposites. Intercalated polythiophene in vanadium pentoxide (V_2O_5) xerogels. J Am Chem Soc 111:4139
93. Wu C-G, Kanatzidis MG, Marcy HO, DeGroot DC, Kannewurf CR (1989) Polym Mater Sci Eng 61:969
94. Liu Y-J, DeGroot DC, Schindler JL, Kannewurf CR, Kanatzidis MG (1993) Stabilization of anilinium in vanadium (V) oxide xerogel and its post-intercalative polymerization to poly (aniline) in air. J Chem Soc Chem Commun 593
95. Wu C-G, DeGroot DC, Marcy HO, Schindler JL, Kannewurf CR, Liu Y-J, Hirpo W, Kanatzidis MG (1996) The postintercalative intralamellar polymer growth in polyaniline/metal oxide nanocomposites is facilitated by molecular oxygen. Chem Mat 8:1992
96. Somani PR, Marimuthu R, Mandale AB (2001) Synthesis, characterization and charge transport mechanism in conducting polyaniline/V_2O_5 composites. Polymer 42:2991
97. Ferreira M, Zucolotto V, Huguenin F, Torresi RM, Oliveira ON Jr (2002) Layer-by-layer nanostructured hybrid films of polyaniline and vanadium oxide. J Nanosci Nanotech 2:29
98. Ferreira M, Huguenin F, Zucolotto V, da Silva JEP, de Torresi SIC, Temperini MLA, Torresi RM, Oliveira ON Jr (2003) Electroactive multilayer films of polyaniline and vanadium pentoxide. J Phys Chem B 107:8351
99. Li ZF, Ruckenstein E (2002) Intercalation of conductive polyaniline in the mesostructured V_2O_5. Langmuir 18:6956
100. Leroux F, Koene BE, Nazar LF (1996) Electrochemical lithium intercalation into a polyaniline/V_2O_5 nanocomposite. J Electrochem Soc 143:L181
101. Leurox F, Goward G, Power WP, Nazar LF (1997) Electrochemical Li insertion into conductive polymer/V_2O_5 nanocomposites. J Electrochem Soc 144:3886

102. Lira-Cantú M, Gómez-Romero P (1999) The organic-inorganic polyaniline/V_2O_5 system—Application as a high-capacity hybrid cathode for rechargeable lithium batteries. J Electrochem Soc 146:2029
103. Kuwabata S, Idzu T, Martin CR, Yoneyama H (1998) Charge-discharge properties of composite films of polyaniline and crystalline V_2O_5 particles. J Electrochem Soc 145:2707
104. Huguenin F, Torresi RM, Buttry DA (2002) Lithium electroinsertion into an inorganic-organic hybrid material composed from V_2O_5 and polyaniline J Electrochem Soc 149:A546
105. Varela H, Torresi RM (2000) Ionic exchange phenomena related to the redox processes of polyaniline in nonaqueous media. J Electrochem Soc 147:665
106. Lira-Cantú M, Gómez-Romero P (1999) The polyaniline-V_2O_5 system: improvement as insertion electrode in lithium batteries. Int J Inorg Mat 1:111
107. Huguenin F, Torresi RM, Buttry DA, da Silva JEP, de Torresi SIC (2001) Electrochemical and Raman studies on a hybrid organic-inorganic nanocomposite of vanadium oxide and a sulfonated polyaniline. Electrochem Acta 46:3555
108. Varela H, Huguenin F, Malta M, Torresi RM (2002) Materials for cathodes of secondary lithium batteries. Quim Nova 25:287
109. McKinnon WR (1995) In: Bruce PG (ed) Solid state electrochemistry. Cambridge University Press, Cambridge, p 163
110. Huguenin F, Ticianelli EA, Torresi RM (2002) XANES study of polyaniline-V_2O_5 and sulfonated polyaniline-V_2O_5 nanocomposites. Electrochim Acta 47:3179
111. Holland GP, Yarger JL, Buttry DA, Huguenin F, Torresi RM (2003) Solid-state NMR study of ion-exchange processes in V_2O_5 xerogel, polyaniline/V_2O_5, and sulfonated polyaniline/V_2O_5 nanocomposites. J Electrochem Soc 150:A1718
112. Holland GP, Buttry DA, Yarger YL (2002) Li-7 NMR studies of electrochemically lithiated V_2O_5 xerogels. Chem Mater 14:3875
113. Huguenin F, Torresi RM (2008) Investigation of the electrical and electrochemical properties of nanocomposites from V_2O_5, polypyrrole, and polyaniline. J Phys Chem C 112:2202
114. Goward GR, Leroux F, Nazar LF (1998) Poly(pyrrole) and poly(thiophene)/vanadium oxide interleaved nanocomposites: positive electrodes for lithium batteries. Electrochim Acta 43:1307
115. Wong HP, Dave BC, Leroux F, Harreld J, Dunn B, Nazar LF (1998) Synthesis and characterization of polypyrrole vanadium pentoxide nanocomposite aerogels. J Mater Chem 8:1019
116. Huguenin F, Girotto EM, Torresi RM, Buttry DA (2002) Transport properties Of V_2O_5/polypyrrole nanocomposite prepared by a sol-gel alkoxide route. J Electroanal Chem 536:37
117. Harreld JH, Dunn B, Nazar LF (1999) Design and synthesis of inorganic-organic hybrid microstructures. Int J Inorg Mater 1:135
118. Demets GJF, Anaissi FJ, Toma HE (2000) Electrochemical properties of assembled polypyrrole/V_2O_5 xerogel films. Electrochim Acta 46:547
119. Posudievsky OY, Kozarenko OA, Dyadyun VS, Jorgensen SW, Spearot JA, Koshechko VG, Pokhodenko VD (2011) Characteristics of mechanochemically prepared host-guest hybrid nanocomposites of vanadium oxide and conducting polymers. J Power Sources 196:3331
120. Murugan AV, Kale BB, Kwon C-W, Campet G, Vijayamohanan K (2001) Synthesis and characterization of a new organo-inorganic poly(3,4-ethylene dioxythiophene) PEDOT/V_2O_5 nanocomposite by intercalation. J Mater Chem 11:2470
121. Kwon C-W, Murugan AV, Campet G, Portier J, Kale BB, Vijaymohanan K, Choy J-H (2002) Poly(3,4-ethylenedioxythiophene) V_2O_5 hybrids for lithium batteries. Electrochem Commun 4:384
122. Kuwubata S, Tomiyori M (2002) Rechargeable lithium battery cells fabricated using poly (methyl methacrylate) gel electrolyte and composite of V_2O_5 and polypyrrole. J Electrochem Soc 149:A988

123. Ponzio EA, Benedetti TM, Torresi RM (2007) Electrochemical and morphological stabilization of V_2O_5 nanofibers by the addition of polyaniline. Electrochim Acta 52:4419
124. Malta M, Torresi RM (2005) Electrochemical and kinetic studies of lithium intercalation in composite nanofibers of vanadium oxide/polyaniline. Electrochim Acta 50:5009
125. Chun-Guey W, Jiunn-Yih H, Shui-Sheng H (2001) Synthesis and characterization of processible conducting polyaniline/V_2O_5 nanocomposites. J Mater Chem 11:2061
126. Huguenin F, Girotto EM, Ruggeri G, Torresi RM (2003) Structural and electrochemical properties of nanocomposites formed by V_2O_5 and poly(3-alkylpyrroles). J Power Sources 114:133
127. Wang GC, Zhao J, Li XW, Li CZ, Yuan WK (2008) Synthesis and characterization of electrically conductive and fluorescent poly(N-[5-(8-hydroxyquinoline)methyl]aniline)/ V_2O_5 xerogel hybrids. Synth Met 159:366
128. Liu M, Visco SJ, De Jonghe LC (1991) Novel solid redox polymerization electrodes—all-solid-state, thin-film, rechargeable lithium batteries. J Electrochem Soc 138:1891
129. Shouji E, Buttry DA (1999) New organic-inorganic nanocomposite materials for energy storage applications. Langmuir 15:669
130. Park N-G, Ryu KS, Park YJ, Kang MG, Kim D-K, Kang S-G, Kim KM, Chang S-H (2002) Synthesis and electrochemical properties of V_2O_5 intercalated with binary polymers. J Power Sources 103:273
131. Liu YJ, Schindler JL, Degroot DC, Kannewurf CR, Hirpo W, Kanatzidis MG (1996) Synthesis, structure, and reactions of poly(ethylene oxide) V_2O_5 intercalative nanocomposites. Chem Mater 8:525
132. Ruiz-Hitzky E, Aranda P, Casal B (1992) New polyoxyethylene intercalation materials in vanadium-oxide xerogel. J Mater Chem 2:581
133. Chen W, Xu Q, Hu YS, May LQ, Zhu QY (2002) Effect of modification by poly(ethylene oxide) on the reversibility of insertion/extraction of Li+ ion in V_2O_5 xerogel films. J Mater Chem 12:1926
134. Jin AP, Zhu QY, Chen W, Volkov VL, Zakharova GS, Liu HX, Zhou J, Xu Q (2007) Electrical and electrochromic characterization of poly (ethylene-oxide)/V_2O_5 xerogel films. Solid State Phenom 124:363
135. Jin A, Chen W, Zhou Q, Yang Y, Volkov VL, Zakharova GS (2008) Electrical and electrochemical characterization of poly (ethylene oxide)/V_2O_5 xerogel electrochromic films. Solid State Ionics 179:1256
136. Lutkenhaus JL, Hammond PT (2007) Electrochemically enabled polyelectrolyte multilayer devices: from fuel cells to sensors. Soft Mater 3:804
137. Galiote NA, Huguenin F (2007) Lithium ion diffusion into self-assembled films composed from WO3 and polyallylamine. J Phys Chem C 111:14911
138. Huguenin F, Ferreira M, Zucolotto V, Nart FC, Torresi RM, Oliveira ON (2004) Molecular-level manipulation of V_2O_5/polyaniline layer-by-layer films to control electrochromogenic and electrochemical properties. Chem Mater 16:2293
139. Huguenin F, Santos DS, Bassi A, Nart FC, Oliveira ON (2004) Charge storage capability in nanoarchitectures of V_2O_5/chitosan/poly (ethylene oxide) produced using the layer-by-layer technique. Adv Func Mater 14:985
140. Huguenin F, Nart FC, Gonzalez ER, Oliveira ON (2004) Using the quadratic logistic equation to analyze intercalation of lithium ions in layer-by-layer V_2O_5 films. J Phys Chem B 108:18919

Prospective on the Use of Nanostructured Magnesium Alloys as Anode Materials for Ni–MH Rechargeable Batteries

Sydney Ferreira Santos, Flavio Ryoichi Nikkuni and Edson Antonio Ticianelli

1 Introduction

The commercialization of Ni–MH batteries started at late 80s. These cells were predominantly used for portable electronics for long time, replacing the NiCd cells. Presently, the use of Ni–MH cells had shifted from portable electronics to power tools, hybrid electrical vehicles and electrical vehicles.

Concerning the market of secondary (rechargeable) cells, Ni–MH batteries have Ni–Cd and Li-ion cells as main competitors. Comparing to Ni–Cd batteries, Ni–MH-type cells have larger volumetric and gravimetric energy densities. Furthermore, the absence of cadmium in Ni–MH cells is advantageous from the environmental point of view due to the high toxicity of this element [1].

Conversely, Ni–MH batteries have been experiencing a hard competition with Li-ion batteries which quickly increased their share market and nowadays are predominant in portable electronic applications such as cell phones, notebooks, tablets, among others. These batteries have larger energy densities (volumetric and gravimetric) than Ni–HM ones.

S.F. Santos (✉)
Centro de Engenharia, Modelagem e Ciências Sociais Aplicadas,
Universidade Federal do ABC, Avenida dos Estados 5001,
CEP 09210-580 Santo André, SP, Brazil
e-mail: sydney.ferreira@ufabc.edu.br

F.R. Nikkuni
Fundação Parque Tecnológico Itaipu, Av. Tancredo Neves, 6731,
C.P 2039, CEP 85.867-900 Foz do Iguaçu, PR, Brazil

E.A. Ticianelli
Instituto de Química de São Carlos, Universidade de São Paulo,
Av. Trab. São-Carlense, 400, CEP 13560-970 São Carlos, SP, Brazil

F.L. de Souza and E. Leite (eds.), *Nanoenergy*,
DOI 10.1007/978-3-319-62800-4_8

Table 1 Comparison between Ni–MH, Ni–Cd, and Li-ion sealed cells

Property	Ni–Cd	Ni–MH	Li-ion
Energy density	Fair	Good	Excellent
Power density	Excellent	Very good	Very good
Low-temperature operation	Excellent	Very good	Very good
Charge retention	Fair	Fair	Very good
Charge acceptance	Very good	Very good	Excellent
Efficiency	Good	Good	Excellent
Life	Good	Good	Excellent

Table 1 compares some key features of Ni–MH, Ni–Cd, and Li-ion cells [2]. In terms of price, Ni–MH cells are situated between Ni–Cd (less expensive) and Li-ion (more expensive) counterparts.

For some applications, Ni–MH batteries have to compete not only with Li-ion batteries, as aforementioned, but also with emerging technologies such as low-temperature fuel cells and super-capacitors.

Although the present predominance of Li-ion cells in portable electronics, Ni–MH cells stand in an important position in the market for a few applications such as electrical vehicles (EVs), hybrid electrical vehicles (HEVs), power tools, stationary applications and so on. The use of Ni–MH cells in these applications is justified by advantageous features, highlighting their superior safety even under abusive operating conditions, flexibility of operation temperature, and environmental friendliness [3, 4]. Moreover, for most of these applications the combination of moderate energy density and high power density of Ni–MH cells is enough to meet the present requirements. But to remain in the market for the future, Ni–MH batteries must continuously enhance their performances, mainly concerning the specific energy densities, which are determined by the hydrogen storage capacities of the hydride electrodes, their power densities, controlled by the high rate dischargeability (HRD) properties of the hydride electrodes, and their cycle life, controlled by corrosion/degradation of the hydride electrodes. For instance, current commercial Ni–MH batteries used in HEVs deliver specific power in the range of 150–400 W/kg, while the Department of Energy of US target is over 650 W/kg, clearly indicating the need of improving the HRD of the hydride electrodes [3].

In addition to the well-established use of metal hydrides electrodes in Ni–MH batteries, some other electrochemical applications of these materials have been investigated in the last years such as anodes in Li-ion and Li-air batteries [4, 5], in metal hydride–air batteries [6, 7] and as anodes in alkaline fuel cells [8–12]. These cells with built-in energy storage (resulted from the charging of the hydride-forming alloy with hydrogen) can work similarly to batteries in situations of scarcity of the hydrogen external source, oxidizing the hydrogen content stored in the form of hydride. A variety of these alkaline fuel cells which can use metal hydride anodes is the direct borohydride fuel cell. In these cells, the hydride-forming alloy electrode promotes the borohydride oxidation reaction (BOR) been charged with hydrogen from the borohydride instead of hydrogen gas fuel [13–15].

In this chapter, the basic concepts and hydrogen storage alloy electrodes used in MH batteries are introduced. Moreover, the potentiality and recent investigations concerning the use of magnesium—based hydrides as anode materials are discussed based on recent results.

2 Principles of Nickel—Metal Hydride Batteries

A Ni–MH cell is an electrochemical device for energy storage. These cells are classified as secondary cells since they can be charged again after the complete exhaustion of the cell (i.e., after all the reactants have been converted in product species, the electrochemical reaction can be reverted). These cells are composed by three main parts: a positive electrode (cathode), a negative electrode (anode), and an electrolyte. The cathode is the NiOOH compound, the anode is a hydrogen storage alloy, and the electrolyte is a concentrated solution of KOH. A Ni–MH cell is schematically represented in Fig. 1 [16–19].

The electrochemical reactions that take place during the Ni–MH cell operation are shown below. The reactions are reversible and the direction from left to right indicates charging, whereas the opposite direction indicates discharging [16–19].

In the cathode side, the reaction that takes place is

$$Ni(OH)_2 + OH^- \leftrightarrow NiOOH + H_2O + e^- \quad E = 0.52\,V \tag{1}$$

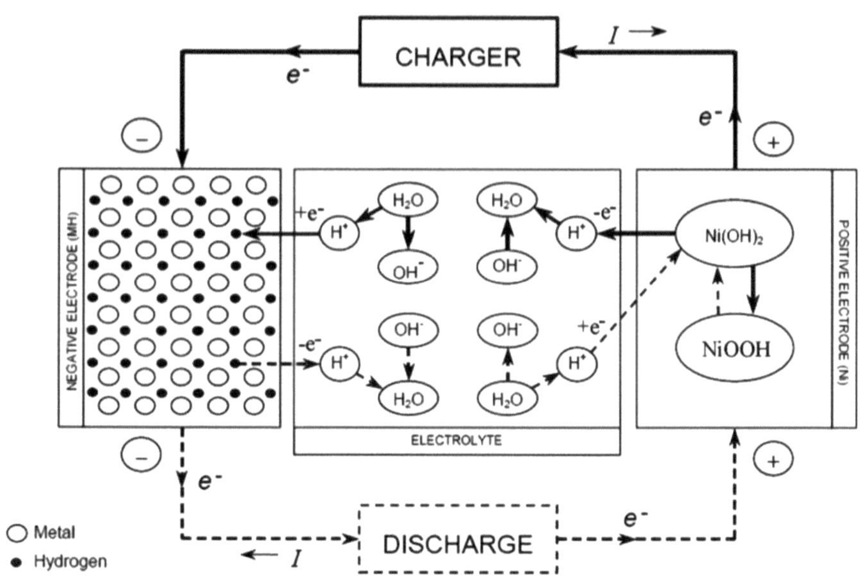

Fig. 1 Schematic representation of a Ni–MH cell. *Full line* represents the electrical circuit during charging while the *dotted line* represents the circuit during discharging [16, 17]

while the reaction in anode side is

$$M + H_2O + e^- \leftrightarrow MH + OH^- \quad E = 0.83\,V \tag{2}$$

Then, the overall cell reaction is

$$Ni(OH)_2 + M \leftrightarrow NiOOH + MH \quad E = 1.35\,V \tag{3}$$

In a Ni–MH cell, the upper limit of the theoretical discharge capacity, i.e., the specific energy density of the cell, is controlled by the reversible hydrogen storage capacity of the alloy electrode. Therefore, the development of novel hydrogen storage materials with larger gravimetric storage capacities is fundamental for attaining cells with higher specific energy densities. Moreover, to increase the power density of a Ni–MH cell, the development of metal hydrides with high rate dischargeability (HRD) properties is needed.

The hydrogen storage materials used in Ni–MH cells are typically intermetallic compounds. Further information on these materials such as their compositions and crystal structures are giving in the following section.

In hydride-forming materials, hydrogen is up taken into their crystal structures, first forming an interstitial solid solution (α phase) for low concentration of hydrogen. In this phase, hydrogen is randomly distributed in the available interstitial sites throughout the crystal structure. When the limit of solubility is exceed a new phase started to nucleate into the microstructure. This phase is a metal hydride (β phase). In this phase hydrogen occupy periodically the available interstitial sites in crystal structure. These two phases are represented schematically in Fig. 2.

Due to its small atomic size, hydrogen exhibits large limits of solubility in metals to form interstitial solid solutions when compared to other light elements (N, C, O, B). Conversely, even the limits of solubility of hydrogen in metals are small when compared to the solubility observed in substitutional solid solutions with extensive solubility (i.e., those solid solutions which follows the so called "Hume-Rotery rules" [20]). The limited solubility of interstitial elements in metals arises from the high elastic strain imposed by solute atoms to the host lattice. Even hydrogen (the smallest interstitial atom) is much larger than the interstitial spaces in host lattice and slightly displaces the neighbor metallic atoms from their ideal equilibrium

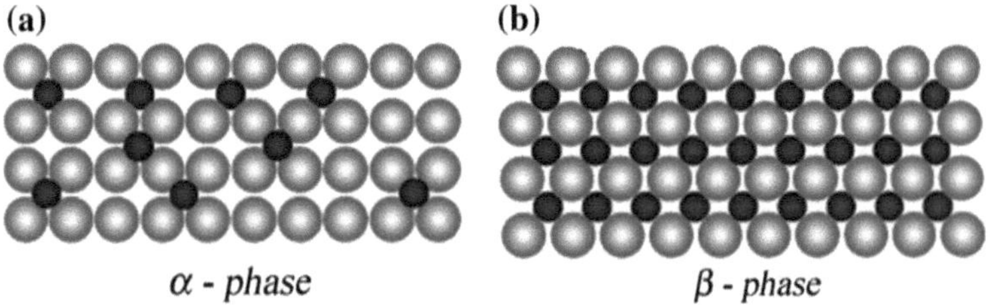

Fig. 2 Schematic representation of the atomic arrangement for **a** interstitial solid solution (α-phase) and **b** metal hydride (β–phase)

positions. This difference in volume size is the origin of the elastic strain. In extensive substitutional solid solutions the elastic strain fields around the solute atoms are much lower.

When a metal with positive heat of mixing with hydrogen is submitted to a hydrogen atmosphere, it absorbs the atoms of this element until reaching the equilibrium with the gaseous atmosphere. If the pressure is increased, more hydrogen will be absorbed until reaching a new equilibrium condition. Thus, for each pressure (p) there is a certain amount of hydrogen which can be absorbed in equilibrium condition (c). This behavior is known as Sievert's law and it can be mathematically represented by the following expression:

$$c = k \cdot \sqrt{p}, \tag{4}$$

where c means the hydrogen concentration in the metal, p is the equilibrium pressure of H_2, and k is a constant which depends on temperature (Sievert's constant).

When the material is hydrogenated under a gas phase atmosphere at constant temperature, pressure–composition isotherms (PCI) can be determined for this material. These isotherms can be measured for different temperatures, as illustrated in Fig. 3 [16].

As aforementioned, in solid solution composition range, hydrogen metal interaction follows the Sievert's Law and the hydrogen content is function of the pressure. When the maximum solubility at a given temperature is attained further hydrogen absorption results on the formation of a new phase named metal hydride

In equilibrium condition, the transformation from interstitial solid solution (usually denominated α-phase) to hydride (β-phase) takes place under constant pressure. This pressure is the equilibrium pressure of hydride formation (or equilibrium pressure of absorption) and corresponds to the plateaus of pressure

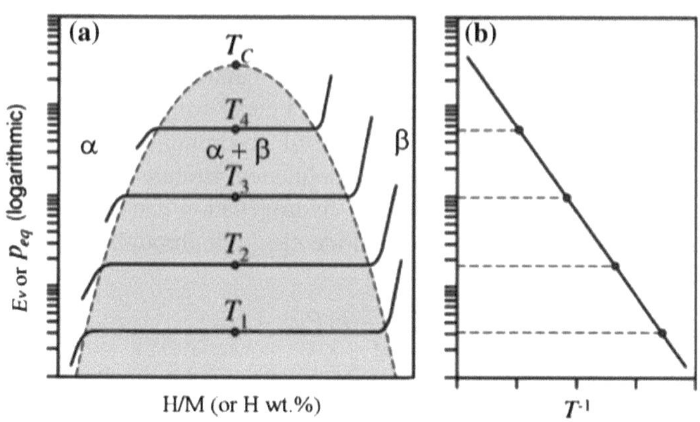

Fig. 3 Schematic representation of some PCI curves (**a**) and the corresponding van't Hoff plot (**b**) [16]

observed for the *pressure–composition* isotherms observed in Fig. 3 for different temperatures. As represented in this Fig. 3, increasing the temperature a higher equilibrium pressure is attained. Under desorption, the same reaction takes place but the equilibrium pressure of desorption is lower than that of absorption. Ideally, the plateau pressure of hydride formation/decomposition is flat. This behavior is assumed to the occupation of energetically equivalent interstitial sites by hydrogen atoms. But, in several cases, a slopping plateau is observed during (de)hydriding. This behavior can by ascribed to a few reasons. One of them is the presence of internal stresses developed from dislocation pile-ups. This accumulation of defects is generated from the sudden volume change associated to the hydride formation/decomposition. Slopping plateau can also be caused by internal oxide precipitates. In the case of alloys and intermetallic compounds, the slopping plateau can be originated from chemical inhomogeneity [21]. Moreover, depending on the amount of defects accumulated and mechanical properties of the hydrogen storage material, hydride particles may have extensive fracturing and pulverization (decrepitation), resulting in loose of the electrical contact between the hydride particle and the conductive binder (typically, nickel fine powder or high surface area carbon) and also increasing the exposition of the fresh surfaces generated during pulverization to the strong alkaline electrolyte which results in corrosion of the anode material. These phenomena are responsible for the gradual decay of the discharge capacity of the hydride anode which will shorten the cycle-life performance of the battery.

From a set of PCI curves, it is possible to obtain the thermodynamic properties of the hydride, through the van't Hoff equation, expressed as

$$\ln P_{\mathrm{eq}} = \frac{\Delta H}{R \cdot T} - \frac{\Delta S}{R} \tag{5}$$

From the linearization of this equation (Fig. 3b), one can obtain the values of enthalpy (ΔH) and entropy (ΔS) of $\alpha \rightarrow \beta$ reaction. The ΔS term mostly corresponds to the change from molecular hydrogen to dissolved solid hydrogen, i.e., it is nearly the standard entropy of hydrogen and does not strongly depend on the nature of the metal. The term ΔH depends on the metal hydrogen bond [16, 22, 23]. These values can be obtained from a set of PCI curves of absorption or desorption. P_{eq}, R, and T are the equilibrium pressure of absorption or desorption (plateau pressure), the ideal gas constant and the absolute temperature, respectively [16, 22].

The equilibrium pressure for gaseous hydrogenation can be correlated to the electrochemical potential (E_v) of the hydride electrode through the equation

$$E_v = -\frac{1}{2} \cdot \frac{RT}{F} \ln P_{H_2} - \delta E_{ref}, \tag{6}$$

where F is the Faraday constant ($F = 9.6487 \times 10^4$ C/mol), PH_2 is the hydride dissociation pressure, and T the absolute temperature. E_v is measured with reference to the standard reference electrode. Thus, if other reference electrode is used,

δE_{ref} is added to the value of E_v (for instance, −0.926 for Hg/HgO reference electrode) [16, 17]. Therefore, it is possible to obtain the *pressure–composition* isotherms from electrochemical tests (electrochemical PCTs) similarly to the determination of *pressure–composition* isotherms from metal–gas reaction. For information regarding the experimental setup and data analysis of electrochemical PCTs, the interested reader may consult the article of Mossvati et al. [24].

3 AB_5 and AB_2 Alloy Electrodes

There are several different types of alloys that form metal hydrides. For hydrogen storage purposes, the metal hydrides should have large reversible hydrogen storage capacities and adequate thermodynamic properties. But, in the case of anode application, the hydride-forming alloy should have additionally

(i) Adequate corrosion resistance in concentrated alkaline electrolyte;
(ii) Stability over repeated cycles of electrochemical charge/discharge;
(iii) High electrocatalytic activity and reversibility of the electrode reaction;
(iv) Easy activation.

Development of materials suitable to fulfill completely this set of requirements is a challenging task. Presently, there are two different types of alloys used as anode materials in commercial Ni–MH batteries: AB_5 and AB_2 alloys.

The AB_5 alloys are based on the $LaNi_5$ intermetallic compound which have $CaCu_5$ compound as prototype structure and space group P6/mmm. A detailed description of the crystallography of these alloys is beyond the scope of this chapter and further useful information on this subject can be obtained from Wronsky [19] and references therein.

The investigations on the use of intermetallic compounds as anode materials started in early 70s but only in late 80s the minimum requirements for practical application were achieved [25]. Few years later, the commercial production of Ni–MH batteries started having a $LaNi_5$-based alloy as anode material. These alloys have been used up to the present days in commercial Ni–MH cells.

In this alloys both elements La and Ni are partially replaced by other metals in order to improve the electrode performance.

In commercial alloys, La is usually replaced by mischmetal, a commercial mixture of rare earths (La, Ce, Nd, and Pr), aiming to decrease the cost of raw material. The hydrogen storage properties of mischmetal—containing AB_5 alloys are strongly affected by the mischmetal composition. For example, mischmetal with high content of Ce causes an increase in plateau pressure and hysteresis in pressure–composition isotherm loops [19]. This later means that the difference between the plateau pressure of absorption and plateau pressure of desorption increased. This is usually an undesired feature for metal hydrides designed for technological applications but it is a more serious problem for those hydrides aimed for hydrogen storage tanks.

Partial substitution of Ni is also performed with the purpose of improving the reaction kinetics, increase the corrosion resistance, and adjust the plateau pressures of hydrogen absorption and desorption of the alloys.

The substitution of Ni by elements such as Sn and mainly Co can improve the kinetic properties and increase both cycle life and discharge capacity [26, 27]. These beneficial effects have been attributed to the increase in the unit cell parameter and reduction in cell volume expansion during hydriding. Both features lead to the decrease in pulverization of the alloy under galvanostatic cycling. This pulverization process takes place due to successive volume expansion and contraction caused by hydride formation and decomposition resulting in generation and accumulation of defects and, consequently, fractures of the particles and formation of fresh surfaces which are more sensitive to corrosion when exposed to electrolyte. This process results in continuous loss of active material by corrosion.

In spite of these positive effects of Co addition, alternatives to this element have been searched due to its high cost.

Replacement of Al for Ni increases the cycle-life performance of the electrodes through the formation of a protective passive film. Other common substitute for Ni is Mn which decreases the plateau pressures of hydride formation and decomposition maintaining the same hydrogen storage capacity [19, 26–28].

The commercial Ni–MH cells make use of multi-component alloys as anode materials, with several substitutes for both A and B elements. As a common feature, the commercial alloys need relatively large amounts of cobalt to present satisfactory electrode performances, despite the inconvenience of its high cost. A typical example of composition of AB_5 alloy electrode is $MmNi_{3.55}Mn_{0.4}Al_{0.3}Co_{0.75}$ (in at. %), where Mm means mischmetal. This alloy contains over 10% of Co which ensure good durability for the electrode. Conversely, Co is an expensive alloying element been responsible by approximately 40% of the cost of raw material [29, 30].

Concerning the AB_5 alloys, the main constraint is related to the low discharge capacity displayed by these alloys, which absorb nearly 1.2–1.4 wt% of H, depending on composition. These hydrogen storage capacities can lead to discharge capacities nearly 370 mA h/g. The experimental results indicate capacities nearly 300 mA h/g. This discrepancy has been attributed to the pulverization.

The AB_2 alloys used in electrode applications are multielement pseudo-binary alloys, having ZrV_2 and $ZrMn_2$ compounds as prototypes for hydrogen storage. The ZrV_2 compound has structure C14 hexagonal while the $ZrMn_2$ compound has structure C15 cubic. These two phases present high hydrogen storage capacities. Other phases that typically appear in such alloys are the C36 hexagonal, with poor hydrogen storage capacity, and a BCC solid solution with high hydrogen storage capacity [19]. The C14, C15, and C36 structures are members of a group of intermetallic compounds denoted Laves phases. Some useful crystallographic information can be obtained in Wronsky [19] and references therein.

The discharge capacity of AB_2 alloys is typically larger than those of the AB_5 counterparts, reaching values around 390 mA h/g. Conversely, AB_2 alloys exhibit lower electrocatalytic activity, poorer activation behavior, and higher costs than AB_5 counterparts, limiting their utilization [1, 19, 31].

Concerning the synthesis of AB_2 and AB_5 alloys, they are typically produced by melting under controlled atmosphere. The preferential processing routes are arc-melting and induction melting. In the first one, chunks of pure elements are placed into the furnace chamber which is repeatedly evacuated and filled with an inert gas (argon, in general) to ensure a low partial pressure of oxygen. Thus an electrical arc between the metal chunks and a nonconsumable tip of tungsten is produced and it allows to reach high temperatures (over 2000 °C) promoting the fusion and intermixing of pure elements. The obtained ingots are usually turned upside down and remelted few times to achieve homogenization of the chemical composition. This arc-melting furnaces usually have a water cooled copper base were the raw material is places and melted avoiding the necessity of crucibles, a potential source of contamination. On the other hand, dendritic segregation is usually observed for arc-melted alloys and in some cases subsequent high temperature annealing is necessary to achieve the desired microstructure. In the case of induction melting, similar procedures for cleaning the atmosphere are employed. In this case, the melting is usually carried out in crucibles and the selection of the crucible material is an important task in order to avoid or minimize the contamination of the cast ingot. Conversely, the magnetic agitation promoted by the magnetic field generated by the induction coil can promote a refining of the microstructure and homogenization at some extent.

4 Mg alloy electrodes

Magnesium-based materials (alloys and composites) are very attractive for hydrogen storage applications due to a set of promising properties of Mg, such as

(i) High volumetric and gravimetric hydrogen storage capacities;
(ii) Low density;
(iii) Availability;
(iv) Relative low cost.

For instance, Mg and Mg_2Ni when fully hydrided (i.e., completely converted to MgH_2 and Mg_2NiH_4, respectively) can reach gravimetric hydrogen storage capacities of 7.6 and 3.6 wt%, respectively. As mentioned in previous section, the commercially used AB_5-type alloys can only reach up to 1.4 wt% of hydrogen.

These features have motivated the study of hydrogen storage in a wide range of Mg-based materials (with polycrystalline, nanocrystalline, amorphous, nanocomposites structures) since long time [32–37].

Notwithstanding the above-mentioned encouraging properties, the slow kinetics of hydrogen absorption/desorption has been a major drawback to apply Mg-based materials in hydrogen storage applications. Moreover, the poor corrosion resistance of magnesium in alkaline solution is a major problem for electrochemical applications.

Polycrystalline magnesium and magnesium alloys can only absorb and release hydrogen at high temperatures and usually with slow kinetics (an exception of this trend is show in Ref. [33]). In pure Mg and Mg–Ni alloys, for example, the room temperature electrochemical (de)hydrogenation is negligible. For this reason, only after the development of highly metastable magnesium alloys (i.e., alloys with amorphous or nanocrystalline structures) these materials started to be considered for anode applications. The main investigated Mg alloys for anode application are those of the Mg–Ni system. Alloys of this system are mostly synthesized by mechanical alloying even though other routes are also possible. For Mg alloys, routes involving fusion introduce a considerable degree of complexity because Mg easily evaporates and it is very reactive with oxygen and moisture, especially in liquid state. In mechanical alloying, powders of pure Mg and Ni are placed together, in appropriate stoichiometry, in a milling vial with a certain number of balls and milled by a predefined period of time. Before milling, the vial is evacuated and filled with argon in order to prevent the oxidation during processing. The impact of the balls against small powder agglomerates during milling promotes repeated events of cold welding and fracture. After certain time, the particles become mixtures of deformed Mg and Ni regions with lamellar-like microstructures. The large amount of Mg/Ni interfaces and structural defects (vacancies, dislocations, staking faults, etc.) generated by plastic deformation create short-circuits for inter-diffusion, promoting atomic mixture of the metallic elements through a solid state reaction which was facilitated since the staking of several small plates of the metals (lamellar-like morphology) can work as a sequence diffusion couples (nanometric in thickness) [38]. This is the typical behavior observed for mechanical alloying of soft elements [16, 39].

Depending on the type of mill, material to be processed, desired microstructure, final average particle size, and materials application, a number of other processing variables need to be settled, such as: (i) ball-to-powder weight ratio, (ii) number of balls, (iii) diameter of the balls, (iv) in some cases, the milling speed, (v) milling atmosphere, etc. A deeper description of the mechanical alloying process is far beyond the scope of the present chapter and there is a vast literature for consulting on this subject [39–41]. One of the most interesting and complete source of information is the classical review article of Suryanarayana [39] which deals with all relevant aspects of mechanical alloying and correlated techniques.

The mechanical alloying of Mg–Ni binary system can generate a variety of different microstructures, depending on the selected stoichiometry, processing parameters, etc. This process is mainly indicated to obtain alloys with metastable microstructures such as: extended solid solutions (i.e., metastable solid solutions supersaturated in solute), fully amorphous alloys, nanocrystalline alloys (single or multi-phase), and mixtures of amorphous and nanocrystalline phases, nanocomposites, and so on. The processing parameters adopted during mechanical alloying will strongly affect the microstructural and morphological features of the synthesized alloys and therefore will greatly influence the electrochemical properties of the final product.

Liu et al. [42] investigated the Mg_xNi_{100-x} alloy series, with x ranging from 10–90 at.%. These authors reported that fully amorphous alloys were obtained between 30 and 70 at.% of Mg and yielded the best electrochemical properties among this series. The highest value of discharge capacity was achieved by the $Mg_{50}Ni_{50}$ alloy (387 mAh/g), but its cycle—life performance was very poor. After nine cycles, the discharge capacity was only 35% of the initial one. Similar behavior has been reported for MgNi binary alloys by other authors [43, 44].

Investigations on the surface chemical composition of Mg–Ni alloys have been performed trying to understand their electrochemical performances. Surface and subsurface analysis of the $Mg_{50}Ni_{50}$ alloy, by combining X-ray photoelectron spectroscopy (XPS), Auger electron spectroscopy (AES), and Ar^+ sputtering, indicated that at top surface Mg prevails in oxidized state and Ni in metallic state [43]. Penetrating the subsurface, there was observed an enrichment of Ni up to about 60 at.%. The good electrocatalytic activity of this alloy was ascribed to this Ni-rich subsurface layer which could act as a catalyst for the electrochemical reactions [43].

In a study of Zhang et al. [44] three different compositions of Mg–Ni amorphous alloys were investigated. The larger value for the maximum discharge capacity was 490 mAh/g for the $Mg_{50}Ni_{50}$ alloy. The maximum discharge capacity decreased for larger amounts of Ni (370 mAh/g for $Mg_{40}Ni_{60}$ and 200 mAh/g for $Mg_{33}Ni_{67}$). Conversely, the increase in Ni content improved the cycling performance of the electrodes. Surface analyses indicated that Mg prevails in oxidized state on top surface for all three compositions while Ni remains mostly in metallic state for the $Mg_{50}Ni_{50}$ composition. Increasing Ni content resulted in a higher Ni oxidized/Ni metallic ratio which can be related to the decrease of the maximum discharge capacity. Moreover, subsurface analyses indicate an increase of the thickness of Ni subsurface layer for the alloys with larger Ni content.

The aforementioned experimental results on Mg–Ni alloys highlight the necessity of improving their electrochemical properties, mainly their cycling performances in order to obtain alloys suitable for practical applications. There are several strategies that have been investigated aiming to optimize the electrochemical properties of Mg–Ni alloys. These strategies can be sub-divided in two main groups:

(i) Modification of surface chemical composition,
(ii) Modification of bulk chemical composition.

The former group comprises all routes where only the surface is modified and the center of the particle remains unchanged. These routes include all types of coatings, lixiviation and etching of Mg–Ni particle surfaces.

The second group of strategies comprises modification of chemical composition in full particles' volume. This is typically the case of addition of alloying elements which takes place during the synthesis of the alloy.

Concerning the modification of the surface chemical composition, the modified surface (usually coated with some type of protective layer) must prevent or minimize the corrosion of the alloy electrode but cannot hinder the absorption and release of hydrogen. Moreover, the modified surface should present high electrocatalytic activity and mechanical stability. One of the most investigated methods of modifying the chemical composition of alloy electrodes is the *electroless deposition* which allows the particles of the hydrogen-absorbing alloy to be completely covered with a metal or compound with good corrosion resistance. In this technique the coating is performed by immersing the active material (hydrogen-absorbing alloy) or the working electrode in a Becker with a chemical solution containing the coating precursor (usually salts of the element to be deposited). Thus the reaction takes place on the surface of the hydrogen storage alloy in the presence of a reducing agent. For instance, a bath containing $CuSO_4$ (0.16 g/ml) and H_2SO_4 (pH = 4–5) has been employed for Cu electroless deposition on hydrogen-absorbing alloy electrodes [45]. The electroless deposition usually presents improvements of the high rate dischargeability, decreasing of cyclic degradation, lowering the charge transfer resistance, and lowering the charge/discharge overpotential in AB_5-type alloys [45–48]. But, in the case of Mg–Ni alloys, the experimental results have not been so promising like those obtained for the AB_5 counterparts. Rongeat et al. [49] performed the electroless coating of the $Mg_{50}Ni_{50}$ alloy with a fine and dense chromate coating and observed a constant discharge capacity of the coated electrode for the first two cycles. After that, a decay of the discharge capacity similar to that observed for the uncoated one was reported. This behavior was attributed to the rupture of the coating layer caused by the huge mechanical stresses on this layer due to the large volume expansion promoted by hydriding. These explanation is not complete satisfactory since AB_5 alloys also experience pronounced volume expansion due to hydride formation and their electrochemical performance is improved in a larger extent by electroless deposition. This highlights that further investigation is still necessary to understand the different performances attained by electroless coated AB_5 and Mg–Ni alloys.

Another technique for modifying the surface chemical composition of the alloy electrodes is *mechanical coating* which consists of dispersing a coating element by short milling times in a parent Mg–Ni alloy to promote the improvement on its corrosion resistance [49, 50]. Differently of electroless deposition, mechanical coating does not form a compact and continuous layer on the particle surface of the Mg–Ni alloy. This feature probably makes the coating obtained by this technique less sensitive to the expansion caused by hydride formation. Conversely, the particles are not fully protected against the contact with the electrolyte and protected from corrosion. Figure 4 shows the scanning electron microscopy (SEM) images of the coating materials used for mechanical coating of a Mg–Ni alloy, the uncoated (bare) Mg–Ni alloy and Mg–Ni coated alloys [50]. Figure 4a shows the morphology of the ball-milled bare alloy. This alloy has a relatively equiaxial particle shape with sizes ranging from 1 to 5 μm. In Figs. 4b, c the microstructures of the coating materials can be observed. There is a clear morphological difference between the coating materials and bare alloy. For the coating materials, two types of morphologies can be observed: (i) small agglomerates of fine and rounded particles, in small amount; and (ii) a large amount of flattened particles. The morphology of Mg–Ni bare alloy is probably due to an equilibrium between cold welding and

Fig. 4 SEM images of **a** ball-milled and uncoated Mg–Ni alloy, **b** Ni coating, prepared by ball-milling pure Ni granulates; **c** Ni-5% Al coating, prepared by ball-milling Ni and Al powders in appropriate stoichiometry; **d** Mg–Ni coated with Ni by mechanical coating; and **e** Mg–Ni coated with Ni-5% Al by mechanical coating [50]

fracture events during ball-milling while the morphology of coating materials suggest a predominance of deformation and cold welding. Figures 4d, e show the microstructures of the Mg–Ni alloys coated by Ni and Ni-5% Al, respectively. The resulting nanocomposites have similar morphologies, composed by fine spherical particles. Flattened particles are no longer observed. This feature can be ascribed to the low fraction of coating materials (5 wt%) and to an effective coating of the Mg–Ni particles.

The X-ray diffraction patterns for the coating materials (not shown here) indicate that Al was only partially solubilized in Ni–Al coating material. The amount of solubilized Al was estimated to about 2.2 at.% [50].

Figure 5 shows curves of discharge capacity as a function of the number of cycles of bare nanostructured $Mg_{50}Ni_{50}$ and two coated $Mg_{50}Ni_{50}$ alloys (one coated with 5 wt% of Ni and the other with 5 wt% of Ni–Al alloy) [50]. The coating materials were dispersed by 1 h of further milling in a planetary ball-mill. It is possible to see that mechanical coating improved both the maximum discharge capacity and the cycle-life performance of the electrodes, but in a limited extent. The results indicate that more extensive investigations on processing/coating parameters are still necessary in order to optimize the electrode performances.

Li et al. [51] also investigated the mechanical coating of a metal hydride electrode with nickel. These authors mechanically coated the Mg_2NiH_4 compound with

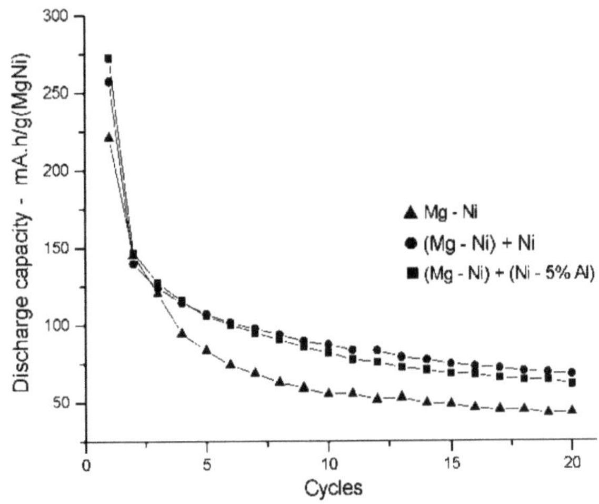

Fig. 5 Discharge capacity versus number of cycles of bare, Ni coated, and Ni-5% Al coated $Mg_{50}Ni_{50}$ alloys [50]

nanometric Ni powder (<50 nm in particle size), adding amounts of Ni ranging from 0 to 3 molar ratio. Mechanical coating was performed by ball-milling the Mg_2NiH_4 sample with nano-sized Ni powder in a planetary ball-mill at a rotating speed of 400 rpm by 5 h using ball-to-powder ratio of 40:1. In these conditions, a solid state reaction between the metal hydride and nano-Ni coating toke place as indicated by observed nanocrystallization and amorphization as function of the Ni content added. It was observed that the maximum discharge capacities of ball-milled samples, observed in the first discharge, increased from 578 mAh/g (0% Ni) to 896 mAh/g (2% nano-Ni) and then decreased to 810 mAh/g (3% nano-Ni). In contrast to the positive effect of this coating on the maximum discharge capacity of the hydride electrodes, it was not observed any beneficial effect on the cycle-life performance of the coated electrodes. All the investigated electrodes showed fast degradation of their discharge capacities.

As aforementioned, modifications of the surface chemical composition can lead to improvements of the electrode performance of Mg–Ni alloys but until now the best electrochemical results were reported for modifications of the bulk chemical composition of the alloy electrodes, i.e., addition of alloying elements to the binary Mg–Ni system. Thus, a number of ternary and quaternary alloys have been investigated and some of than have showed remarkable improvements on the electrochemical properties when compared to the binary counterparts. Some of these metallic elements improve the cycle-life performance of the electrodes while reducing their maximum discharge capacities. This behavior is observed for instance for Co, Al, Si, Cu, W, Ti, Al, Mn, among others [49–53]. In a previous report [53] we investigated the effect of Cr, Co, Nb, Ti, and V additions on the electrochemical properties of the nanostructured $Mg_{50}Ni_{50}$ alloy. It was observed increases in maximum discharge capacities for the additions of Zr, Nb, and Cr metals, given rise to the ternary $Mg_{45}Ni_{45}M_{10}$ alloys (where M is the added metal). Additions of Zr and Nb decreased the cycle-life performance when compared to the binary alloy while Cr slightly increased this property. These results of discharge capacity versus cycling number are shown in Figs. 6 and 7.

Some promising results have been reported for amorphous alloys of the Mg–Ni–Ti and Mg–Ni–Al systems, which presented a limited decrease of the maximum discharge capacity and improvements of cycle-life performance [49]. Mg–Ni–V system also presented interesting electrode performances [54]. In alloys of this system the same capacity of $Mg_{50}Ni_{50}$ alloy and improved cycle-life performance were achieved [54]. In the case of the Ti addition, it was observed by surface analysis the formation TiO_2 on top surface of the alloy particles and also decrease in $Mg(OH)_2$ formation rate [55]. These results indicate a preferential oxidation of Ti on the surface of the particles creating a thin layer of TiO_2. This passive film might protect the bulk particle against corrosion improving the cycle-life performance of the electrodes.

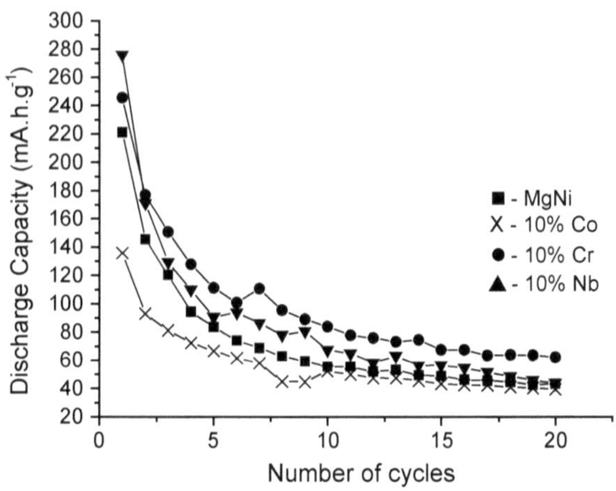

Fig. 6 Discharge capacity as a function of the number of cycles of the Mg–Ni, Mg–Ni–Co, Mg–Ni–Cr and Mg–Ni–Nb alloys [53]

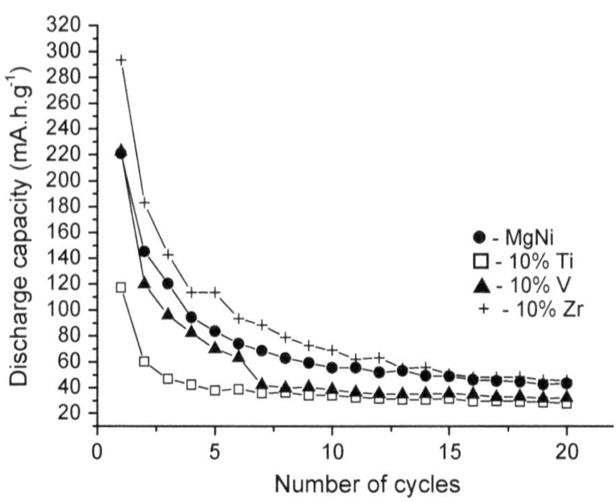

Fig. 7 Discharge capacity as a function of the number of cycles of the Mg–Ni, Mg–Ni–Ti, Mg–Ni–V and Mg–Ni–Zr alloys [53]

Rare earths have been investigated too as alloying elements for Mg–Ni alloy electrodes. Huang et al. [56] investigated the effect of Nd addition and reported that the maximum discharge capacity increased for the alloys with larger contents of this element. When the amount of Nd was in the range of 10–15 mol%, the maximum discharge capacity was close to 580 mAh/g. Moreover, the maximum discharge capacity after 20 cycles of charge/discharge was 80% of the initial one. This retained capacity is larger than those observed for Mg–Ni binary alloys. The effect of La addition on the structure and electrode performance of the Mg_2Ni alloy synthesized by melting-spinning technique (i.e., a technique of rapid solidification which allows obtaining thin metallic ribbons, in the range of tens of micrometers of thickness, with metastable microstructures) was reported by Ren et al. [57]. Using the same processing parameters, these authors observed that Mg_2Ni alloy presented a nanocrystalline structure and the addition of La favored the formation of amorphous phase. This behavior indicates that La addition increased the glass—forming ability of the Mg–Ni alloy. It was also observed that the increase in La content improved the maximum discharge capacity and cycle—life performance of the electrodes.

Noble metals also have attracting attention as alloying elements for the Mg–Ni alloys. The most investigated of these metals is Pd [58–60]. As a general trend, it has been reported that Pd additions promoted increase in cycle—life performances. To avoid a decay of the maximum discharge capacity, the amount of Pd in the alloy should be low. As an example, Ma et al. [59] reported improvement in cycle-life performance by ball-milling $(Mg_{50}Ni_{50})$ + 6 mol% of Pd. The maximum discharge capacities reported for both unalloyed and Pd-added alloys was almost the same. Similar results were reported by Souza et al. [61] for the $Mg_{49.5}Ni_{49.5}Pt_1$ (in at.%) ball-milled alloy.

Recently, Mg–Ni-based quaternary alloys have been investigated. Depending on the alloying elements, further improvements can be obtained by these quaternary systems when compared with the ternary ones. Figure 8 shows the curves of discharge capacity *vs.* number of cycles for some quaternary $Mg_{51}Ti_4Ni_{43}M_2$ alloys

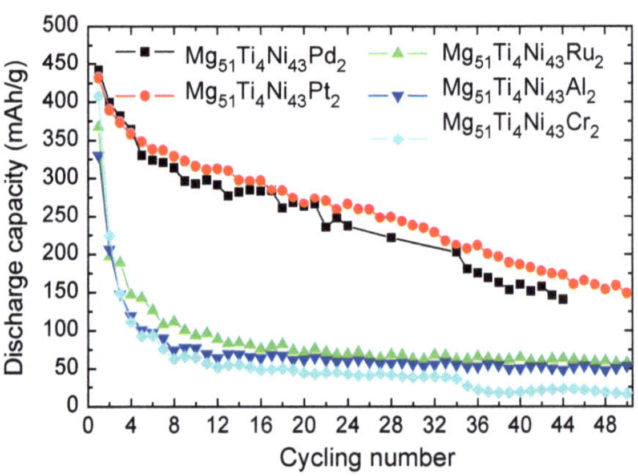

Fig. 8 Discharge capacities *vs.* cycling number of $Mg_{51}Ti_4Ni_{43}M_2$ alloys [62]

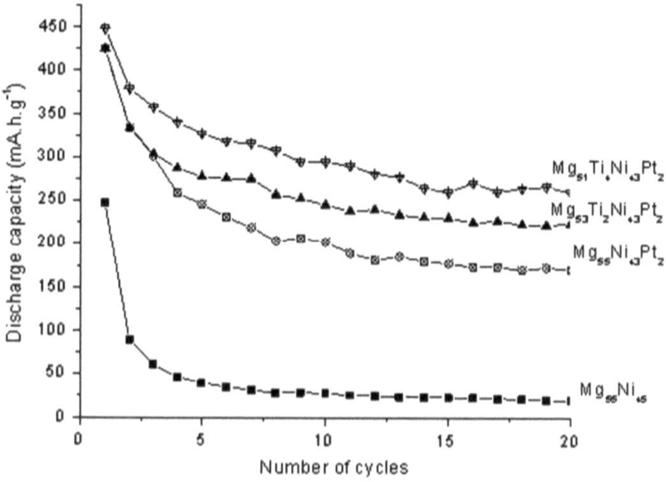

Fig. 9 Discharge capacities versus cycling number of $Mg_{55}Ni_{45}$ and Pt-containing alloys [63]

(in at.%) where M is the alloying element [62]. From these results, it is possible to observe that the additions of Pd and Pt presented much better results than the additions of Al, Ru, or Cr. The elaboration of multielement alloys (i.e., at least four components in the system) for electrode applications open a wide range of possibilities concerning the designing of microstructures in order to obtain a material with optimized electrochemical properties.

In Fig. 9 the effects of Ti and Pt on the electrochemical properties of the $Mg_{55}Ni_{45}$ alloy can be observed [63]. The addition of Pt significantly increased the maximum discharge capacity and improved the cycle life of the alloy. A partial replacement of Ti for Mg resulted in further improvements of these electrochemical properties. From these results, one can see that the addition of both Ti and Pt simultaneous resulted in the best electrode performance. Concerning the cycling stability, the major target of this investigation, the decrease in degradation rate for these quaternary alloys can be ascribed to different protection mechanisms acting simultaneously. As mentioned above, Ti decreases the corrosion rate of Mg through the formation of a protective layer of TiO_2 on the particle's surface. In the case of Pt, the protective behavior can be ascribed to a displacement of the corrosion potential reinforcing the cathodic character of the alloy in a similar way of that reported for Ti–Pd and Ti–Pt alloys [64, 65].

In other investigation of our group, the effect of partial substitution of noble metals for Ni in the $Mg_{49}Ti_6Ni_{45}$ was carried out [66]. The alloys were synthesized by mechanical alloying with nominal composition of $Mg_{49}Ti_6Ni_{45-X}NM_X$, where NM means Pd or Pt and assume values of 2 and 4 at.%. The structure of the alloys was characterized by X-ray diffraction (XRD) and transmission electron microscopy (TEM). Figure 10 shows the XRD patterns of the abovementioned alloys [66]. It is possible to observe the presence of broadened peaks in all samples due to

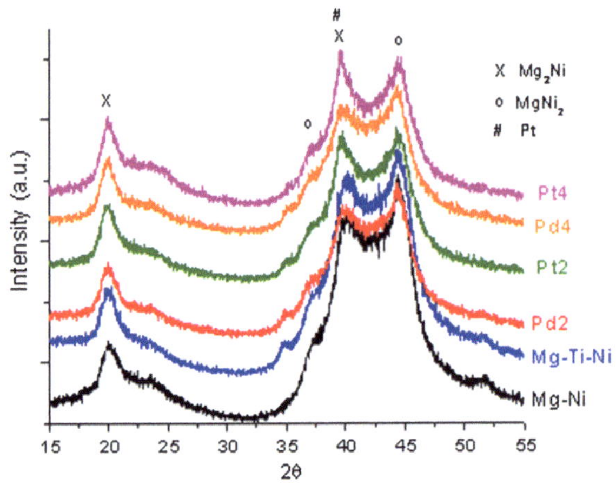

Fig. 10 XRD patterns of $Mg_{55}Ni_{45}$ and $Mg_{49}Ti_6Ni_{45-x}MN_x$ alloys [66]

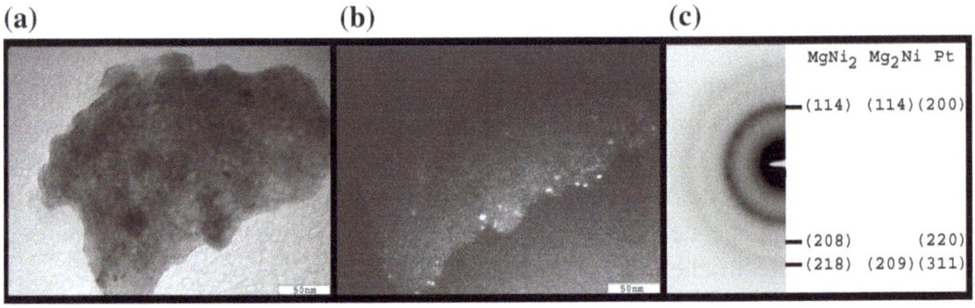

Fig. 11 Bright field image (**a**), dark field image (**b**), and selected area electron diffraction pattern (**c**) of the $Mg_{49}Ti_6Ni_{43}Pt_2$ alloy [66]

reduced crystallite sizes and strain generated by ball-milling. The Mg_2Ni and $MgNi_2$ intermediate phases were identified in all samples while there is evidence of Pt phase in the alloys containing this element and also of an amorphous band overlapped by some diffraction peaks.

Figures 11, 12, 13 and 14 show the results of transmission electron microscopy (TEM) for $Mg_{49}Ti_6Ni_{45-x}MN_x$ alloys [66]. All TEM bright field and dark field images indicate that the alloys composed by sub-micrometric particles presented microstructures formed by nanocrystalline phases dispersed on an amorphous matrix (Figs. 11, 12, 13 and 14). The Mg_2Ni and $MgNi_2$ phases were identified in all the synthesized alloys in selected area electron diffraction patterns (SAEDP) in good agreement to the results of X-ray diffractometry (Fig. 10). In the SAEDP, there is some indication of diffraction rings which may be related to Pd and Pt phases, but these rings are overlapped with those of Mg_2Ni and $MgNi_2$ phases. In the case of Pt-containing alloys, the fraction of amorphous phase is apparently larger than that of Pd-containing alloys. There was not observed the presence of Pt and Pd-based intermediate phases, however evidences of these elements in unalloyed state were indicated by XRD and selected area electron diffraction (SAED).

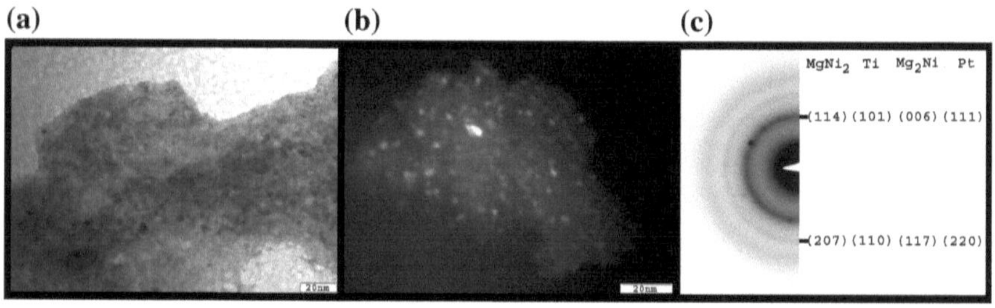

Fig. 12 Bright field image (**a**), dark field image (**b**), and selected area electron diffraction pattern (**c**) of the $Mg_{49}Ti_6Ni_{41}Pt_4$ alloy [66]

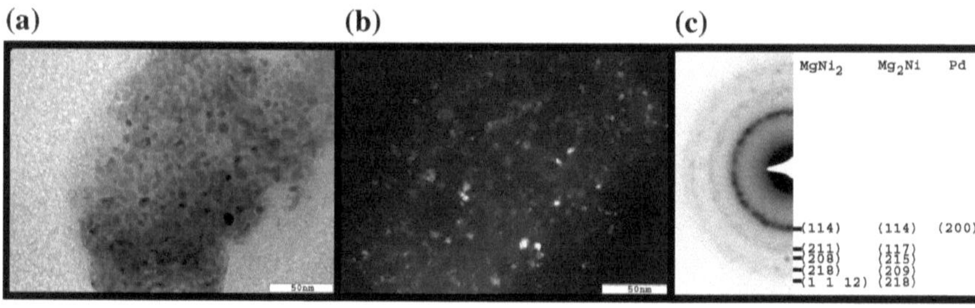

Fig. 13 Bright field image (**a**), dark field image (**b**), and selected area electron diffraction pattern (**c**) of the $Mg_{49}Ti_6Ni_{43}Pd_2$ alloy [66]

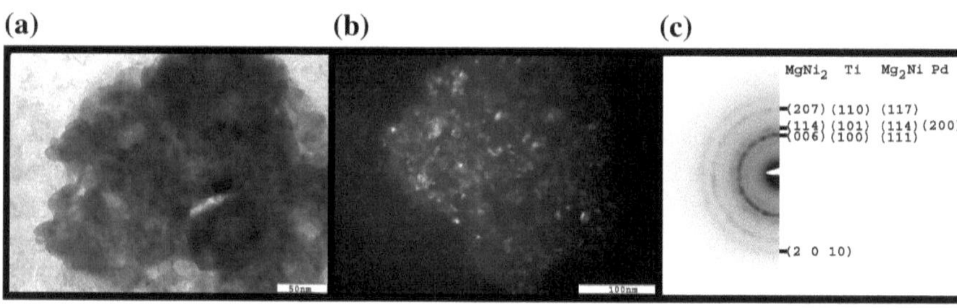

Fig. 14 Bright field image (**a**), dark field image (**b**), and selected area electron diffraction pattern (**c**) of the $Mg_{49}Ti_6Ni_{41}Pd_4$ alloy [66]

The average crystallite sizes with respective standard deviations for the $Mg_{49}Ti_6Ni_{45-x}NM_x$ alloys were obtained from the TEM dark field images. These results are shown in Table 2 and indicate the nanocrystalline nature of the samples which helps the uptake and release of hydrogen since the resulting interfacial regions in these particles accelerate hydrogen diffusivity.

The curves of discharge capacity versus cycling number for the $Mg_{55}Ni_{45}$ and $Mg_{49}Ti_6Ni_{45-x}NM_x$ alloys (with $x = 0$, 2, and 4 at.%) are presented in Fig. 15 and quantitative values are shown in Table 3 [66]. From these results, it is possible to

Table 2 Average crystallite sizes with respective standard deviations for $Mg_{49}Ti_6Ni_{45-x}NM_x$ alloys

Alloy	Crystallite size (nm)	Std. dev. (nm)
$Mg_{49}Ti_6Ni_{43}Pd_2$	7.3	2.5
$Mg_{49}Ti_6Ni_{41}Pd_4$	9.4	5.2
$Mg_{49}Ti_6Ni_{43}Pt_2$	8.9	4.7
$Mg_{49}Ti_6Ni_{41}Pt_4$	10.5	13.6

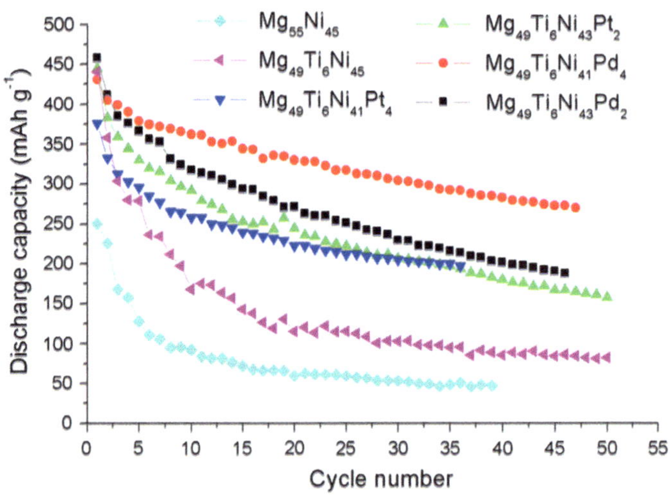

Fig. 15 Discharge capacities versus cycling number of $Mg_{55}Ni_{45}$ and $Mg_{49}Ti_6Ni_{45-x}NM_x$ [66]

observe that the partial substitution of Mg by Ti increased the discharge capacity of the alloy electrode from 250 mA h/g for $Mg_{55}Ni_{45}$ to 441 mA h/g for $Mg_{55}Ti_6Ni_{45}$ alloy. It is also observed that the addition of 2 at.% of noble metal further increased the discharge capacity of the electrodes. For the alloy with 2% of Pt, the maximum discharge capacity increased from 250 (binary alloy) to 445 mA h/g while the alloy with 2 at.% Pd showed a maximum discharge capacity of 457 mA h/g. Further increase in the amount of noble metal (up to 4 at.%) resulted in a slight decrease of the maximum discharge capacity for the quaternary alloys, especially for the Pt-containing alloy. This trend is related to the decrease in relative amount of Mg (hydride-forming element) in quaternary alloys, when the amount of noble metal is increased from 2 to 4 at.% of noble metal, recalling that Pt is more heavy than Pd.

Concerning the cycle life of the alloy electrodes, the electrochemical tests (Fig. 15; Table 3) show the maximum discharge capacity and the relative discharge capacity (discharge capacity in the *n*th cycle) after 10 and 20 cycles. $Mg_{55}Ni_{45}$ and $Mg_{49}Ti_6Ni_{45}$ alloys exhibited very low cycling stability, which indicates that the addition of Ti to the binary alloy was only effective to improve the maximum discharge capacity but not the cycle life of the electrode. The addition of a noble

Table 3 Maximum discharge capacity (mA h/g) and the relative discharge capacity (RDC, which is calculated by the ratio between the discharge capacity in the *n*th cycle and the maximum discharge capacity as a percentage) in the 10th and 20th cycles

Alloy composition (at.%)	Max. discharge capacity (mA h/g)	RDC after 10 cycles (%)	RDC after 20 cycles (%)
$Mg_{55}Ni_{45}$	250	37	24
$Mg_{49}Ti_6Ni_{45}$	441	38	26
$Mg_{49}Ti_6Ni_{43}Pd_2$	457	69	59
$Mg_{49}Ti_6Ni_{41}Pd_4$	431	84	77
$Mg_{49}Ti_6Ni_{43}Pt_2$	445	66	55
$Mg_{49}Ti_6Ni_{41}Pt_4$	376	69	59

metal (Pd or Pt) strongly improved the cycling stability of the alloy electrodes. $Mg_{49}Ti_6Ni_{43}Pd_2$ and $Mg_{49}Ti_6Ni_{43}Pt_2$ presented similar degradation rates. When the amount of noble metal is increased up to 4 at.%, the degradation rate for the Pd-containing alloy electrode was further reduced, but the same trend was not observed for the alloy with 4 at.% of Pt.

Considering both maximum discharge capacities and cycle-life performances, the best electrochemical results were attained by the $Mg_{49}Ti_6Ni_{41}Pd_4$ alloy electrode. In addition, Pd is a less expensive alloying element than Pt.

Another alloy system which has been extensively investigated recently is the La–Mg–Ni. These alloys are usually represented as A_nB_m where A is La (or other rare-earth element) and Mg; and B is Ni (which can be partially substituted by other elements such as Al, Si, Co, etc.) [4, 67]. The most important series of these alloys are $La_{3-x}Mg_xNi_9$ (AB_3), $La_{4-x}Mg_xNi_{14}$ (A_2B_7) and $La_{5-x}Mg_xNi_{19}$ (A_5B_{19}) with, respectively, trigonal $PuNi_3$, hexagonal Ce_2Ni_7 or trigonal Gd_2Co_7, hexagonal Pr_5Co_{19}, or trigonal Ce_5Co_{19} types of crystal structures. All these alloys belong to a family of layered structures, into which a single Laves-type $La_{2-x}Mg_xNi_4$ layer and several Haucke $CaCu_5$-type $LaNi_5$ layers are stacked along the trigonal/hexagonal axes. For instance, the AB_3 type structure is obtained by ($AB_5 + 2\ AB_2 = A_3B_9 = 3\ AB_3$). For further details, on the complex crystal structures of these layered phases (also known as long-period stacking ordered phases or super-lattice alloys), the interested reader can consult, among others, Refs. [4, 67–69]. These layered La–Mg–Ni alloys have been shown improved electrochemical properties when compared to $LaNi_5$ alloys such as higher discharge capacities (about 25% superior) and more important very low self-discharge behavior which is highly desirable for metal hydride electrodes. Because of these interesting features, Sanyo introduced in the market a commercial batteries (Sanyo Eneloop family) using this type of hydride anode [4].

5 Concluding Remarks

Despite the fact that Ni–MH battery is a mature technology which has been replaced by other technologies (mainly, Li-ion batteries) in several applications, it remains to be used in a number of important applications, especially those demanding high power capacity. The continuous research activity in novel alloy electrodes for Ni–MH batteries has been performed and improvements on the electrochemical properties of these materials have been attained and introduced in the market as can be observed La-Ni-Mg super-lattice compounds. Several alloy systems have been investigated in laboratory and the results obtained by quaternary systems having a transition metal and a noble metal as alloying elements for Mg–Ni-based alloys can be considered very promising but further improvements are necessary to allow the technological use of such materials. Until now, the number of quaternary alloys investigated is very small. Moreover, a more detailed understanding on the correlation between the electrochemical properties, surface states and microstructures is necessary to obtain alloys with optimized electrode performances. This goal can be accomplished only with systematic investigation of these multielement alloys and their mechanisms of degradation. Moreover, in quaternary alloys other transition metals different from Ti need to be investigated, such as Al, Cr, and V, among others. One approach to drive the directions of the investigations in quaternary alloys is to select alloying elements from those which presented interesting results in ternary alloys. Furthermore, the study of quaternary alloys with rare earths as alloying elements instead of transition metals was not explored so far.

There is also a lack of investigations concerning the processing routes of hydrogen storage alloys for anode applications. The investigated alloys are mostly synthesized by mechanical alloying and the processing parameters of this route have not been sufficiently investigated. Moreover, other processing techniques are much less investigated. For instance, rapid solidification techniques, such as melt-spinning, are promising routes to obtain nanocrystalline or amorphous Mg alloys and are much less studied than ball-milling. Furthermore, new processing techniques of severe plastic deformation started to be investigated for hydrogen storage alloys applied in solid state hydrogen storage but almost nothing have been done for electrode applications.

Finally, bottom up approaches, which are widely used to produce nanomaterials in several energy conversion and storage technologies (for instance, low-temperature fuel cell electrocatalysts and Li-ion electrode materials), are rarely investigated for Mg-based hydride electrodes.

The above-mentioned points are only few examples on the potentiality of nanotechnology to create new hydrogen storage materials suitable for anode applications.

References

1. Fetcenko MA, Ovshinsky SR, Reichman B, Young K, Fierro C, Zallen A, Mays W, Ouchi T (2007) Recent advances in NiMH battery technology. J Power Sources 165:544–551
2. Linden D, Reddy TB (2002) Handbook of batteries, 3rd edn. McGraw-Hill, USA
3. Li MM, Yang CC, Wang CC, Wen Z, Zhu YF, Zhao M, Li JC, Zheng WT, Lian JS, Jiang Q (2016) Design of hydrogen storage alloys/nanoporous metals hybrid electrodes for nickel-metal hydride batteries. Sci Rep 6:27601
4. Young KH, Nei J (2013) The current status of hydrogen storage alloy development for electrochemical applications. Materials 6:4574–4608
5. Oumellal Y, Rougier A, Nazri GA, Tarascon J-M, Aymard L (2008) Metal hydrides for lithium-ion batteries. Nat Mater 7:916–921
6. Sakai T, Iwaki T, Ye Z, Noréus D, Lindstrom O (1995) Air-metal hydride battery construction and evaluation. J Electrochem Soc 142:4040–4045
7. Hu W-K, Ye Z, Noréus D (2001) Influence of MH electrode thickness and packing density on the electrochemical performance of air-MH batteries. J Power Sources 102:35–40
8. Wang XH, Chen Y, Pan HG, Xu RG, Li SQ, Chen LX, Chen CP, Wang QD (1999) Electrochemical properties of $Ml(NiCoMnCu)_5$ used as an alkaline fuel cell anode. J Alloys Compd 293:833–837
9. Hu W-K, Gao X-P, Kiros Y, Middelman E, Noréus D (2004) Zr-based AB_2-type hydrogen storage alloys as dual catalysts of gas-diffusion electrodesin an alkaline fuel cell. J Phys Chem B 108:8756–8758
10. Chartouni D, Kuriyama N, Kiyobayashi T, Chen J (2002) Int J Hydrogen Energy 27:945–952
11. Hu W-K, Noréus D (2003) Rare-earth-based AB_5-type hydrogen storage alloys as hydrogen electrode catalysts in alkaline fuel cells. J Alloys Compd 356–357:734–737
12. Chen Y, Santos DMF, Sequeira CAC, Lobo RFM (2012) Studies of modified hydrogen storage intermetallic compounds used as fuel cell anodes. Crystals 2:22–33
13. Lee S-M, Kim J-H, Lee H-H, Lee PS, Lee J-Y (2002) The characterization of an alkaline fuel cell that uses hydrogen storage alloys. J Electrochem Soc 149:A603–A606
14. Paschoalino WJ, Ticianelli EA (2013) An investigation of the borohydride oxidation reaction on LaNi-based hydrogen storage alloys. Int J Hydrogen Energy 38:7344–7352
15. Paschoalino WJ, Thompson SJ, Russell AE, Ticianelli EA (2014) The borohydride oxidation reaction on La-Ni-based hydrogen-storage alloys. ChemPhysChem 15:2170–2176
16. Santos SF, Ishikawa TT, Ticianelli EA (2009) Fundamentals and technological applications of hydrogen—absorbing Mg amorphous alloys. In: Telle JR, Pearstine NA (eds) Amorphous materials: research, technology and applications. Nova Science Publishers, USA, pp 219–237
17. Harding Battery Handbook for Quest® Rechargeable Cells and Battery Packs (2004) One energy centre shores, MI. www.hardingenergy.com
18. Notten PHL, Ouwerkerk M, van Hal H, Beelen D, Keur D, Zhuo J, Feil H (2004) High energy density strategies: from hydride-forming materials research to battery integration. J Power Sources 129:45–54
19. Wronski ZS (2001) Materials for rechargeable batteries and clean hydrogen energy sources. Int Mater Rev 46:1–49
20. Smallman RE, Bishop RJ (1999) Modern physical metallurgy and engineering materials: science, processing and applications, 6th edn. Butterworth Heinemann, UK
21. Park CN, Luo S, Flanagan TB (2004) Analysis of sloping plateaux in alloys and intermetallic hydrides I: diagnostic features. J Alloys Compd 384:203–207
22. Huot J (2010) Metal hydrides. In: Hirscher M (ed) Handbook of hydrogen storage, Willey-VCH, Weinheim
23. Schlapbach L, Zuttel A (2001) Hydrogen-storage materials for mobile applications. Nature 414:253–258

24. Mosavati N, Young K-H, Meng T, Simon Ng KY (2016) Electrochemical open-circuit voltage and pressure-concentration-temperature isotherm comparison for metal hydride alloys. Batteries 2:1–11
25. Reilly JJ, Adzic GD, Johnson JR, Vogt T, Mukerjee S, McBreen J (1999) The correlation between composition and electrochemical properties of metal hydride electrodes. J Alloys Compd. 293–295:569–582
26. Ambrosio RC, Ticianelli EA (2002) Effect of cobalt on the physicochemical properties of a simple LaB_5 metal hydride alloy. J Power Sources 110:73–79
27. Lin Q, Zhao S, Zhu DJ, Song B, Mei Z (2003) Investigation of hydriding properties and structure of $MlNi_{5-x}Sn_x$ system. J Alloys Compd 351:91–94
28. Liu J, Yang Y, Li Y, Yu P, He Y, Shao H (2007) Comparative study of $LaNi_{4.7}M_{0.3}$ by powder microelectrode technique. Int J Hydrogen Energy 32:1905–1910
29. Ayari M, Paul-Boncour V, Lamloumi J, Percheron-Guégan A, Guillot M (2005) Study of the aging of $LaNi_{3.55}Mn_{0.4}Al_{0.3}(Co_{1-x}Fe_x)_{0.75}$ ($0 \leq x \leq 1$) compounds in Ni–MH batteries by SEM and magnetic measurements. J Magn Magn Mater 288:374–383
30. Ben Moussa M, Abdellaoui M, Mathlouthi H, Lamloumi J, Percheron-Guégan A (2008) Electrochemical properties of the $MmNi_{3.55}Mn_{0.4}Al_{0.3}Co_{0.75-x}Fe_x$ ($x = 0.55$ and 0.75) compounds. J Alloys Compd 458:410–414
31. dos Santos AR, Ambrosio RC, Ticianelli EA (2004) Electrochemical and structural studies on nonstoichiometric AB_2-type metal hydride alloys. Int J Hydrogen Energy 29:1253–1261
32. Awad AS, Nakhl M, Zakhour M, Santos SF, Souza FL, Bobet J-L (2016) Effect of microwave irradiation on hydrogen sorption properties of hand mixed MgH_2—10 wt.% carbon fibers. J Alloys Compd 676:1–8
33. Asselli AAC, Santos SF, Huot J (2016) Hydrogen storage in filed magnesium. J Alloys Compd 687:586–594
34. Santos SF, Ishikawa TT, Botta Filho WJ, Huot J (2014) MgH_2 + FeNb nanocomposites for hydrogen storage. Mater Chem Phys 147:557–562
35. Jain P, Lang J, Skryabina N, Fruchart D, Santos SF, Binder K, Klassen T, Huot J (2013) MgH_2 as dopant for improved activation of commercial Mg ingot. J Alloys Compd 575:364–369
36. Liang G, Huot J, Boily S, Van Neste A, Schulz R (1999) Catalytic effect of transition metals on hydrogen sorption in nanocrystalline ball milled MgH_2–Tm (Tm=Ti, V, Mn, Fe and Ni) systems. J Alloys Compd 292:247–252
37. Dornheim M, Doppiu S, Barkhordarian G, Boesenberg U, Klassen T, Gutfleisch O, Bormann R (2007) Hydrogen storage in magnesium-based hydrides and hydride composites. Scripta Mater 56:841–846
38. Zhang DL, Ying DY (2001) Solid state reactions in nanometer scaled diffusion couples prepared using high energy ball milling. Mater Sci Eng, A 301:90–96
39. Suryanarayana C (2001) Mechanical alloying and milling. Prog Mat Sci 46:1–184
40. Murty BS, Ranganathan S (1998) Novel materials synthesis by mechanical alloying/milling. Int Mater Rev 43:101–141
41. El-Eskandarany MS (2001) Mechanical alloying for fabrication of advanced engineering materials. Noyes Publications, Norwich
42. Liu W, Wu H, Lei Y, Wang Q, Wu J (1997) Amorphization and electrochemical hydrogen storage properties of mechanically alloyed Mg–Ni. J Alloys Comp 252:234–237
43. Zhang SG, Yorimitsu K, Nohara S, Morikawa T, Inoue I, Iwakura C (1998) Surface analysis of an amorphous MgNi alloy prepared by mechanical alloying for use in nickel–metal hydride batteries. J Alloys Compd 270:123–126
44. Zhang SG, Hara Y, Morikawa T, Inoue I, Iwakura C (1999) Electrochemical and structural characteristics of amorphous $MgNi_x$ ($x \geq 1$) alloys prepared by mechanical alloying. J Alloys Compd 293:552–555
45. Feng F, Northwood DO (2004) Effect of surface modification on the performance of negative electrodes in Ni/MH batteries. Int J Hydrogen Energy 29:955–960

46. Ambrosio RC, Ticianelli EA (2005) Studies on the influence of palladium coatings on the electrochemical and structural properties of a metal hydride alloy. Surf Coating Tech 197:215–222
47. Deng C, Shi P, Zhang S (2006) Effect of surface modification on the electrochemical performances of $LaNi_5$ hydrogen storage alloy in Ni/MH batteries. Mater Chem Phys 98: 514–518
48. Ambrosio RC, Ticianelli EA (2005) Electrochemical and X-ray absorption spectroscopy studies of copper coatings on a hydrogen storage alloy. J Electroanalytical Chem 574: 251–260
49. Rongeat C, Grosjean MH, Ruggeri S, Dehmas M, Bourlot S, Marcotte S, Roué L (2006) Evaluation of different approaches for improving the cycle life of MgNi-based electrodes for Ni–MH batteries. J Power Sources 158:747–753
50. Santos SF, de Castro JFR, Ishikawa TT, Ticianelli EA (2008) Effect of Mechanical Coating with Ni and Ni-5% Al on the structure and electrochemical properties of the Mg-50% Ni Alloy. J Mater Sci 43:2889–2894
51. Li M, Zhu Y, Yang C, Zhang J, Chen W, Li L (2015) Enhanced electrochemical hydrogen storage properties of Mg_2NiH_4 by coating with nano-nickel. Int J Hydrogen Energy 40:13949–13956
52. Liu W, Wu H, Lei Y, Wang Q, Wu J (1997) Effects of substitution of other elements for nickel in mechanically alloyed $Mg_{50}Ni_{50}$ amorphous alloys used for nickel—metal hydride batteries. J Alloys Compd 261:289–294
53. Santos SF, de Castro JFR, Ishikawa TT, Ticianelli EA (2007) Effect of transition metal additions on the electrochemical properties of a MgNi-based alloy. J Alloys Compd 434: 756–759
54. Iwakura C, Shyn-ya R, Miyanohara K, Nohara S, Inoue H (2001) Effects of Ti–V substitution on electrochemical and structural characteristics of MgNi alloy prepared by mechanical alloying. Electrochem Acta 46:2781–2786
55. Han SC, Lee PS, Lee JY, Zuttel A, Schlapbach L (2000) Effects of Ti on the cycle life of amorphous MgNi-based alloy prepared by ball milling. J Alloys Compd 306:219–226
56. Huang L-J, Tang J-G, Wang Y, Liu J-X, Wu DC (2009) Effects of microstructure on the electrode properties of melt–spun Mg-based amorphous alloys. J Alloys Compd 485:186–191
57. Ren HP, Zhang YH, Li BW, Zhao DL, Guo SH, Wang XL (2009) Influence of the substitution of La for Mg on the microstructure and hydrogen storage characteristics of $Mg_{20-x}La_xNi_{10}$ (x = 0–6) alloys. Int J Hydrogen Energy 34:1429–1436
58. Yamaura SI, Kim HY, Kimura H, Inoue A, Arata Y (2002) Electrode properties of rapidly solidified $Mg_{67}Ni_{23}Pd_{10}$ amorphous alloy. J Alloys Compd 347:239–243
59. Ma T, Hatano Y, Abe T, Watanabe K (2004) Effects of Pd addition on electrochemical properties of MgNi. J Alloys Compd 372:251–258
60. Santos SF, Castro JFR, Ticianelli EA (2013) Microstructures and electrode performances of $Mg_{50}Ni_{(50\text{-}X)}Pd_x$ alloys. Cent Eur J Chem 11:485–491
61. Souza EC, de Castro JFR, Ticianelli EA (2006) A new electrode material for nickel-metal hydride batteries: MgNiPt alloy prepared by ball-milling. J Power Sources 160:1425–1430
62. Nikkuni FR, Santos SF, Ticianelli EA (2010) Effects of Pd and Pt additions on the structure and electrochemical performance of Mg-Ti-Ni alloy electrodes. Proceedings of the Brazil MRS Meeting, Ouro Preto, Brazil
63. de Castro JFR, Santos SF, Nikkuni FR, Ishikawa TT, Ticianelli EA (2010) Structural and electrochemical characteristics of $Mg_{(55-x)}Ti_xNi_{(45-y)}Pt_y$ metal hydride electrodes. J Alloys Compd 498:57–61
64. Nakagawa M, Matsuya S, Udoh K (2001) Corrosion behavior of pure titanium and titanium alloys in fluoride-containing solutions. Dental Mater J 20:305–314
65. Nakagawa M, Matsuya S, Udoh K (2002) Effects of fluoride and dissolved oxygen concentrations on the corrosion behavior of pure titanium and titanium alloys. Dental Mater J 21:83–92

66. Nikkuni FR, Santos SF, Ticianelli EA (2013) Microstructures and electrochemical properties of $Mg_{49}Ti_6Ni_{(45-x)}M_x$ (M=Pd and Pt) alloy electrodes. Int J Energy Res 37:706–712
67. Yartys V, Noreus D, Latroche M (2016) Metal hydrides as negative electrode materials for Ni–MH batteries. Appl Phys A 122:1–11
68. Xie L, Li J, Zhang T, Song L, Kou H (2017) Microstructure and hydrogen storage properties of Mg-Ni-Ce alloys with a long-period stacking ordered phase. J Power Sources 338:91–102
69. Liu Y, CaoY, Li Huang L, Gao M, Pan H (2011) Rare Earth–Mg–Ni-based hydrogen storage alloys as negative electrode materials for Ni/MH batteries. J Alloys Compd 509:675–686

Neutralization Batteries

William G. Morais, Gilberto Lima and Fritz Huguenin

1 Introduction

In a few years from now, the worldwide energy production is estimated to become inefficient to provide the ever-rising energy demand per capita, especially due rhythms of the population growth. Emission of greenhouse gases generated by fuel fossil combustion tends to increase the damage on the environment and the well-being of the population [1]. In this context, research for enhancing the participation of the renewable energy into the world energy matrix has become most relevant.

Currently, the world energy consumption is estimated at 19.2% based upon renewable sources [2]. In order to increase this perceptual, improvements in the energy conversion efficiency from solar, wind, biomass and geothermal into electrical energy is necessary.

Another useful possibility is to explore the potential of the ionic gradient with the purpose to convert it into electrical power [3]. Since the studies developed by Pattle in 1954 [3], technologies with low environmental impact were proposed to acquire energy from the mixture of solutions with different electrolyte concentration. For example, we can mention the hydroelectric pile based on membranes acid and basic [3], osmotic pump [4], dialytic battery [5], devices based on the differences in the vapour pressure [6], electrochemical concentration cells [7], CAPMIX [8] and mixing entropy battery [9].

The technologies to obtain energy with selective insertion electrodes [9], in solutions containing different electrolyte concentration, could be advanced to

W.G. Morais · G. Lima · F. Huguenin (✉)
Departamento de Química, Faculdade de Filosofia, Ciências E Letras de Ribeirão Preto, Universidade de São Paulo, Ribeirão Preto (SP) 14040-901, Brazil
e-mail: fritz@ffclrp.usp.br

F.L. de Souza and E. Leite (eds.), *Nanoenergy*,
DOI 10.1007/978-3-319-62800-4_9

harvest energy during the neutralization of acid solutions [10–13]. Our group has developed a system that can store energy after a cycle of proton insertion/deinsertion into a selective host matrix. This study was initially presented in the article entitled "Proton Electroinsertion in Self-Assembled Material for Neutralization Pseudocapacitors" [10].

The addition of bases or acids in water treatment might emerge as a viable strategy to convert chemical energy into an electric and sustainable work, if this system could reduce treatment costs of this kind of effluent, it might become commercially valuable. Further, this technology could be applied to extract energy by treating industrial wastewater, sewage, swimming pools and water reservoirs.

These Neutralization Batteries will be discussed as an idealized acid–base machine, whether the maximum work comes from the flow of matter between an acid and a basic reservoir. The method description as well the working principle and formalism concerning this idealized machine are detailed below.

2 Working Principle and Thermodynamic Formalism

The method applied to harvest energy after a full cycle of charge/discharge of the battery with different pHs solutions, is shown in the Fig. 1. This machine can produce work by operating between an acid reservoir and a basic reservoir in four steps. Two of these steps are reversible buffered stages and isothermal, the other two steps are open for the acid and the base entering into the system. The machine also presents two insertion electrodes; one of them is selective for the electroinsertion of protons while the other one does not enable it.

The thermodynamic cycle can be understood as described forwards. Step 1: entrance of an acid solution into the electrochemical cell; Step 2: electroinsertion of the protons in the selective electrode, and electrodeinsertion of the other ionic species in the second (positive or negative) electrode; Step 3: entrance of the base into the electrochemical cell; Step 4: electrodeinsertion of the protons in the selective electrode and electrodeinsertion of the other ionic species in the second electrode; finally, there is an exchange of the electrolytic solutions with low activity of protons for another acid solution. During the electrochemical cycle, protons are removed from the more concentrated solution (step 2) and added in the less concentrated solution (step 4). Thus, after the thermodynamic cycle, the change in partial entropy associated with the variation in the proton concentration is converted into electrical work.

Although the electrode selective for proton can be positive or negative, the processes of the insertion of proton (reduction) and its deinsertion (oxidation) must take place in more and less acid medium, respectively. Moreover, in the acid solution, the step associated with the electroinsertion of proton in the positive electrode is spontaneous, while the step associated with the electrodeinsertion corresponds to a non spontaneous process in less acid medium. Similarly, when the host matrix selective for proton is placed as the negative electrode, the

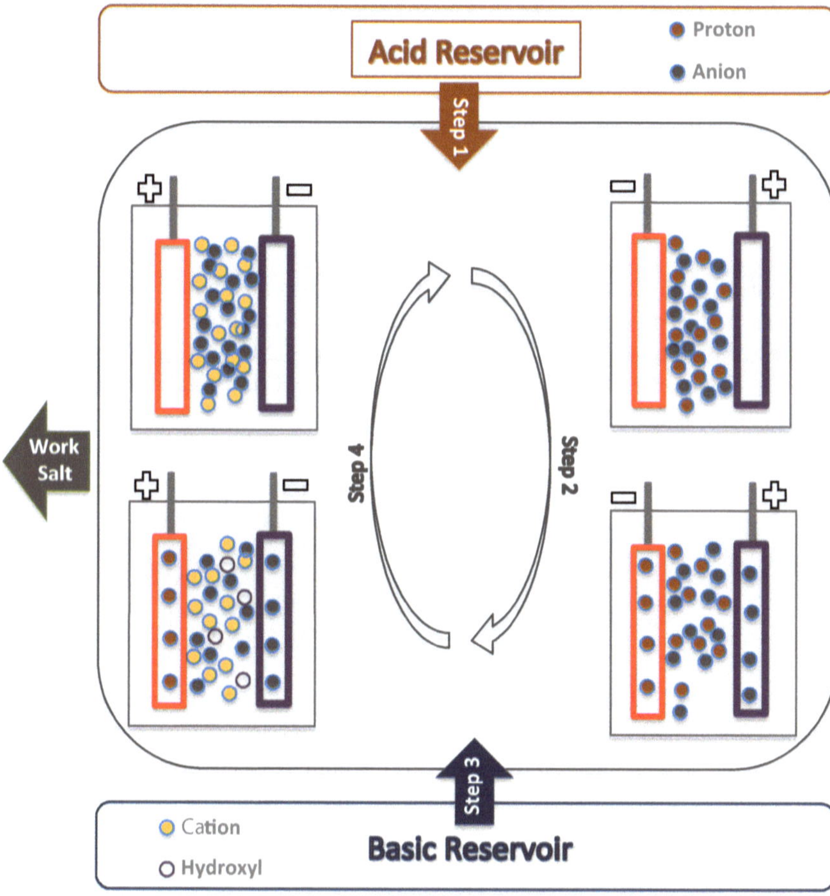

Fig. 1 Scheme of the working principle of the acid–base machine

nonspontaneous electroinsertion and the spontaneous electrodeinsertion occur in acid and less acid medium, respectively.

It is possible to demonstrate the maximum work, as well the efficiency, produced by the machine, for this purpose we shall consider the electrochemical cell configuration described onward. Selective insertion/deinsertion of protons (H^+) and anions (X^-) occurs in the negative (M_1) and positive (M_2) electrodes, respectively. First, using an external power source, ions are removed from the acid solution to promote the following nonspontaneous processes:

$$M_1 + H^+ + e^- \rightarrow M_1H \quad (1)$$

$$M_2 + X^- \rightarrow M_2X + e^- \qquad (2)$$

However, with the addition of the base, the following spontaneous processes take place:

$$M_1H \rightarrow M_1 + H^+ + e^- \qquad (3)$$

$$M_2X + e^- \rightarrow M_2 + X^- \qquad (4)$$

We intend to proceed, based on these reactions, by equating the electrochemical potentials of the products and reagents, for the cell reaction, one can relate the activities of the proton (a_{H^+}), anions (a_{X^-}), M_1 (a_{M_1}), M_1H (a_{M_1H}), M_2 (a_{M_2}), and M_2X (a_{M_2X}) with the electromotive force (E_{rxn}) in acid (Eq. 5) and neutral (or less acid) medium of the resulting solution (Eq. 6),

$$E_{rxn} = E^o_{rxn}(T_1) - \frac{RT_1}{F}\ln\left(\frac{a^a_{M_1H}}{a^a_{M_1}}\frac{a^a_{M_2X}}{a^a_{M_2}}\frac{1}{a^a_{H^+}a^a_{X^-}}\right) \qquad (5)$$

$$E_{rxn} = -E^o_{rxn}(T_1) + \frac{RT_1}{F}\ln\left(\frac{a^n_{M_1H}}{a^n_{M_1}}\frac{a^n_{M_2X}}{a^n_{M_2}}\frac{1}{a^n_{H^+}a^n_{X^-}}\right) \qquad (6)$$

in which R is the gas constant, F is the Faraday constant, T_1 is the temperature and E^o_{rxn} is the standard electromotive force. The superscripts "a" and "n" stand for activity in the acid and neutral (or less acid) medium, respectively. The total sum of the electromotive forces ($E_{rxn,neut}$) during the spontaneous, in acid medium, and nonspontaneous, in less acid medium, reversible buffered, and isothermal process corresponds to:

$$E_{rxn,neut} = \frac{RT_1}{F}\ln\left(\frac{a^a_{H^+}\,a^a_{X^-}}{a^n_{H^+}\,a^n_{X^-}}\right) \qquad (7)$$

In this case, the temperature (T_1) is equal for both spontaneous and nonspontaneous processes. Considering that during the neutralization the ionic force remains constant and, according to Deby–Hückel law, the average activity coefficients depend exclusively on the ionic force, one can cancel the anion activities in acid and neutral mediums from Eq. 7:

$$E_{rxn,neut} = \frac{RT_1}{F}\ln\left(\frac{a^a_{H^+}}{a^n_{H^+}}\right) \qquad (8)$$

The efficiency associated with the neutralization ($\varepsilon_{\mathrm{neut}}$) of the idealized machine can be determinate using the work (w) and the enthalpy of neutralization (ΔH_{r}):

$$\varepsilon_{\mathrm{neut}} = \frac{-w}{\Delta H_{\mathrm{r}}} = \frac{-\Delta G}{\Delta H_{\mathrm{r}}} = \frac{FE_{\mathrm{rxn}}}{\Delta H_{\mathrm{r}}} = \frac{RT_1 \ln\left(\frac{\mathrm{a}^{\mathrm{a}}_{\mathrm{H}^+}}{\mathrm{a}^{\mathrm{n}}_{\mathrm{H}^+}}\right)}{\Delta H_{\mathrm{r}}} \quad (9)$$

Finally, admitting the acid reservoir has a unitary activity, the efficiency is a function of the pH of the resulting solution from the following neutralization process:

$$\varepsilon_{\mathrm{neut}} = \frac{2.3RT_1\mathrm{pH}}{\Delta H_{\mathrm{r}}} \quad (10)$$

The efficiency is now defined as a function of the hydrogen potential, as seen in Eq. 10. Since this acid–base machine operates according to an idealized cycle, the hydrogen potential has no dependence on electrodes or the electrolyte. Therefore, the hydrogen potential is determined on a purely electrical and mechanical basis. This is possible because the resulting electrical work can be used to raise the masses of the neighbourhood.

Some small deviations in the pHs can take place, for practical systems, during the non-buffered charge/discharge processes, these alterations depend on the relation between the number of electroactive sites in the host matrixes and the amount of protons in the electrolyte solution. Nevertheless, the maximum work produced by the system does not depend on these pH oscillations. Indeed, this work is a function of the state for these conditions, so it depends only on the initial and the final states, and it is not taken into consideration the heat released during the reversible electrochemical processes (overpotential and null ohmic drop in Steps 2 and 4).

The heat released from the neutralization can increase the efficiency of the system if the process takes place adiabatically (step 3). For this case, the temperature for the nonspontaneous, T_1, and the spontaneous, T_2, processes will be different, in which $T_2 > T_1$ when E_{rxn} $(T_2) > E_{\mathrm{rxn}}$ (T_1). The total efficiency is given by Eq. 11:

$$\varepsilon_{\mathrm{total}} = \frac{\left(E^{\mathrm{o}}_{\mathrm{rxn}}(T_1) - E^{\mathrm{o}}_{\mathrm{rxn}}(T_2)\right) + \frac{R}{F}\left(T_2 \ln\left(\frac{\mathrm{a}^{\mathrm{n}}_{\mathrm{M_1H}}}{\mathrm{a}^{\mathrm{n}}_{\mathrm{M_1}}}\frac{\mathrm{a}^{\mathrm{n}}_{\mathrm{M_2X}}}{\mathrm{a}^{\mathrm{n}}_{\mathrm{M_2}}}\frac{1}{\mathrm{a}^{\mathrm{n}}_{H^+}\,a^{\mathrm{n}}_{\mathrm{X}^-}}\right)_{T_2} - T_1 \ln\left(\frac{\mathrm{a}^{\mathrm{a}}_{\mathrm{M_1H}}}{\mathrm{a}^{\mathrm{a}}_{\mathrm{M_1}}}\frac{\mathrm{a}^{\mathrm{a}}_{\mathrm{M_2X}}}{\mathrm{a}^{\mathrm{a}}_{\mathrm{M_2}}}\frac{1}{\mathrm{a}^{\mathrm{a}}_{\mathrm{H}^+}\,a^{\mathrm{a}}_{\mathrm{X}^-}}\right)_{T_1}\right)}{\Delta H_{\mathrm{r}}} \quad (11)$$

3 Electrochemical Properties Required to Maximize the Efficiency in the Energy Conversion

The voltage curves of the neutralization battery (Fig. 2) have a similar shape of the usual batteries, during the charge/discharge processes. However, the voltage during the discharge of the neutralization battery is higher than in the charge process, differently from usual secondary batteries, since the processes take place in different electrolyte concentration. Thereby, the neutralization batteries produce electrical work equivalent to the integral along the cycle.

The materials, used to establish the neutralization batteries, have no need of showing the same electrochemical properties likewise the secondary battery case. The values of the voltage, during the discharge, are not vital for the good performance of these systems, since the electrical work depends on potential involved in both charge and discharge steps.

When it comes to choosing the electrodes, inexistence of parallel reactions, small loss of energy due the overpotentials and high voltage variation, as function of the concentration of the electrolyte solution in the equilibrium states are factors that must be considered. Thus, intercalation materials that due low voltage and high molar mass, do not show high densities of energy could present suitable properties to electrochemically convert entropic energy. Therefore, these materials can provide an alternative for these neutralization batteries, once the modified electrodes contain higher charge capacity and practical reversibility.

The basic properties concerning the material to convert entropic energy are as follow:

(a) The host matrix should be electroactive and with a structure that enables the practical insertion/deinsertion, selectively, of ions and chemically reversible. The smaller the overpotentials, the greater the converted energy from the system, after a full charge/discharge cycle.
(b) The electroinsertion/electrodeinsertion processes require selectivity with the charge of the ion. The positive and negative species present in the solution,

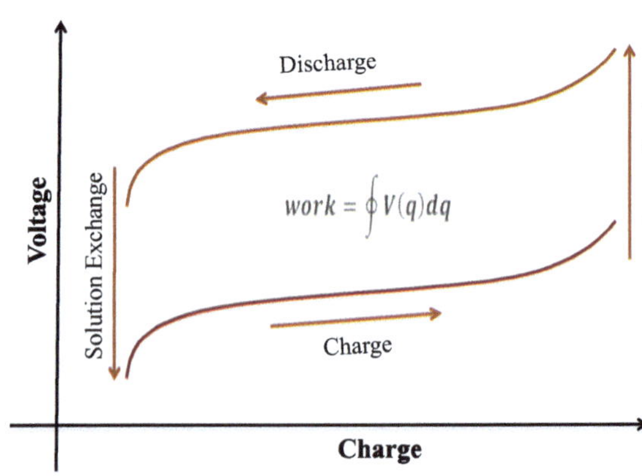

Fig. 2 Voltage variation as function of the charge for a neutralization battery. The integral of the cycle corresponds to the electrical work produced by the battery

which the concentration varies during the charge/discharge processes, should intercalate in different electrodes in the battery. This selectivity is essential to assure the equilibrium voltage variation during the exchange of the electrolytic solution.

(c) In order to increase the durability of the battery, minimal change in the structure of the electrode is required.
(d) In order to achieve higher energy conversion per cycle, the electrodes should present high charge capacity. Since, amount of entropic energy associates the full charge/discharge cycle, there is no dependence on the transference of the energy (or energy density) during the discharge step.
(e) In order to achieve low overpotentials related to the mass transport, high mobility is required for the inserted ions.
(f) In order to reduce the ohmic drop in the electrodes, high conductivity or alternative paths for the electronic transport is required for the host matrix.
(g) A significant voltage variation, associated with the changes in concentration of the electrolyte (or the ionic species that take part of the electrochemical reaction), is required. With the variation of the concentration of the ionic species in solutions, the ones that participate in the electroinsertion/electrodeinsertion processes, the positive and negative electrode should vary in the opposite direction.
(h) Low cost.
(i) Low toxicity.

4 Electrochemical Results for Neutralization Batteries

Some examples of nanomaterials that have been investigated as host matrices will be discussed ahead, as well the electrochemical studies for determining parameters associated to the electroinsertion processes involved in the neutralization batteries.

Phosphomolybdic acid (PMA) and poly(3,4-ethylenedioxythiophene)-poly (styrenesulfonate) (PEDOT-PSS) equally can be used as a host matrix because they have intrinsic pseudocapacitive properties such as high rate of proton insertion/deinsertion. Furthermore, the high proton conductivity for the PMA and charge carriers (i.e. polaron, bipolaron and free carriers) for the PEDOT-PSS contribute to reduce the overpotentials during the electroreduction and electrooxidation [14–19].

We were particularly interested in preventing the high solubility in aqueous medium presented by the PMA, then the self-assembled films were prepared with poly(allylamine) (PAH) and PEDOT-PSS, using the layer-by-layer method (LbL), due the electrostatic interactions between protonated amine groups of the PAH, oxygen atoms of the PMA and sulfonic groups of the poly(styrenesulfonate) [20–22]. An additional vantage in the implementation of the LbL method is associated with the control in the thickness of the film, which allows maximizing the harvested

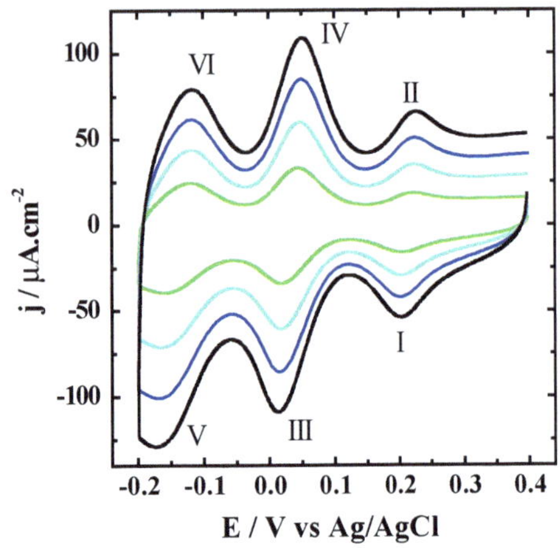

Fig. 3 Cyclic voltammograms for PMA/PAH/PEDOT-PSS film in HCl (pH = 1) with (*green dash line*) 50, (*light blue dash line*) 100, (*dark blue dash line*) 150 and (*black dash line*) 200 mV s^{-1}

charge with low diffusional and ohmic overpotentials by dimensioning of the electrodes. Conclusively, synergistic effects can be achieved by the self-assembled method due the intimate contact between the components in nanocomposites [23], promoting enhancements on the specific capacity and assuring the practical reversibility of the electrochemical processes.

Figure 3 shows the cyclic voltammogram for the LbL composed by PMA/PAH/PEDOT in HCl (pH = 1) with several scan rate (v = 50, 100, 150 and 200 mV s^{-1}). Low overpotentials are observed in the cyclic voltammograms. Additionally, a small change in the peak potential and the peak currents have a linear variation with the scan rate. The volumetric specific capacity is 20.3 C cm^{-3} for the film with 120 ± 4 nm of thickness. PEDOT-PSS can offer alternative electronic paths to the electroactive sites isolated by the PAH [17], also the charge transfer resistance is reduced, and it might be associated to the electrostatic interactions between the activated complex and the sulfonic group from the PSS during the insertion/deinsertion processes.

The determination of the stoichiometry of the global reaction, for each voltammetric peak, was achieved with the variation in the pH of the medium with and without the addition of KOH, aiming to preserve the ionic force constant. It was able to perceive a shift in the average peak potential of 70 mV per unity of pH, between 0.5 and 2.5. These values are close to the variation in the equilibrium potential for two electrons and two protons reactions, which are associated to the disproportionation reactions involving oxidation states I, III and V of the PMA [14]. Thus, the three redox pairs (I–II, III–IV and V–VI in Fig. 3) correspond to the reactions 12, 13 and 14, respectively.

$$PMo_{12}O_{40}^{-3} + 2e^{-} + 2H^{+} \leftrightarrows H_2PMo_{10}{}^{VI}Mo_2{}^{V}O_{40}^{-3} \quad (12)$$

$$H_2PMo_{10}^{VI}Mo_2^{V}O_{40}^{-3} + 2e^{-} + 2H^{+} \leftrightarrows H_4PMo_8^{VI}Mo_4^{V}O_{40}^{-3} \quad (13)$$

$$H_4PMo_8^{VI}Mo_4^{V}O_{40}^{-3} + 2e^- + 2H^+ \leftrightarrows H_6PMo_6^{VI}Mo_6^{V}O_{40}^{-3} \quad (14)$$

The presence of PSS tends to block the electroinsertion/electrodeinsertion of the chloride ions, though anions play a part in the charge compensation mechanism [24]. Moreover, covalent bonds C–Cl, present in the PEDOT polymeric chain, contribute to decrease the anionic transport [25]. The reaction associated to the electroreduction/electrooxidation of the conducting polymer is suggested to be as:

$$[PEDOT - PSS]\,Cl + e^- + H^+ \leftrightarrows H\,[PEDOT - PSS]\,Cl \quad (15)$$

Electrochemical measurements in the frequency domain were carried out in order to analyze the steps involved in the electrochemical reactions. Thus, measurements from electrochemical impedance spectroscopy were performed for the LbL films formed from (a) PMA/PAH, (b) PEDOT-PSS/PAH and (c) PMA/PAH/PEDOT-PSS. These kinetic studies can investigate the contribution of the PMA and PEDOT-PSS components in the PMA/PAH/PEDOT-PSS electrode. Figure 4 shows the Nyquist plots for (a) PMA/PAH, (b) PEDOT-PSS/PAH and (c) PMA/PAH/PEDOT-PSS at 0.2 V, with theoretical data obtained from the model presented together.

The beginning of the semicircles at high frequencies is associated with the charge transfer resistance (R_{ct}) and the electric double layer capacitance (C_{dl}). At lower frequencies, a capacitive process related to the finite diffusion of the electroinserted protons, is observed. The semi-infinite diffusion is not observed as a consequence of the small thickness and/or high proton mobility inside the host matrix.

On the basis of the Eqs. 12–15, the following elementary steps for the electroreduction mechanism were proposed:

$$2H_i^+ + 2e^- \rightarrow 2(H^+, e^-)_i \quad \text{(for the sites of the PMA)} \quad (16)$$

$$H_i^+ + e^- \rightarrow (H^+, e^-)_i \quad \text{(for the sites of the PEDOT)} \quad (17)$$

$$(H^+, e^-)_i \rightarrow (H^+, e^-)_s \quad (18)$$

Equations 16 and 17 relate the proton transfer into the host matrix, which is linked with the reduction of the PMA and PEDOT-PSS sites, respectively. This distinction is made due the molecularity of the elementary steps. The chemical specie $(H^+, e^-)_i$ represents the presence of protons and electrons at solution/electrode interface. Due to the concentration gradient and electroneutrality, these species diffuse together into the matrix, which is represented by $(H^+, e^-)_i$. On the basis of the reaction mechanism and the Butler–Volmer equation, the current density oscillatory component ($\tilde{i}$) is given by [26]:

$$\tilde{i} = nFkb\bar{c}_{P,0}\exp\left(\alpha nF\overline{V}/RT\right)\tilde{V} + nF\,\exp\left(\alpha nF\overline{V}/RT\right)\tilde{c}_{P,0} \quad (19)$$

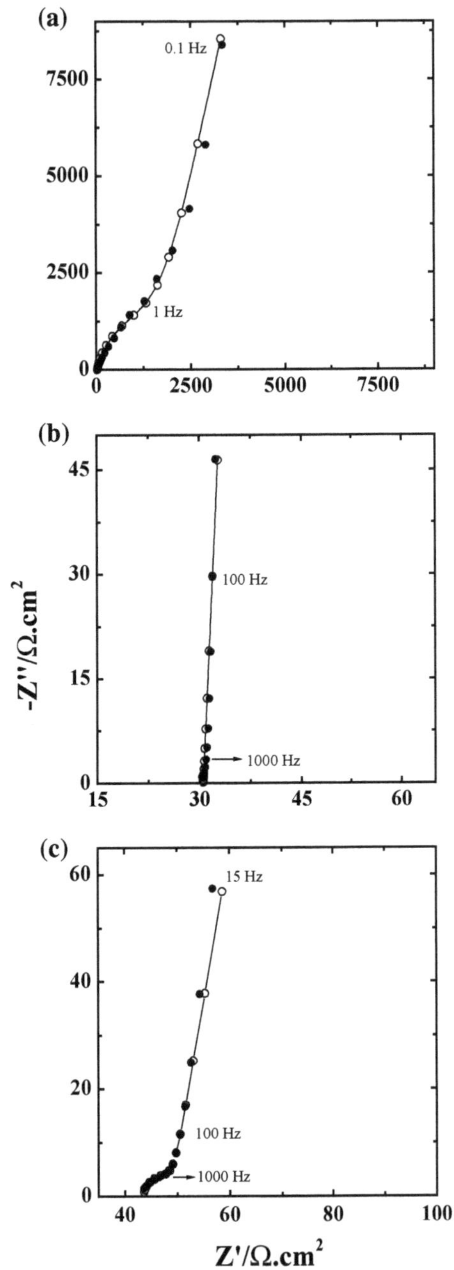

Fig. 4 Nyquist diagram for the **a** PMA/PAH, **b** PEDOT-PSS/PAH and **c** PMA/PAH/PEDOT-PSS films in HCl (pH = 1) at (*filled circle*) 0.2 V. (*Circle*) Theoretical impedance data

$$\tilde{i} = -nFD_{\mathrm{P}}\left(\mathrm{d}\tilde{c}_{\mathrm{p}}/\mathrm{d}y\right)_{y=0} \tag{20}$$

where n is the number of electrons per mol, F is the Faraday constant, k is the velocity constant, $b(=\alpha nF/RT)$ is a constant, α is the transfer coefficient, $\bar{c}_{\mathrm{P},0}$ is the average proton concentration on the electrode surface in the stationary state, $\tilde{c}_{\mathrm{P},0}$ is the oscillating component of this concentration, $\overline{V}$ is the potential at the stationary state, and D_{p} is the proton diffusion coefficient inside the host matrix and y is the position. By taking into consideration the oscillatory component of the applied potential, the impedance as a function of the frequency is obtained:

$$Z(\omega) = R_{\Omega} + \frac{R_{\mathrm{ct}} + Z_{\mathrm{D}}(\omega)}{1 + (j\omega)^{\beta}(R_{\mathrm{ct}} + Z_{\mathrm{D}}(\omega))C_{\mathrm{dl}}} \tag{21}$$

where $Z_{\mathrm{D}}(\omega)$ is the diffusional impedance and R_{Ω} is the resistance of the film and the electrolyte solution. Based on these initial and boundary conditions, proposed by Ho et al. [27], one can express this transfer function as

$$Z_{\mathrm{D}}(\omega) = \frac{1}{(i\omega)^{\gamma}C_{\mathrm{L}}} = \frac{(\mathrm{d}E/\mathrm{d}q)}{(i\omega)^{\gamma}} \tag{22}$$

In which C_{L} is limit capacitance and corresponds to the variation of the injected charge as a function of the equilibrium potential (dq/dE). It can be demonstrated, by the comparative analysis of the C_{L} values, that the presence of the PEDOT-PSS in the PMA/PAH self-assembled matrix enhances the proton storage capacity. For the PMA/PAH films, the values of C_{L} were 105, 165 and 125 μF cm^{-2} (at 0.2, 0.0 and −0.2 V, respectively), while for the PMA/PAH/PEDOT-PSS film they were 270, 390 and 270 μF cm^{-2} (at 0.2, 0.0 and −0.2 V, respectively). Owing to the dispersions frequency, associated with the surface roughness of the electrode and the anomalous diffusion, the terms β and γ are included for a better adjustment [28].

Equations 23 and 24 are obtained on the basis of the proposed reactional mechanism and Eqs. 21 and 22, which allow the determination of the rate constants. The elementary steps of proton transfer (Eqs. 16 and 17) were considered in parallel, once the proton transfer at the interface between the electrolyte solution and the electrodes, in the case of the PMA/PAH/PEDOT-PSS film, involves two distinct sites.

$$R_{\mathrm{ct}} = \frac{RT}{4F^2\alpha k_1'\left(\bar{c}_{\mathrm{P},0}\right)^2} = \frac{RT}{4F^2\alpha\bar{c}_{\mathrm{P},0}k_1} \tag{23}$$

$$R_{\mathrm{ct}} = \frac{RT}{F^2\alpha\bar{c}_{\mathrm{P},0}\left(k_2 + 4k_1'\bar{c}_{\mathrm{P},0}\right)} = \frac{RT}{F^2\alpha\bar{c}_{\mathrm{P},0}k_{\mathrm{ap2}}} \approx \frac{RT}{F^2\alpha\bar{c}_{\mathrm{P},0}k_2} \tag{24}$$

The fact that the apparent constant $k_{\mathrm{ap}\,2}\left(= k_2 + 4k_1'\bar{c}_{\mathrm{P},0}\right)$ to be independent of the proton concentration allowed its substitution for the constant k_2, which is related to the proton transfer rate to compensate the charges of the PEDOT-PSS sites electrochemically reduced. The values of the rate constant $k_1(=\bar{c}_{P,0}k'_1)$, which is associated with PMA sites, are much smaller than the values of the constant k_2. The obtained values for k_1 were 4.3×10^{-7} cm s^{-1} and 1.1×10^{-5} cm s^{-1} (at 0.2 and 0.0 V, respectively), while for k_1 they were 7.6×10^{-4} cm s^{-1} and 9.7×10^{-4} cm s^{-1} (at 0.2 and 0.0 V, respectively). These results state low practical irreversibility for the PMA/PAH/PEDOT-PSS film, and indicate the influence of PEDOT-PSS in the self-assembled host matrix for increasing the proton electroinsertion rate at electrolytic solution/electrode interface.

An experimental simulation of an acid–base machine (Neutralization Battery) is described further. This procedure provides a proof of concept for the idealized machine. The curves of the proton electroinsertion and electrodeinsertion for the PMA/PAH/PEDOT-PSS were obtained with the aid of a three electrode cell, with a platinum sheet and a silver wire, as counter and quasi-reference electrodes, respectively. These conditions were used in order to achieve an experimental approximation of an acid–base machine operating between the acid (HCl) and the basic (NaOH) reservoirs, the PMA/PAH/PEDOT-PSS served as the proton selective electrode, while the silver electrode was considered as the second electrode having null overpotential. However, the specifications of this second electrode can be disconsidered, because its contribution to the work produced should be null, due the fact that the chloride concentration is considered to be constant during the neutralization. Also, minimal overpontentials, associated with the electrical resistance and proton diffusion, can be achieved with low current applied (5 μA cm^{-2}) and small thickness of the film (108 ± 4 nm). This experimental simulation allowed us to obtain 22 kJ per mol of electroinserted proton (Fig. 5).

Therefore, the change in the potential for the PMA/PAH/PEDOT-PSS film, as a function of the pH, is associated with proton intercalation and it can be used to capture energy in a neutralization battery.

Another experimental simulation of a acid–base machine was performed between the acid (H_2SO_4) and the basic (KOH) reservoirs. The curves of the proton electroinsertion and electrodeinsertion for the PMA/PAH/PEDOT-PSS were obtained with the aid of a three electrode cell, with a platinum sheet and Hg/Hg_2SO_4, in saturated H_2SO_4, as counter and reference electrodes, respectively.

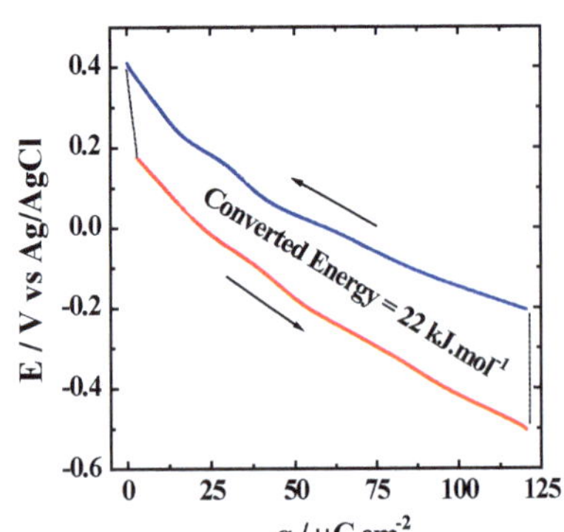

Fig. 5 Electroinsertion at (*red dash line*) NaCl 0.02 M (HCl 0.1 M) and electrodeinsertion (*blue dash line*) NaCl 3 M (pH = 6) curves for the PMA/PAH/PEDOT-PSS film with j = 5 μA cm^{-2}

Moreover, selectivity for proton electroinsertion in this electrochemical system was investigated with the use of electrochemical quartz crystal microbalance (EQCM). The slope (272 μg C^{-1}) of the curves associated the mass change as a function of the charge for the PMA/PAH/PEDOT-PSS, in H_2SO_4 at pH = 1, suggested that the hydrated proton was electroinserted/electrodeinserted in the host matrix, while no anions participated in the charge compensation. The charge compensation mechanism plays a very important role in neutralization battery, because one of the electrodes must only allow the insertion of protons, whereas the other one should allow specifically the insertion of the anion.

The areas between electroreduction, at pH = 0, and electrooxidation, at several pHs, curves (Fig. 6a) for the PMA/PAH/PEDOT-PSS film were determined in order to obtain the efficiency (ε_{neut}) of the machine, which was calculated with the aid of the Eq. 9 and plotted in the Fig. 6b.

The values obtained experimentally were close to the ones predicted by the thermodynamic formalism previously discussed. With the aid of the neutralization enthalpy at 298 K (55.815 kJ mol^{-1}), it was possible to determinate the work (w) produced by the machine as a function of the proton activity of the resulting solution. The neutralization of a H_2SO_4 solution (pH = 0 to pH = 6) made it

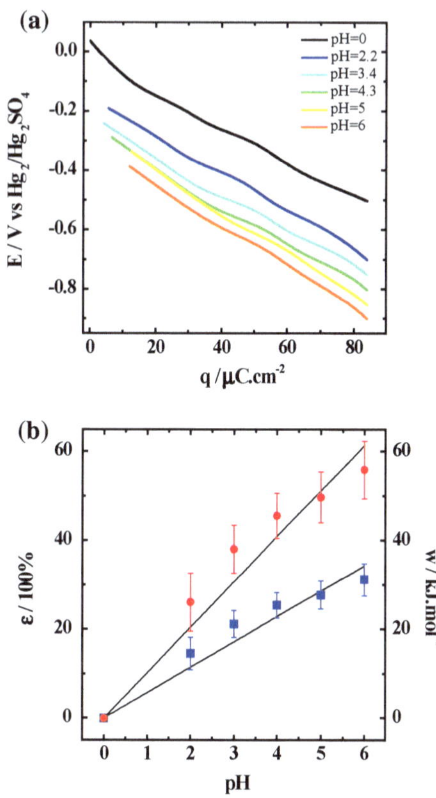

Fig. 6 **a** Eletroreduction, at pH = 0, and electrooxidation, at several values of pH, of the self-assembled PMA/PAH/PEDOT-PSS film with $j = 5$ μA cm^{-2}. **b** Theoretical (*black dash line*) and experimental efficiency (*filled circle*) and work (*filled square*) as a function of pH of the resulting solution after neutralization

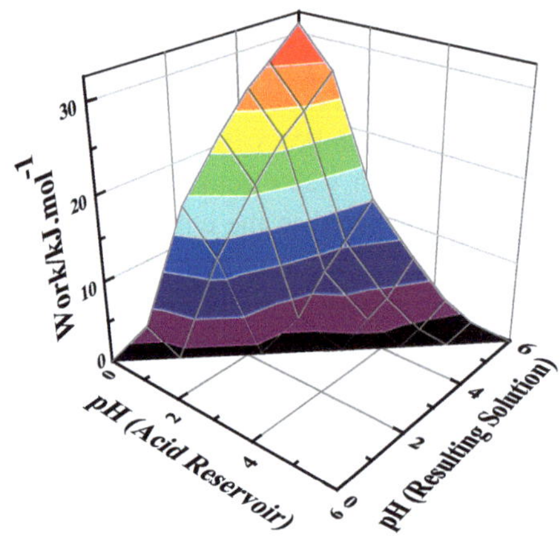

Fig. 7 Work produced by the acid–base machine, in the experimental conditions, as a function of the pH of the acid reservoir and of the resulting solution

possible the conversion of 31.2 kJ per mol of proton electroinserted, which corresponds to an efficiency of 56.3%.

The redox curves, for the PMA/PAH/PEDO-PSS film, shown in the Fig. 6a–b were used to produce a diagram for all the operational combinations of the machine, at the studied conditions. The work values, as a function of the pH of the acid and the resulting solutions, are close to the ones predicted by the formalism (Eq. 9) (Fig. 7).

Experimental simulation of Neutralization Batteries was also realized using positive and negative electrodes. Thus, we investigated other host matrices as positive electrodes for these batteries. Variation in the voltage can be achieved using a second electrode that does not involve protons in the redox processes, due the change in the potential as a function of the pH for PMA/PAH/PEDOT-PSS film. On this basis, an electrosynthesized PEDOT film was prepared and studied in order to use it as the positive electrode in the neutralization battery. During the reduction/oxidation processes, this conducting polymer inserts/deinserts sulphate anions, respectively, presenting broad voltammetric waves at more negative potentials, as well, a pseudocapacitive behaviour at more positive potentials. The minor change in the peak potential as a function of the pH states a major anion participation in the charge compensation mechanism. The electrochemical properties discussed above enable the utilization of the electrossynthesyzed PEDOT film in the assembling of the neutralization battery [24, 29]. The electrochemical properties of this film were analyzed with aid of the electrochemical impedance spectroscopy in H_2SO_4 at pH = 1, and in K_2SO_4 50 mM at pH = 6. The values of the charge transfer resistance were 15.7 and 12.7 Ω cm^{-2} at pH = 1 and pH = 6, respectively, which indicates the ion electroinsertion rate is high. Thus, this electrode presents suitable properties as positive electrode for neutralization batteries, once demonstrates low practical irreversibility. Moreover, the semi-infinite diffusion was not observed at low frequency, due to high mobility of charge carries into the host matrix.

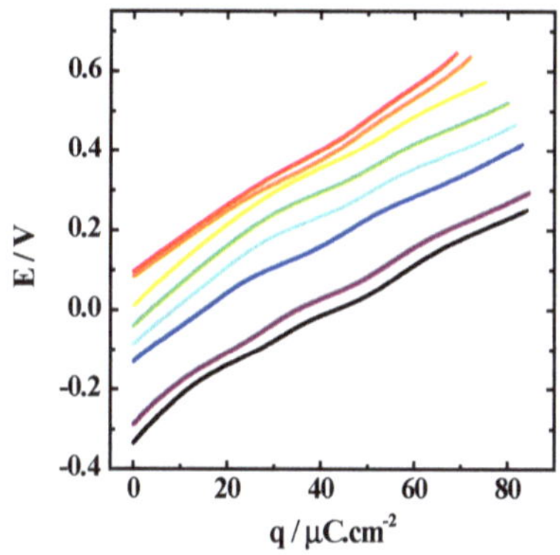

Fig. 8 The charge/discharge curves for the acid–base machine composed by self-assembled PMA/PAH/PEDOT-PSS film and electrosynthesized PEDOT: charge at pH = 0 (*black dash line*), pH = 1 (*purple dash line*), pH = 2.2 (*dark blue dash line*), pH = 3.4 (*light blue dash line*), pH = 4.3 (*green dash line*), pH = 5 (*yellow dash line*) at 298 K, and discharge at pH = 6 at 298 K (*orange dash line*) and 318 K (*red dash line*). $j = 5\ \mu A\ cm^{-2}$

Table 1 Values of work, efficiency and work per concentration of base as a function of the pH of the acid reservoir performed by the machine during the neutralization

pH	0	1	2.2	3.4	4.3	5
w/kJ mol^{-1}	31.167	26.0	12.979	8.219	3.977	1.495
ε	55.84	46.58	23.25	14.73	7.12	2.68
w/C_b (kJ dm^3 mol^{-2})	1.95	8.67	5.57×10^2	8.30×10^3	3.90×10^4	6.01×10^4

In order to reproduce a system configuration closer to a practical one, an experiment was carried out in which an acid–base machine composed by a PMA/PAH/PEDOT-PSS film, as the negative electrode, and an electrosynthesized PEDOT film, as the positive electrode, were utilized to produce work. In this configuration, the charging process was performed in H_2SO_4 solutions, with pH lower than 6, followed by the exchange of the acid solution for the K_2SO_4 solution at pH = 6, where the discharge process took place (Fig. 8).

This electrochemical configuration provided values, of efficiency and work per mol of proton electroinserted, close to the ones expected from the thermodynamic formalism, these results are presented at the Table 1. Some small changes in the values, when compared to those obtained from the three electrodes cell configuration, are due the overpotentials associated with the electroinsertion/electrodeinsertion of sulphate anions into/from the PEDOT electrode, in more and less acid medium, respectively.

In order to estimate the production cost of a practical device, the ratio w/C_b should be taken into consideration; it represents the work (w), performed by the machine, normalized by the concentration of the used base (C_b). On the basis of

w/C_b, one can infer that the treatment of acid solutions, with a high pH, is more profitable per volume of solution. Nevertheless, by treating solutions with very low proton activity, the work produced will not be sufficient to compensate the necessary investment in the production of the neutralization battery. Therefore, a suitable pH range is essential in order to apply these devices in wastewater treatment with a satisfying cost-benefit.

The heat released during the neutralization can be used to enhance the performance of the machine; in this case, the procedure should be carried out adiabatically so the temperature in the discharge process will be higher than in the charge. For instance, the neutralization of 1.5 mol L^{-1} protons in solution provokes an increase of 20.1 K in the temperature. Thereby, a discharge process, at pH = 6 and 318 K (Fig. 8), was performed with the full cell configuration.

On the basis of the area between the discharge (at pH = 6 and 318 K) and charge (pH = 6 and 298 K) curves, it was possible to obtain 812 J mol^{-1}, which corresponds to 34.48% of work expected for the Carnot's Cycle and stands a high value for practical thermal machines. Though, this comes from the assumption that the host matrix would intercalate all the protons in the solution, what is an experimental difficulty.

Using the neutral, or slightly acid, solutions resulting from the neutralization reactions that occur in this machine, it is still possible to obtain energy with the insertion/deinsertion of the alkali cations present in these solutions. A different class of intercalation electrodes named polycyanometalates might be used for this purpose. Unlike polyoxometalates, in which the charge compensation mechanism involves protons, polycyanometalates have a structure which allows the alkaline ions selective intercalation and when used together with polyoxometalates the result is increased energy storage in these electrochemical systems.

Prussian blue analogues (PBA) belong to the class of polycyanometalates and have been extensively studied as intercalation electrodes of alkaline ions [30–32]. When suitable electrolyte is used, PBAs exhibit a long life cycle, fast intercalation rate and high load storage capacity, meeting the requirements for neutralization batteries. These analogues are represented by the general formula $A_xM_y[M'(CN)_6]_z$, where A is an alkaline cation and M and M′ are transition metals [33–37]. Transition metals cations are octahedrally coordinated by hexacyanometalates (HCM) groups and this arrangement produces an open-channel network, along the three crystallographic axes, which allows the electroinsertion/electrodeinsertion of cations [38].

These compounds may display redox activity, of one or more transition metals in their structure, and their electroinsertion/electrodeinsertion processes can be represented as in the Eq. 25 [32, 39].

$$A_xM_y\left[M'^{(III)}(CN)_6\right]_z + a(A^+ + e^-) \rightleftharpoons A_{(x+a)}M_y\left[M'^{(II)}(CN)_6\right]_a\left[M'^{(III)}(CN)_6\right]_{(z-a)} \tag{25}$$

Dissimilar from Prussian blue, $Fe_4[Fe(CN)_6]_3$, which only allows insertion/desinsertion of potassium ions, PBAs enable the effective accommodation of others alkaline cations such as Li^+, Na^+, Mg^{2+} and NH_4^+ during the electroinsertion process [37, 38, 40–43]. Nickel hexacyanoferrate (NPBA) is a Prussian blue analogue in which nickel ions occupy some sites of the cubic structure face centered, occupied by the iron ions, connected by CN bridges, creating channels in the direction <100>, which allow the insertion/deinsertion of hydrated sodium ions in a chemically reversible way [38].

Cyclic voltammograms of self-assembled films, prepared from solutions of Ni $(NO_3)_2.6H_2O$ (20 mM) and a mixed aqueous solution of $K_3[Fe(CN)_6]$ (20 mM) and $NaNO_3$ (15 mM) [11], presented high stability, in aqueous NaCl solutions, and reversibility, with relation to the intercalation of sodium ions, in solutions with different pHs, according to Eq. 26:

$$Na^+Ni^{(II)}\left[Fe^{III}(CN)_6\right] + x(Na^+ + e^-) \rightleftharpoons Na^+_{1+x}Ni^{(II)}\left[Fe^{III}(CN)_6\right]\left[Fe^{II}(CN)_6\right]_{(1-x)} \quad (26)$$

The NPBA cyclic voltammograms in NaCl solutions reveal the change in peak potentials as a function of the concentration of sodium ions, and showed that the electrode potential changed up to 120 mV when the sodium ion concentration varied from 20 mM to 5.1 M. During the neutralization, the concentrated salt solution (neutral or less acid) can be discharged into a suitable environment, and the acid solution, with low concentration of salt, can be added to the full electrochemical cell. Thus, using alkaline ion selective electrodes, in the neutralization batteries, it is possible to store energy due the partial change in entropy associated with the variation of alkaline ions concentration [9, 11, 44] (Fig. 9).

Following our previously reported approach, in these kinds of batteries, the discharge potential is higher than the charge potential. Therefore, it is possible to

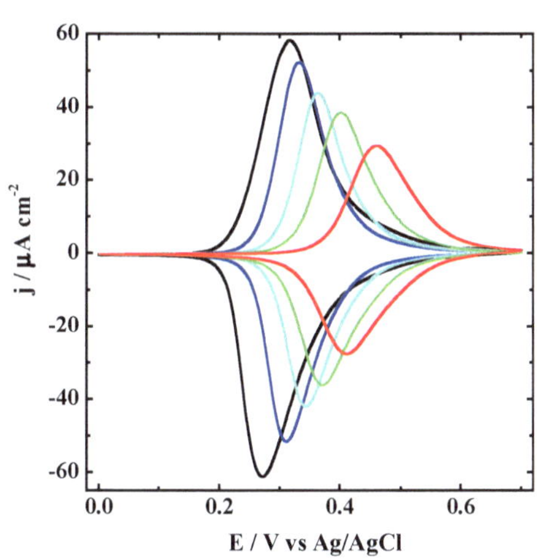

Fig. 9 Potentiodynamic profile for the NPBA film in NaCl at (*black dash line*) 20 mM, (*dark blue dash line*) 80 mM, (*light blue dash line*) 320 mM, (*green dash line*) 1.28 M, and (*red dash line*) 5.12 M and pH = 4 with $v = 10$ mV s^{-1}

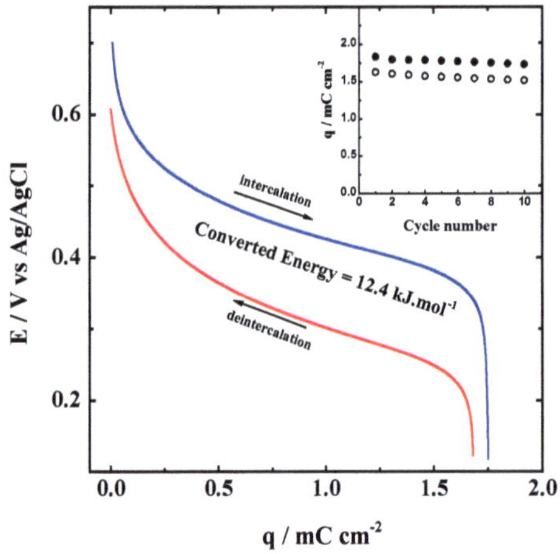

Fig. 10 Electrochemical energy extraction cycle for the NPBA film: electroreduction at (*dark blue dash line*) 3.0 M NaCl and electrooxidation curves at (*red dash line*) 20 mM NaCl and pH = 4 with j = 10 μA cm^{-2}. *Inset* (*circle*) Oxidation and (*filled circle*) reduction charge as a function of number of cycles

harvest energy in these experimental conditions, which is essential. Figure 10 shows that it was possible to convert 12.4 kJ per mole of intercalated sodium ions.

The kinetic analysis of the sodium ion electroinsertion/electrodeinsertion was performed by electrochemical impedance spectroscopy, using the Nyquist diagrams for the NPBA electrode in NaCl 0.6 M electrolytic solution at *dc* potentials of 0.55, 0.43, 0.37 and 0.32 V, Fig. 11. A better comprehension of the system is possible with the aid of the frequency domain models and, as well, with the application of the Eqs. 27 and 28, which represent the ion transfer from the electrolytic solution to the electrode/electrolytic solution interface and how the pair $(Na^+, e^-)_i$ diffuses into the host matrix, respectively.

$$Na_i^+ + e^- \rightarrow (Na^+, e^-)_i \tag{27}$$

$$(Na^+, e^-)_i \rightarrow (Na^+, e^-)_s \tag{28}$$

The impedance of these processes can be expressed by Eq. 21 and helped to determine the rate constant (k), where the charge transfer resistance corresponds to $\frac{RT}{F^2 \alpha k}$ [26]. The sodium ion diffusion impedance can be represented (between 200 and 40 Hz) in two different limiting cases [27]. First, limiting case, semifinite diffusion ($\omega \gg 2D_c/L^2$), shown in Eq. 29, and second limiting case, finite diffusion ($\omega \ll 2D_c/L^2$) at low frequencies, shown in Eq. 30, which furnishes the D_c values of the electron-ion pair where V_M and L are the molar volume and thickness of the host matrix, x is the stoichiometric ratio in Eq. 26 [27].

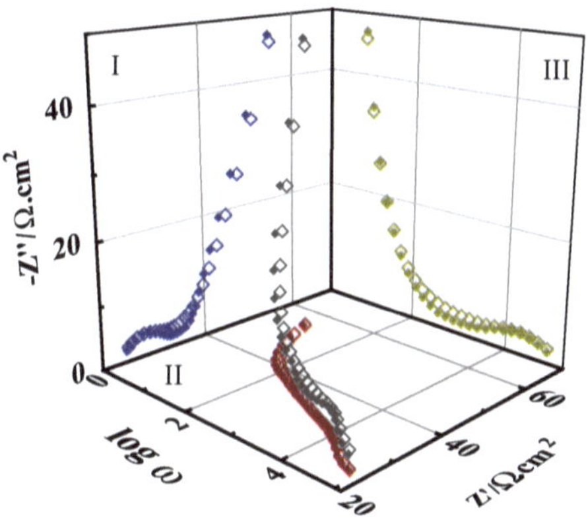

Fig. 11 Three-dimensional Impedance diagrams for the NPBA film at 0.32 V in NaCl 0.6 M at pH = 4: (*I*) (*blue diamond*) real part versus imaginary part of impedance, (*dark filled diamond*) corresponding theoretical data; (*II*) (*red diamond*) real part of impedance versus logarithm of the frequency, (*dark filled diamond*) corresponding theoretical data; and (*III*) (*yellow diamond*) imaginary part of impedance versus logarithm of the frequency, (*dark filled diamond*) corresponding theoretical data

$$Z_D(\omega) = \left|\left(\frac{V_M}{(2D_c)^{0.5}F}\right)\left(\frac{dE}{dx}\right)\omega^{-0.5}\right|(1-j) = \left|\left(\frac{L}{(2D_c)^{0.5}}\right)\left(\frac{dE}{dq}\right)\omega^{-0.5}\right|(1-j) \tag{29}$$

$$Z_D(\omega) = \left|\frac{V_M}{F}\left(\frac{dE}{dx}\right)\left(\frac{L}{3D_c}\right)\right| - j\left|\frac{V_M}{F\omega L}\left(\frac{dE}{dx}\right)\right| = \left|\left(\frac{L^2}{3D_c}\right)\left(\frac{dE}{dq}\right)\right| + \frac{\left(\frac{dE}{dq}\right)}{j\omega} = \left|\left(\frac{L^2}{3D_c}\right)\left(\frac{dE}{dq}\right)\right| + \frac{1}{j\omega C_L} \tag{30}$$

In fact, we can also use the model of anomalous diffusion and a dimensionless parameter, γ (lower than 1), which accounts for frequency dispersions observed at low frequencies, should be included in the Eq. 30 as shown in the Eq. 31. In this case, the mean square displacement of the diffusing sodium ion follows a power law dependence on time indicating that the number of sodium ions diffusing in the self-assembled materials are not conserved [27, 44].

$$Z_D(\omega) = \left|\left(\frac{L^2}{3D_c}\right)\left(\frac{dE}{dq}\right)\right| + \frac{1}{(j\omega)^{\gamma}C_L} \tag{31}$$

Table 2 Values of the parameters obtained from electrochemical impedance data at different *dc* potentials

E (V)	R_{Ω} (Ω cm^2)	C_{dl} (μF cm^{-2})	R_{ct} (Ω cm^2)	k (10^{-7} mol cm^{-2} s^{-1})	D_c (10^{-8} cm^{-2} s^{-1})	C_L (mF cm^{-2})	γ
0.55	23.8	7	5	1.6	1	1.8	0.8
0.43	23.8	7	2	4.0		10.0	0.85
0.37	23.8	7	3	2.6		14.2	0.8
0.32	23.8	7	10	0.8	1	5.7	0.76

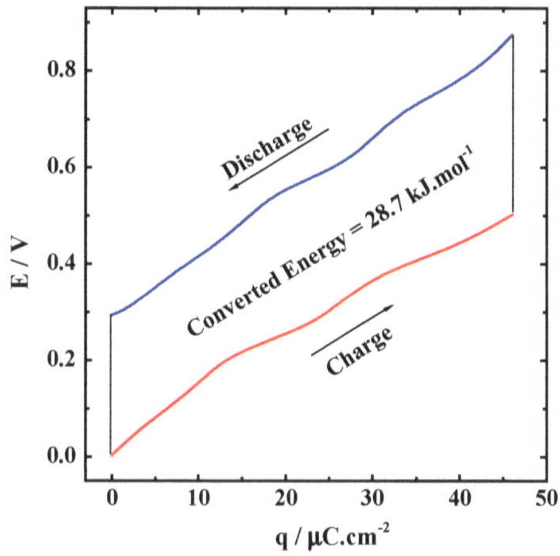

Fig. 12 The charge/discharge curves for the acid–base machine composed by self-assembled PMA/PAH/PEDOT-PSS film and NBPA film: charge at (*red dash line*) pH = 1 and NaCl 20 mM and discharge at (*blue dash line*) pH = 6 and NaCl 3 M. $j = 10\ \mu A\ cm^{-2}$

The k and D_c as well as other parameters obtained from fit the impedance diagrams using these electrochemical models in the frequency domain are shown in Table 2. These values obtained for k and D_c, indicate fast interfacial transfer of sodium ions from the electrolytic solution to the host matrix. Electroinsertion of hydrated Na^+ ions leads to low values of R_{ct} due the minimum electrostatic repulsions at the electrode/solution interface and low values of activation energy (E_a) for the intercalation process. Furthermore, there is no significant sign of semi-infinite diffusion, because the modified electrode had a small thickness (100 nm) and high chemical diffusion coefficient [11].

On the basis of the electrochemical investigation on NPBA film in acid and neutral medium, this host matrix presents suitable properties to be used in Neutralization Batteries. So, we have simulated a battery formed from PMA/PAH/PEDOT-PSS film and NPBA film as negative and positive electrodes, respectively. Neutralizing 0.1 M H_2SO_4 with KOH solution saturated with NaCl, 28.7 kJ per mol of ion electroinserted is obtained, as is shown in Fig. 12.

5 Final Considerations

Electrochemical systems are noteworthy by high energy conversion efficiency and, in the particular case of the neutralization batteries, using selected electrodes and specific experimental conditions, we have demonstrated high efficiency for conversion of entropic energy into electrical work. However, we should overcome challenges to make feasible this technology, encouraging the treatment of acid (and/or alkaline) wastewaters and contributing to sustainable growth. For a practical neutralization battery, the conversion efficiency decreases significantly as a function of the magnitude of overpotentials, once entropic energy is converted to useful work after charge/discharge cycles. Thus, the development of insertion electrodes with low irreversibility at different concentrations of proton and salt must be encouraged. Additionally, these host matrices should present high charge capacity, selectivity, ciclability, low cost and low toxicity. It is important to mention that the performed work depends on the logarithmic variation of activity, while the neutralization process is a linear function of the concentration of ionic species. Henceforward, one should manufacture nanomaterials with the suitable properties at the pH of wastewaters, aiming to achieve a profitable treatment and the environmental preservation.

Acknowledgments We are grateful to FAPESP (Project 2015/ 16867-9) for financial support.

References

1. Yang ZG, Zhang JL, Kintner-Meyer MCW, Lu XC, Choi DW, Lemmon JP, Liu J (2011) Chem Rev 111:3577
2. REN21, Renewables (2016) Global status report. REN21 Secretariat, Paris
3. Pattle RE (1954) Nature 174:660
4. Levenspiel O, de Nevers N (1974) Science 183:157
5. Weinstien JN, Leitz FB (1976) Science 191:557
6. Olsson M, Wick GL, Isaacs JD (1979) Science 206:452
7. Kiviat FE (1976) Science 194:719
8. Brogioli D (2009) Phys Rev Lett 103:58501
9. La Mantia F, Pasta M, Deshazer HD, Logan BE, Cui Y (1810) Nano Lett 2011:11
10. Facci T, Gomes WJAS, Bravin B, Araujo DM, Huguenin F (2014) Langmuir 30:426
11. Gomes WJAS, de Oliveira C, Huguenin F (2015) Langmuir 31:8710
12. Bravin B, Gomes WJAS, Huguenin F (2016) J Electroanal Chem 765:52
13. Morais WG, Gomes WJAS, Huguenin F (2016) J Phys Chem C 120:17872
14. Sadakane M, Steckhan E (1998) Chem Rev 98:219
15. Ruiz V, Suárez-Guevara J, Gomez-Romero P (2012) Electrochem Commun 24:35
16. Suppes GM, Cameron CG, Freund MS (2010) J Electrochem Soc 157:A1030
17. Lei C, Wilson P, Lekakou C (2011) J Power Sources 196:7823
18. Pandey K, Lakshmi N (1999) J Mater Sci 34:1749
19. Ahonen HJ, Lukkari J, Kankare J (2000) Macromolecules 33:6787
20. Smith RR, Smith AP, Stricker JT, Taylor BE, Durstock MF (2006) Macromolecules 39:6071
21. Wang YH, Wang XL, Hu CW (2002) J Colloid Interface Sci 249:307
22. Kurth DG, Volkmer D, Ruttorf M, Richter B, Muller A (2000) Chem Mater 12:2829
23. Vaillant J, Lira-Cantu M, Cuentas-Gallegos K, Casan-Pastor N, Gomez-Romero P (2006) Prog Solid State Chem 34:147
24. Plieth W, Bund A, Rammelt U, Neudeck S, Duc L (2006) Electrochim Acta 51:2366
25. Toth PS, Janaky C, Berkesi O, Tamm T, Visy C (2012) J Phys Chem B 116:5491
26. Orazem ME, Tribollet B (2008) John Wiley & Sons, Inc.
27. Ho C, Raistrick ID, Huggins RA (1980) J Electrochem Soc 127:343
28. Bisquert J, Compte A (2001) J Electroanal Chem 499:112
29. Syritski V, Idla K, Öpik A (2004) Synth Met 144:235
30. Liu N, Li W, Pasta M, Cui Y (2014) Front Phys 9:323
31. Pasta M, Wessells CD, Huggins RA, Cui Y (2012) Nat Commun 3:1149
32. Jia Z, Wang J, Wang Y (2014) RSC Advances 4:22768
33. Wessells CD, McDowell MT, Peddada SV, Pasta M, Huggins RA, Cui Y (2012) ACS Nano 6:1688

34. Yue Y, Binder AJ, Guo B, Zhang Z, Qiao ZA, Tian C, Dai S (2014) Angew Chem Int Ed 53:3134
35. Wu X, Cao Y, Ai X, Qian J, Yang H (2013) Electrochem Commun 31:145
36. Wu X-Y, Sun M-Y, Shen Y-F, Qian J-F, Cao Y-L, Ai X-P, Yang H-X (2014) Chemsuschem 7:407
37. Nie P, Shen L, Luo H, Ding B, Xu G, Wang J, Zhang XJ (2014) Mater Chem A 2:5852
38. Wessells CD, Peddada SV, Huggins RA, Cui Y (2011) Nano Lett 11:5421
39. Baioni AP, Vidotti M, Fiorito PA, Ponzio E, Córdoba de Torresi SI (2007) Langmuir 23:6796
40. Lu Y, Wang L, Cheng J, Goodenough JB (2012) Chem Commun 48:6544
41. Lee H, Kim Y-I, Park J-K, Choi JW (2012) Chem Commun 48:8416
42. Itaya KUI, Neff VD (1986) Acc Chem Res 19:162
43. Wessells CD, Peddada SV, McDowell MT, Huggins RA, Cui Y (2012) J Electrochem Soc 159:A98
44. Rica R, Ziano R, Salerno D, Mantegazza F, van Roij R, Brogioli D (2013) Entropy 15:1388

New Ternary Intermetallics Based on Magnesium for Hydrogen Storage: The Fishing Approach

J.-L. Bobet, E. Gaudin and S. Couillaud

1 Introduction

Magnesium allows obtaining a good hydrogen storage capacity in terms of weight percentage but its use is limited by high stability of the hydride and slow kinetics. The kinetics can be improved by (i) mechanical grinding, cold rolling (or other severe plastic deformation) and (ii) addition of various elements (catalysts or activators). However, to change the stability of the hydride, it is necessary to modify the structure of the Mg–Mg bond. De Jongh et al. [1] have demonstrated by ab initio calculations that nanometer magnesium could reversibly absorb hydrogen at room temperature. Vajo et al. [2] also demonstrated the interest of nano-confinement on the thermodynamics and dehydrogenation kinetics of metal hydrides.

Many authors have since managed the synthesis of nano Mg and sorption results seem to be encouraging [3–14]. It is also possible to deposit nano Mg in nano-porous carbon structures or equivalents [6–14]. However, the production of large quantities is not possible at the moment and the extreme sensitivity of these materials towards oxygen makes its use rather complex.

Another way to replace the Mg–Mg bonds is to create binary compounds (or ternary, quaternary, etc.) based on magnesium. Thus, new Mg–TM bonds (TM = transition metal and/or inner transition element) take the place of Mg–Mg metallic bonds. Binary compounds have been extensively studied for the last 50 years, and many binary compounds were already identified.

The $REMg_2$ [15], $REMg_3$ [16], $REMg_{12}$ [17], $REMg_5$ [18], RE_2Mg_{17} [19] and Mg_2X (where X = Ni, Si, Ge, Sn, and RE = rare earth) [20] compounds may be cited. To date, the binary compound which has the best properties of hydrogen

J.-L. Bobet (✉) · E. Gaudin · S. Couillaud
CNRS, ICMCB, Université de Bordeaux, 87 avenue du Docteur Albert Schweitzer, 33600 Pessac, France
e-mail: Jean-Louis.Bobet@icmcb.cnrs.fr

F.L. de Souza and E. Leite (eds.), *Nanoenergy*,
DOI 10.1007/978-3-319-62800-4_10

storage is Mg_2Ni (absorption at 1 bar and 267 °C, maximum capacity = 3.6 wt%). However, many of these ternary compounds decompose under hydrogen to form binary hydrides (i.e., MgH_2, REH_3, etc.) [17, 20].

The combined addition of transition element and internal transition (i.e. rare earths) could open new ways of research. Transition elements form unstable hydrides while rare earths form very stable binary hydrides (even more stable than magnesium). Thus, from a thermodynamic point of view [21] mixing of rare earths and transition elements could lead to obtaining hydrides with intermediate stability. This demonstrates the interest to synthesize and study new intermetallic having potentially good reversible storage capacity under moderate temperature and pressure.

Then, there are two different methods: (i) the development of solid solution derived from already known binary compounds or (ii) the development of new ternary compounds. The first method is certainly the easiest one but it is limited to the original properties of binary compounds which will be more or less preserved (e.g. changes of the thermodynamic properties of solid solutions $TR_{1-x}TR'_xNi_{5-y}TM_y$ which have been widely studied [22, 23]. It is the second method that we are going to develop in the rest of this chapter. However, this method is much more exploratory and therefore has a much higher risk of failure. Despite everything, this exploratory research will lead to the identification of new compounds (and potentially new crystallographic structures) which in itself should already be regarded as a positive result.

In addition, it must be considered that because of their light weight and high stiffness, the magnesium alloys look interesting for mechanical applications. Few ones are already used in automotive and aerospace industries (e.g. AZ and AM families). However, their commercial applications are limited by poor creep resistance and poor tensile properties. Moreover, mechanical properties are quickly decreased at a temperature above 120 °C [24, 25]. Thus, it could be paramount to investigate new magnesium-rich systems. It was already shown that: (i) mixing RE in magnesium matrix can improve the mechanical properties by solid solution strengthening and precipitation strengthening, the corrosion resistance, and the stability in temperature [26] and (ii) use of amorphous or nanocrystalline alloys leads to an increase of mechanical properties [27]. Moreover, such systems are complicated to obtain at large scale.

A lot of compositions have been studied. Usually, the alloys are composed of 10–15 at.% of RE, 10–15 at.% of a transition metal, and 70–80 at.% of magnesium. Such compositions can be amorphized easily because the mixing enthalpies are very different (which is the thermodynamic condition to obtain an amorphous alloy).

Thus, we will focus our study on those used to obtain amorphous materials but we will tend to get defined compounds (if possible with an original crystal structure) and not mixed alloys.

First, the method (so called fishing approach) to obtain new compounds will be described. Then we will give a full description of the synthesis, structure, and hydrogen sorption properties of a new compound ($LaCuMg_8$). Finally, we will describe briefly two other new compounds which are still understudy. Some physical properties (mechanical, electrical, and magnetic) will also be presented in order to demonstrate that other application than hydrogen storage can also be of interest.

2 Experimental Details for the Fishing Approach

Starting materials for the preparation of the sample were rare earth pieces (Stream Chemicals, >99.9%), transition metal rod (Stream Chemicals, >99.9%), and magnesium rod (alpha Aesar, >99.8%). To avoid oxides impurities, the surfaces of the magnesium and rare earth pieces were cleaned in the glove box. Elemental pieces were then weighed in the appropriate amounts and sealed in tantalum ampoules under an argon pressure of ca. 800 mbar. The argon used was purified before with magnesium sponge (673 K). Then, the ampoules were then placed in a quartz tube and sealed under vacuum to be heated at about 1373 K and kept at this temperature for 10 min. After, cooling from 1300 K was done (i) quickly (i.e. air quenched) or (ii) slowly (e.g. 2 K/h). No reaction with the tantalum tubes was observed. The polycrystalline samples were found to be stable in air.

The bulk samples were investigated by electron probe microanalyses (EPMA) with metallic RE (L_α), TM (K_α) and Mg (K_α) as standards. The bulk samples were embedded in a methylmethacrylate matrix and the surface was polished with different silica and diamond pastes. The surface remained unetched for the EPMA measurements. The global quantitative analyses were in good agreement with the initial compositions.

First, we prepare a sample with an exact composition. As we are looking for new magnesium-rich phases, the starting composition is 10 RE + 10 TM + 80 Mg or 15 RE + 15 TM + 70 Mg. In the following, two types of notation were used: (i) the one used in metallurgy (but with mole instead of weight percentages), which provides direct information about the composition of each element in the starting mixture (e.g. $La_{10}Ni_{10}Mg_{80}$ = 10 at.% of La + 10 at.% of Ni + 80 at.% of Mg; Such writing does not provide any information on the nature of the possible existing phases) and (ii) the one used in solid state chemistry to illustrate the composition of the phase (using integer as in $LaCuMg_8$). This last writing indicates the existence of the phase and provides information on its exact composition.

The EPMA analysis allows to highlight (sometimes) some new phases in relatively large abundance. In such case, it is possible to analyze precisely the composition of the new phase and then to start again a new synthesis with the new nominal composition. The same method is again used to optimize the composition of the new phase. It is worth pointing out that in a few cases, it can be difficult to achieve a single phase even after complete optimization of the process. Moreover, in order to improve the crystallinity of some new ternary phases, a low cooling rate (e.g. 2 K/h) was used instead of the usual cooling rate (i.e, air quenched). A crystal suitable for single-crystal X-ray diffraction can be selected on the basis of the size and sharpness of the diffraction spots. Data collection was carried out on an Enraf-Nonius Kappa CCD diffractometer using Mo Kα radiation. Data processing and all of the refinements were performed with the JANA2006 program package [28].

A Gaussian-type absorption correction was applied, and the shape was determined with the video microscope of the Kappa CCD. Details of data collections can be founded in Ref. [29].

In the following, the fishing approach has been used to highlight three new phases: $LaCuMg_8$ (derived from a known structure), $Gd_{13}Ni_9Mg_{78}$ or $Gd_{1+x}Ni_{1-y}Mg_{8-(x-y)}$ (with a new modulated structure but derived from an FCC stacking), and $NdNiMg_5$ (completely new structure with an interesting 2D network).

SEM micrographs were acquired using a Tescan VEGA II-SBH working at 10 and 20 kV with a current beam of 90 μA. TEM Images were performed using JEOL 22200 FS. It was operated at an accelerating voltage of 200 kV. The powder was ground in ethanol and a droplet of this suspension was then deposited on a carbon formvar copper grid. Hydrogen sorption kinetics were investigated by the use of an automatic Sievert-type volumetric apparatus (HERA, Hydrogen Storage System) in the temperature range from room temperature to 330 °C and with 10 bar of H_2 [30].

3 State of the Art

3.1 The Binary Mg_2Ni Compound

This compound crystallizes in a hexagonal lattice (S.G. $P6_222$) with lattice parameters $a = 5.216$ (6) Å and $c = 13.20$ (6) Å [31]. Under hydrogen pressure, Mg_2Ni absorbs up to four hydrogen atoms per formula unit leading to a weight capacity of 3.6% [32, 33].

Various studies have shown that the hydrogenation mechanism consists in two steps: (i) hydrogen, with a maximum H/M $\approx$ 0.3, fits into the hexagonal structure. This induces an increase in cell volume of about 2.2% [33, 34], (ii) then, the hydride Mg_2NiH_4 is formed inducing a structural change (below 250 °C, Mg_2NiH_4 crystallizes in a monoclinic cell (S.G. Cm) with parameters $a = 6.497$ (2) Å, b = 6.414 (1) Å, $c = 6.601$ (2) Å, and $\beta = 93.23$ (2)°; above this temperature, it crystallizes in a cubic lattice of antifluorite type (S.G. Fm-3m) with parameters $a = 6.508$ (1) Å). The study of the positions of hydrogen in the hydride shows the formation of a complex transition metal—hydrogen (i.e. strong bond between Ni and H) [32].

From a kinetic point of view, Mg_2NiH_4 is formed in about 20 min at temperatures between 250 and 350 °C [35]. However, a long activation period at 330 °C is necessary in advance to allow the hydrogen absorption [36]. Many works propose solutions to overcome this problem [37]. For example, Zaluski et al. show a simplified activation and faster absorption (less than 10 min at 300 °C) when Mg_2Ni is prepared by mechanical alloying (i.e. small particle size).

The kinetic can be improved but the thermodynamic properties remained unchanged and relatively closed to the ones of magnesium even if various solid solutions have been proposed [38].

3.2 Ternary Compounds with Magnesium

The goal of this part is not to give an exhaustive list of all the ternary compounds in the RE–TM–Mg system but just to highlight the fact that a high number of phases have already been reported in the ternary systems RE–TM–Mg but also that all the phases are predominantly phases rich in RE or TM.

3.2.1 RE_2TM_2Mg

To date, more than 40 intermetallic of the RE_2TM_2Mg type (RE = rare earth, TM = Cu, Ni) have been identified [39–42]. These compounds crystallize in the tetragonal structure type Mo_2FeB_2 (S.G. P4/mbm) [43]. This structure, shown in Fig. 1, for La_2Ni_2Mg compound, can be described as an assembly of trigonal prism of RE occupied by transition metal Ni (AlB_2-type block) and rare earth cubes occupied by magnesium (CsCl-type blocks).

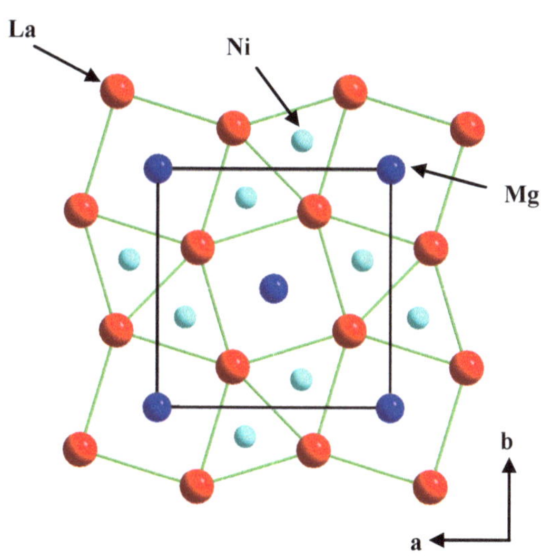

Fig. 1 Structure of La_2Ni_2Mg compound, in projection in the plane (*a*, *b*) [91]

It is worth pointing out that the magnetic behavior depends on the nature of both the RE and the TM. For example, Gd_2Ni_2Mg is antiferromagnetic (T_N = 49 K) [44], while La_2Cu_2Mg shows Pauli paramagnetism and Pr_2Cu_2Mg is ferromagnetic (T_C = 43 K) [45]. The absorption of hydrogen by some compounds was also studied. In 2006, Chotard et al. [46] showed that La_2Ni_2Mg absorbs 8 H/f.u. (i.e. 1.89 wt%) under 30 bar of hydrogen at 100 °C. The hydride formed (i.e. $La_2Ni_2MgH_8$) is very stable and no reversibility was observed.

During hydrogenation, the initial hexagonal lattice becomes monoclinic (GE $P2_1/c$, $\Delta V/V \approx 20\%$). The study of the hydride $La_2Ni_2MgD_8$ by neutron diffraction locates the sites occupied by deuterium. Deuterium (or hydrogen) form complexes with nickel $[Ni_2D_7]_7$ and $[Ni_4D_{12}]_{12}$—which ensure the stability of the hydride [46]. Such structural change has also been observed by Chevalier et al. [47] during the hydrogenation of Ce_2Ni_2Mg leading to the stable hydride $Ce_2Ni_2MgH_{7.7}$.

3.2.2 $TRMMg_2$

The compound $LaNiMg_2$ was identified by Renaudin et al. in 2003 [48]. It is the main representative of the $TRMMg_2$ family. This ternary compound crystallizes in an orthorhombic structure type $MgAl_2Cu$ (S.G. Cmcm) with lattice parameters a = 4.2266 (6) Å, b = 10.303 (1) Å and c = 8.360 (1) Å. At 200 °C and under 5 bar of hydrogen it irreversibly absorbs 7 H/f.u. (i.e. $LaNiMg_2H_7 \rightarrow$ 2.8 wt%). As previously, the absorption of hydrogen results in the formation of a monoclinic structure presented in Fig. 2 (S.G. $P2_1/c$, $\Delta V/V \approx 50\%$). Deuterium atoms are located both in isolated interstitial sites or form tetrahedral complexes $[NiD_4]_4$ (Fig. 2).

To conclude, it is interesting to note that compared to RE_2TM_2Mg systems, the compound $LaNiMg_2$, lighter, allows obtaining a hydrogen storage capacity almost twice larger but still without reversibility.

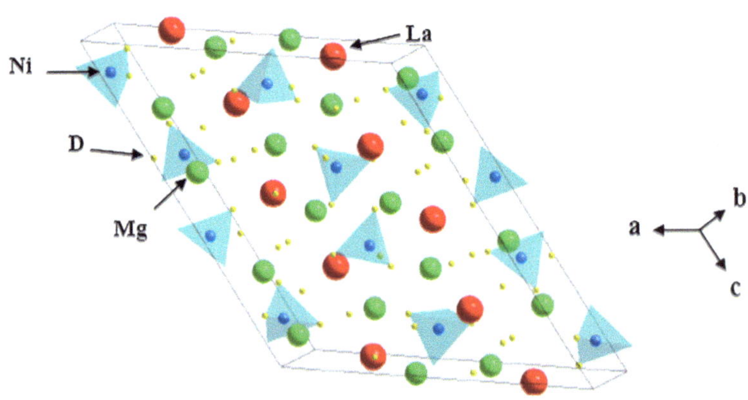

Fig. 2 Structure of the hydride $LaNiMg_2D_7$. Tetrahedra $[NiD_4]_4$ are shown in blue [48]

3.2.3 $RENi_9Mg_2$

In 1997, Kadir et al. [49] revealed this new family of compounds for RE = La, Ce, Pr, Nd, Sm, and Gd). These compounds crystallize in a rhombohedral structure type $PuNi_3$ (S.G. R-3m). This structure can be described as a stacking along the *c*-axis of AB_5 ($CaCu_5$) and AB_2 ($MgCu_2$, $MgZn_2$, $MgNi_2$)-type subunits.

At room temperature and under 30 bar of hydrogen, it absorbs hydrogen to form the hydride $LaNi_9Mg_2H_{\sim 2.3}$. The rhombohedral structure is retained with a volume expansion of about 20% [50]. The hydride formation is reversible and a specific capacity of 0.33% (H/M $\approx$ 0.2) can be reached. This low capacity was improved by replacing the rare earth or the magnesium with calcium [50, 51]. As an example, the compound $La_{0.65}Ca_{1.03}Mg_{1.32}Ni_9$ can absorb reversibly up to 1.87 wt% hydrogen (H/M $\approx$ 1.1) at room temperature. The heat of hydride formation of about −24 kJ/mol H_2 corresponds to an equilibrium pressure of about 2 bar at room temperature.

3.2.4 $RENi_4Mg$

The search for new compounds derived from AB_2–$AuBe_5$ structure type led to the discovery of compounds $RENi_4Mg$. The first compound of this family, $CeNi_4Mg$, was highlighted in 1997 by Geibel et al. [52]. Since then, numerous studies show that these compounds are stabilized for different rare earths (i.e., Gd, Y, Nd, Pr) [53–56]. They crystallize in a cubic lattice type $MgCu_4Sn$ (S.G. F-43m, C15b, $a \approx 7$ Å) [57].

Regarding the reactivity toward hydrogen, different behaviors were observed depending on the rare earth:

- $CeNi_4Mg$ does not react with hydrogen even at temperatures above 300 °C and a hydrogen pressure of 50 bar [54],
- $NdNi_4Mg$ and $LaNi_4Mg$ absorb approximately 4 H/f.u. (i.e. about 1 wt%) at room temperature and with good kinetics [53, 54]. The hydride formation induces a structural modification as the hydride crystallizes in an orthorhombic lattice (for $NdNi_4MgH_4 \rightarrow a = 5.0788$ (2) Å, $b = 5.4887$ (2) Å and $c = 7.3846$ (2) Å). The hydrogen atoms occupy a tetrahedral site [$NdNi_3$] and two bipyramidal sites [Nd_2Ni_2Mg] [53]. High temperatures are needed to destabilize the hydride,
- $GdNi_4Mg$ and YNi_4Mg absorb, respectively, 0.6% and 1.05 wt% of hydrogen reversibly at room temperature [54, 55]. During the formation of the hydride (Gd, Y)$Ni_4MgH_{\sim 2.5}$, the cubic structure is maintained with a volume expansion of about 5%. The enthalpy of formation of the hydride YNi_4Mg is −35.8 kJ/mol H_2 [55].

Within this family, $GdNi_4Mg$ and YNi_4Mg compounds appear as the most interesting for hydrogen storage. Many substitutions were tested to modify the enthalpy of formation of the hydride and/or to improve the storage capacity [58–60]. By substituting nickel by aluminum (i.e., $YNi_{4-x}Al_xMg$), the equilibrium pressure can be adjusted [59].

3.2.5 Ternary Compounds Rich in Magnesium

Considering all the ternary RE–TM–Mg, only a few rich magnesium ternary compounds are known. This is why many research teams are interested in compositions rich in magnesium ternary systems TR–M–Mg [61–67].

Most of the compositions tested lead to the production of mixtures of known phases or amorphous compounds [64–66]. For example, the compositions $Y_{20}Ni_{30}Mg_{50}$, $Y_7Ni_{30}Mg_{63}$, and $Y_{10}Cu_{25}Mg_{65}$ synthesized by Gebert et al. [66] lead to an amorphous compound and/or to a binary mixture of known compounds. Considering all the rare earth and transition metals, only a few magnesium-rich compounds have been identified. For example, we can mention: $RECuMg_4$ [68], $TRCuMg_3$ [68], $Ce_2Fe_2Mg_{15}$ [69], $YZnMg_{12}$ [67]. To date, the reactivity of these systems with hydrogen has not been studied yet.

The increase of magnesium content alleviates the compound and could, therefore, improve the maximum hydrogen sorption capacity. For example, the $LaCuMg_4$ compound (S.G. P62m, a = 10.3911 (1) Å and c = 4.5126 (1) Å) [63] which can be considered as a variant of LaCuMg with inserted $[Mg_6]$ blocks, offer many open spaces (Fig. 3) in which hydrogen could be inserted.

3.2.6 Summary of the State of the Art

To get, just in one snap, the summary of the state of the art, the typical Gibbs triangle RE–Ni–Mg with the 14 defined compounds is presented in Fig. 4.

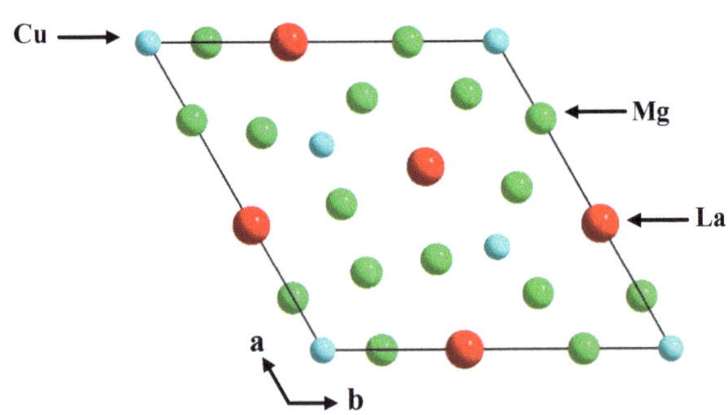

Fig. 3 Crystal Structure $LaCuMg_4$ compound [63]

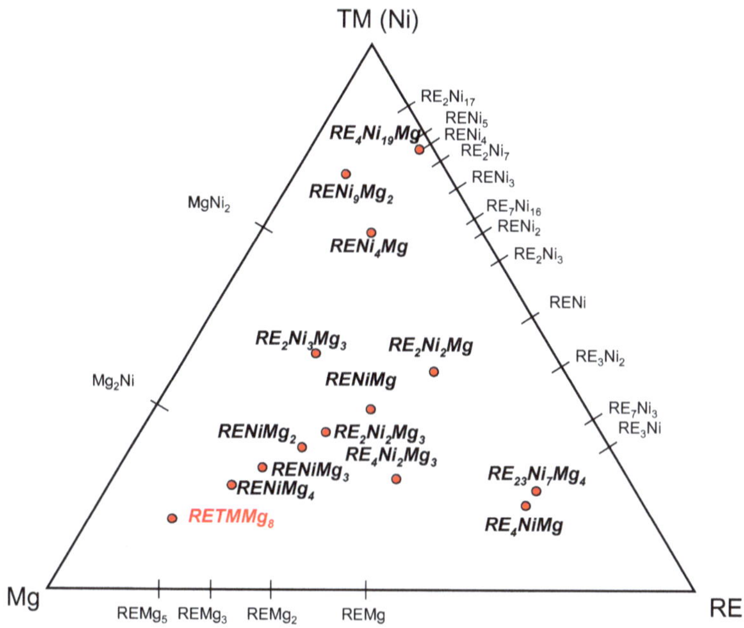

Fig. 4 Gibbs Triangle of RE–Ni–Mg, listing the various existing binary and ternary compounds. However, some compounds do not exist with all rare earths and all transition metal (e.g. $RETMMg_8$ exist for RE = La and TM = Cu)

4 The Study of a New Ternary Compound: $LaCuMg_8$

4.1 Structural Study

Numerous tests were conducted using different RE and various transition metal elements. For example, the $La_{10}\mathbf{Ni}_{10}Mg_{80}$ composition leads to a mixture of three binary compounds (La_2Mg_{17}, $LaMg_3$ and Mg_2Ni) and a ternary compound ($LaNiMg_2$). Using Cu as transition metal leads to a different result as the product is not a mixture but almost a single phase. The X-ray diffractogram of the initial composition $La_{10}\mathbf{Cu}_{10}Mg_{80}$ after fusion is given in Fig. 5.

All the diffraction peaks can be indexed with La_2Mg_{17} structure type, but with a slight shift of the peaks toward larger values of 2θ. The lattice parameters (i.e. a = 10.13 (1) Å and c = 10.10 (4) Å) show a slight contraction toward thus of La_2Mg_{17} (i.e. 2.22% along a and 1.56% along c).

The EPMA analysis of this sample allows us to highlight the existence of a single phase (but with some traces of binary Mg–Cu and magnesium and iron oxides). It is then possible to attribute the difference in lattice parameter between $LaCuMg_8$ and La_2Mg_{17} to a steric effect (R_{Cu} = 1.28 Å while R_{La} = 1.88 Å and R_{Mg} = 1.6 Å).

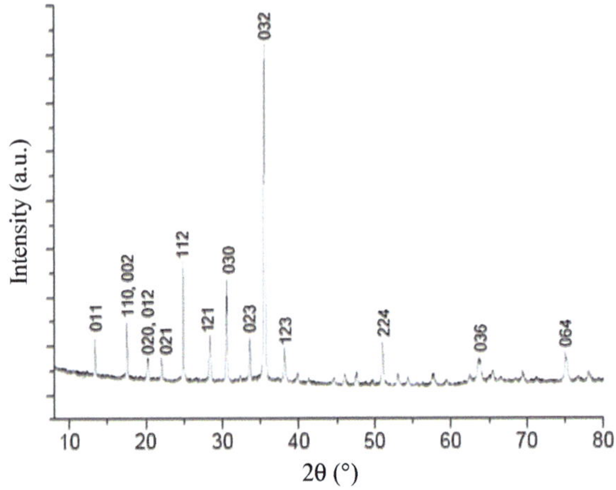

Fig. 5 X-ray diffractogram of the sample $La_{10}Cu_{10}Mg_{80}$ (then called $LaCuMg_8$) after melting. Only the most intense peaks have been indexed

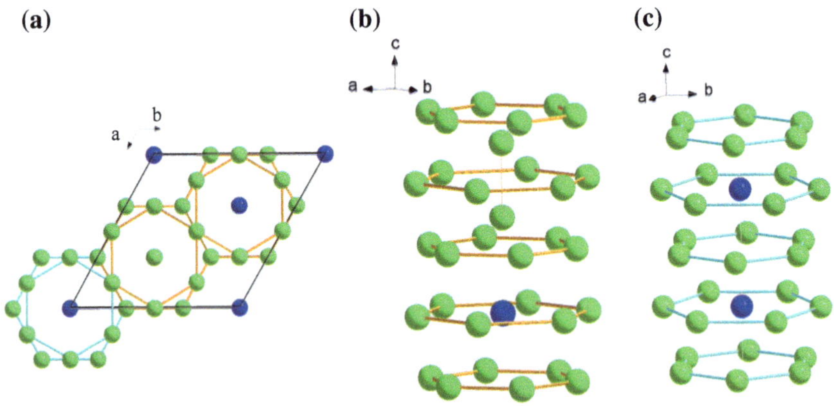

Fig. 6 La_2Mg_{17} structure type in the plane (*a*, *b*) (**a**) and along the c-axis (*b*, *c*). Two types of channels are observed along the *c*-axis: (i) one centered on (1/3, 2/3, *z*) (**b**) and (ii) the other centered on (0, 0, *z*) channel (**c**)

The structure of $LaCuMg_8$ compound was determined by X-ray single-crystal analysis. The extinction conditions observed are consistent with the space group $P6_3/mmc$ of La_2Mg_{17} structure. In this structure, the magnesium atoms occupy four crystallographic sites (i.e. Mg1, Mg2, Mg3, Mg4) and lanthanum atoms occupy two crystallographic sites (i.e. La1, La2). It can be described as a hexagonal assembly of magnesium (Mg2, Mg3, Mg4) forming channels along the *c*-axis (Fig. 6). The lanthanum atoms (La1 and La2) and Mg1 atoms are placed at the center of these channels.

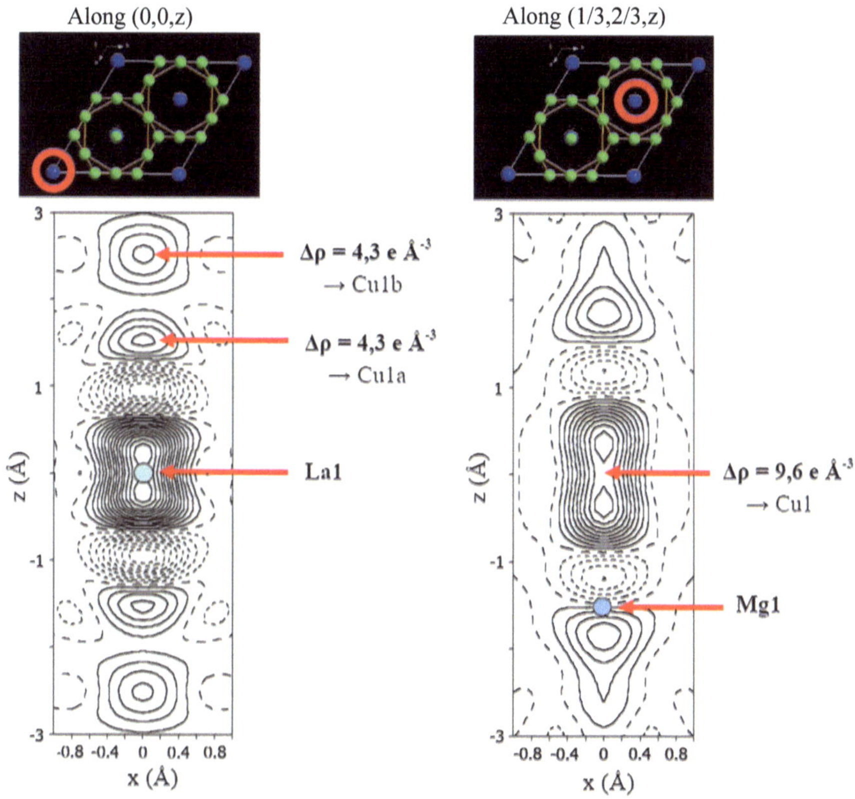

Fig. 7 Fourier Difference maps along the chain (0, 0, *z*) and (1/3, 2/3, *z*) obtained from the XRD measurement on single crystal with the initial model based on the structure type La_2Mg_{17}

The first Rietveld refinement, carried out with the atomic positions La_2Mg_{17} [13], lead to negatives isotropic atomic displacement parameters. The Fourier difference maps were studied along the *z* axis (Fig. 7) and electron densities observed on the cards (0, 0, z) and (1/3, 2/3, z) show:

- a split and a partial occupancy of the La1 position (Fig. 7 left),
- Two new sites on (0, 0, *z*), called hereafter the Cu1a and Cu1b sites (Fig. 7 left),
- a new split site in (1/3, 2/3, *z*), called hereafter the Cu1 site (Fig. 7 right).

Residual negative electron density on La1 site (i.e. new Cu1a, Cu1b and Cu1 site) was attributed to copper atoms considering the stoichiometry deduced from the EPMA analysis. The one on the Mg2 position was corrected by considering a copper–magnesium exchange (called Mg2–Cu2). The refined atomic positions, occupancy, and equivalent atomic displacement parameters of the structure $LaCuMg_8$ are given in Table 1.

Table 1 Atomic position, occupancy rate and equivalent atomic displacement for $LaCuMg_8$

Position	Wyck.	Occupation	*x*	*y*	*z*	U_{eq} ($Å^2$)
La1	4*e*	0.372 (2)	0	0	0.2196 (2)	0.0186 (7)
Cu1a	4*e*	0.081 (3)	0	0	0.098 (2)	0.026 (4)[a]
Cu1b	2*a*	0.093	0	0	0	0.026[a]
La2	2*d*	1	1/3	2/3	3/4	0.0157 (3)
Mg1	4*f*	0.911 (7)	1/3	2/3	0.1000 (4)	0.0219 (12)
Cu1	4*f*	0.089	1/3	2/3	0.2156 (15)	0.027 (7)
Mg2	6*g*	0.634 (7)	1/2	0	0	0.0218 (9)[b]
Cu2	6*g*	0.366	1/2	0	0	0.218[b]
Mg3	12*j*	1	0.3319 (3)	0.9676 (3)	1/4	0.0330 (11)
Mg4	12*k*	1	0.1687 (2)	0.3375 (3)	0.9816 (2)	0.0233 (9)

[a, b]Equivalent atomic displacement constrain at the same value

The La1 position is split and the occupancy rate decreases (according to the results of difference Fourier maps). The occupancy rate is only 37.2% (against 50% initially). Missing La1 atoms are replaced by copper atoms that are placed in Cu1a and Cu1b sites with the respective occupancy rates of 8.1 and 9.3%. In addition, much of the copper atoms are substituted for magnesium on Mg2 site. Finally, other copper atoms are placed in Cu1 position, located between two Mg1 positions around the large hexagon formed by Mg3 atoms. The occupancy rate of Cu1 positions is equal to 8.9% and the one of Mg1 position decreases to 91.1%.

From these refined positions, the composition can be rewritten as $La_{1.74}Cu_{1.53}Mg_{15.73}$ to obtain a total number of atoms of 19 (e.g. 1.74 + 1.53 + 15.73 = 19) as in La_2Mg_{17}. This new structure is, in fact, a variant of the standard structure Th_2Ni_{17} [17, 18, 70, 71]. The structural relationship between the ideal structure Th_2Ni_{17} and the $ThNi_5$ one ($CaCu_5$ type) was highlighted by Givord et al. [71]. It shows a $ThNi_5$–$ThNi_5$–Ni_2Ni_5 stacking sequence (i.e. Th_2Ni_{17}). Then, the chemical formula of the compound $La_{1.74}Cu_{1.53}Mg_{15.73}$ can be rewritten as $(La_{1.74}Cu_{0.25})$ $(Cu_{1.28}Mg_{15.73})$ that is to say A_2B_{17}.

The structure of $LaCuMg_8$ compound is given in Fig. 8. Its comparison with the structure La_2Mg_{17} shows a crystallographic disorder induced by replacement of lanthanum and magnesium by copper. A similar effect was observed by Johnson and Smith replacing the lanthanum by cerium in La_2Mg_{17} compound, to form the compound $(Ce_{1.71}Mg_{0.61})$ $Mg_{16.97}$ [72].

For some compounds crystallizing in Th_2Ni_{17} structure type, it is possible to insert light elements such as carbon and hydrogen [73–75]. For all the compounds, the insert element goes into the 6 h site (x, $2x$, ¼), where x is approximately equal to 0.82. It means that the atom inserted in this position is located in distorted octahedra formed by a square of two Mg3 atoms and two Mg4 atoms with La1 and La2 atom in the apical position.

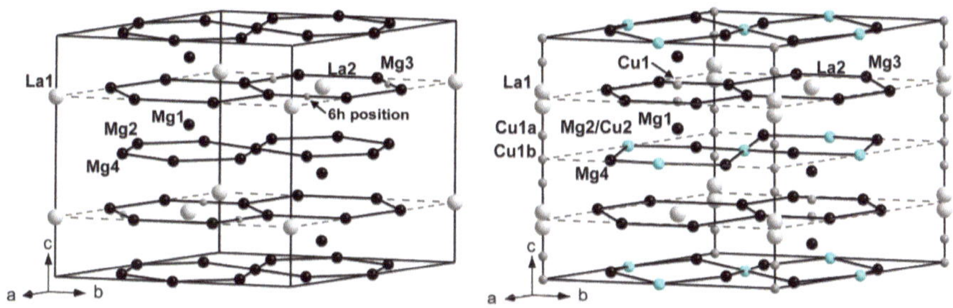

Fig. 8 La_2Mg_{17} (*left*) and $LaCuMg_8$ (*right*) crystal structures. The magnesium network, similar to that of graphite, is highlighted on the figure by links between magnesium atoms

For $Ho_2Fe_{17}H_x$ and $Dy_2Fe_{17}H_x$ compounds, the average distances Fe–H are respectively equal to 1.88 and 1.9 Å and the distances Ho–H and Dy–H are equal to 2.46 and 2.47 Å [73, 74]. If hydrogen is placed hypothetically on the 6 h position in La_2Mg_{17}, the distances Mg–H and La–H would be respectively 2.15 and 2.99 Å. By considering a slight increase of the volume of the structure caused by hydrogenation, these distances are consistent with the metallic radii of the atoms. Moreover, it is also consistent with the distances observed in the binary hydrides of lanthanum and magnesium [76, 77].

From a structural point of view, this approach shows that the inclusion of small heteroatoms such as hydrogen could be possible in La_2Mg_{17} structure and therefore, possible in $La_{1.74}Cu_{1.53}Mg_{15.73}$ structure. To validate this, hydrogenation of the compound $LaCuMg_8$ is studied in detail later.

4.2 *Hydrogen Sorption Study*

The compound La_2Mg_{17} absorbs 17 H/f.u. [17, 78–81] reversibly. Absorption occurs at relatively low temperature (i.e. minimum of 50 °C), while the desorption occurs at higher temperatures (about 300 °C).

Despite some disagreements on the values of the reversible capacity, the absorption mechanism is always the same: during the first absorption, La_2Mg_{17} decomposes irreversibly to form LaH_3 and MgH_2. Then, reversibility is observed for the couple Mg/MgH_2 [17, 78–82].

Gross et al. [31] showed that the absorption kinetics can be improved by a factor of 10, with the addition of $LaNi_5$ (40 wt%). La_2Mg_{17} compound has also been studied for its electrochemical properties (Ni–MH battery) [83, 84]. A discharge capacity of 850 mAh/g (La_2Mg_{17}) was obtained for an electrode composed of La_2Mg_{17} + 200 wt% Ni (e.g. 300 mAh/g for AB_5-type electrode). However, this capacity decreases sharply upon cycling and, after 10 cycles, the capacity decrease to 500 mAh/g and remained stable thereafter.

During the first absorption, the $LaCuMg_8$ compound absorbs slowly (e.g. 27 h for full absorption) hydrogen from 300 °C and reaches a capacity of around 3.2 wt%. This can be considered as an activation step. The analysis by X-ray diffraction reveals that as for La_2Mg_{17}, this activation involves decomposition. The first hydrogenation reaction is:

$$LaCuMg_8 + 9H_2 \rightarrow LaH_3 + 1/2MgCu_2 + 15/2MgH_2$$

Thereafter, the mixture of the three phases [i.e. LaH_3, $MgCu_2$ (or Mg_2Cu) and MgH_2 (or Mg)] will be named "$LaCuMg_8$ activated powder". A Rietveld refinement of the XRD diffractogram allows us to check that the weight percentages of each phase are in good agreement with theory. This also shows that the reaction is complete.

During the first desorption, three phases appear: LaH_{3-x}, Mg, and Mg_2Cu. Subsequent cycles will result in the absorption/desorption of hydrogen by magnesium as in the case of La_2Mg_{17}. Analysis by X-ray diffraction performed after each absorption cycle allows us to highlight a significant difference with the case of La_2Mg_{17}: depending on the absorption temperature, Mg_2Cu or $MgCu_2$ phases are formed. Thus, two reactions can appear upon cycling:

At $T > 200$ °C

$$LaH_{3-x} + Mg_2Cu + 6Mg + (15 + x)/2H_2 \leftrightarrow LaH_3 + 1/2MgCu_2 + 15/2MgH_2$$

And, at $T < 200$ °C

$$LaH_{3-x} + Mg_2Cu + 6Mg + (6 + x)H_2 \leftrightarrow LaH_3 + Mg_2Cu + 6MgH_2$$

Absorption curves at various temperatures are presented in Fig. 9.

This transformation $Mg_2Cu/MgCu_2$ is consistent with the work of Reilly [85], which show that Mg_2Cu reacts with hydrogen at high temperature (≈300 °C) to form MgH_2 and $MgCu_2$ phase.

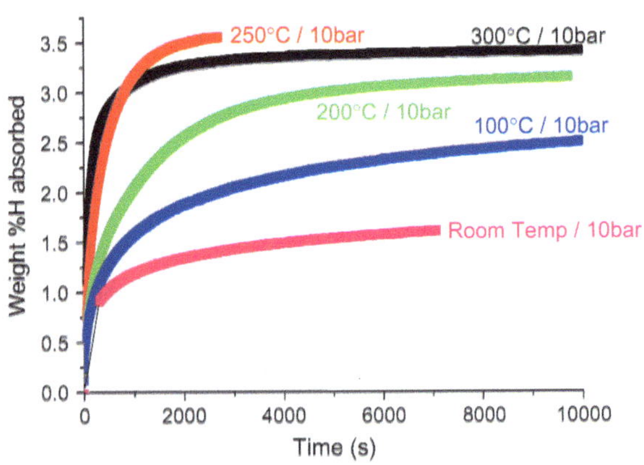

Fig. 9 Absorption kinetics at different temperatures under 10 bar of hydrogen of "$LaCuMg_8$ activated powder"

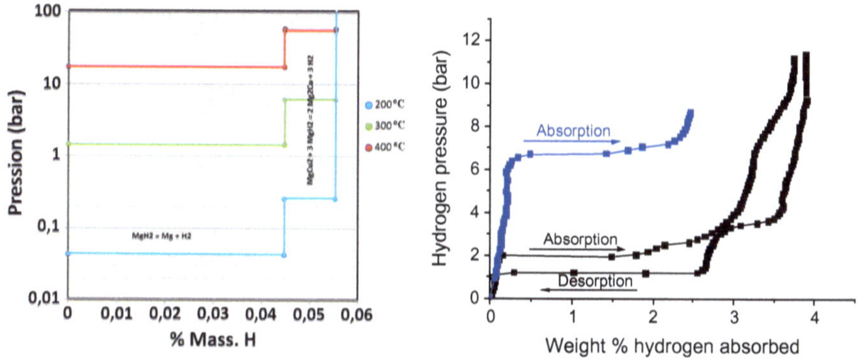

Fig. 10 Theoretical PCT curves (*left*) of the reaction with hydrogen of a mixture 6 Mg + Mg_2Cu at 200, 300 and 400 °C and experimental PCT (*right*) obtained in the case of "LaCuMg8 activated powder"

Also, note that the desorption rates are much higher than those obtained for pure magnesium and are comparable to those obtained for the ground magnesium [86].

Thus, the phase mixture obtained by decomposition of $LaCuMg_8$ makes possible the absorption and the desorption of hydrogen by magnesium with good kinetics and lowered temperatures.

Thermodynamic calculations (Redlich–Kister model and Muggianu formalism) made from the literature data of Cu–Mg [87] and Mg–H systems [88] show two successive plateaus (Fig. 10): (i) the first one at low pressure (e.g. ≈1 bar at 300 °C) corresponding to the hydrogenation reaction of magnesium and (ii) the second one at higher pressure (e.g. ≈6 bar at 300 °C), corresponding to the reaction of Mg_2Cu with hydrogen ($3Mg_2Cu + 3H_2 \rightarrow 3MgH_2 + MgCu_2$). Experimental PCT curves (Fig. 10) performed at 300 and 350 °C confirm this result by highlighting two plateaus (especially at 350 °C).

As shown in Fig. 10, there is a good agreement between experiment and theory. The difference can be attributed to a competition between kinetics and thermodynamics. Indeed, it should be emphasized the catalytic role of Mg_2Cu and/or $MgCu_2$ and also the role played by the transformation $Mg_2Cu/MgCu_2$. The catalytic effect of Mg_2Cu phase (or $MgCu_2$ for the desorption) has already been demonstrated in other studies [89–91]. This phase is assumed to limit the surface oxidation of Mg facilitating the dissociation and hydrogen diffusion.

In situ analysis by synchrotron of hydrogenation and dehydrogenation reactions showed the importance of the transformation reaction $Mg_2Cu/MgCu_2$. By releasing or consuming energy, it facilitates the heat exchange and reduces the reaction temperatures [44].

It is important to note that the absorption kinetics of "$LaCuMg_8$ activated powder" are superior to those of the mixture Mg + Mg_2Cu reported by Andreasen et al. [89]. The improvements of the kinetics are attributable to (i) the decomposition which allows to obtain a more reactive mixture towards H_2, (ii) the presence of LaH_3 phase (H trap) and (iii) to the morphology of the powder. Indeed, $LaCuMg_8$ activated powder consists of an intimate mixture at the micron-scale of three compounds (MgH_2, Mg_2Cu or $MgCu_2$, and LaH_3). Au et al. [92] and Karty et al. [91] suggest that the atomic hydrogen diffusion is better at the interphase Mg/Mg_2Cu. Then, the intimate mixture will increase the Mg/Mg_2Cu interfaces and thus will facilitate the diffusion of hydrogen.

To get a better understanding of the absorption mechanisms, the absorption kinetics was studied using kinetic laws [93].

The kinetic study of hydrogenation and dehydrogenation of magnesium has been the subject of several publications that show that the mechanism involved is relatively simple and can be described using the Avrami-Erofeev equation [94, 95]:

$$F = 1 - \exp\left(-kt^n\right),$$

where F is the fraction of hydrogen absorbed, k the reaction constant at the considered temperature, and n a constant linked to the mechanism of the reaction. This equation can be rewritten as:

$$\ln[-\ln(1 - F)] = \ln k + n \ln t$$

The plot $\ln(-\ln(1 - F)) = f(\ln(t))$ corresponds to a line with a slope n whose intercept depends on the rate constant k. If k is determined at various temperatures, the activation energy of the sample can be obtained (from an Arrhenius plot). To obtain the best fitting, only the domain $0.2 \le F \le 0.8$ is considered [94, 95].

For each absorption temperature, a straight line was obtained. For example, the lines drawn from the experimental data at 300 and 100 °C are shown in Fig. 11. The values of k and n determined by considering a linear regression are shown in Table 2.

The activation energy for absorption determined (using the Arrhenius law with different values of k at different T) is about 65 kJ/mol H_2 (Fig. 12).

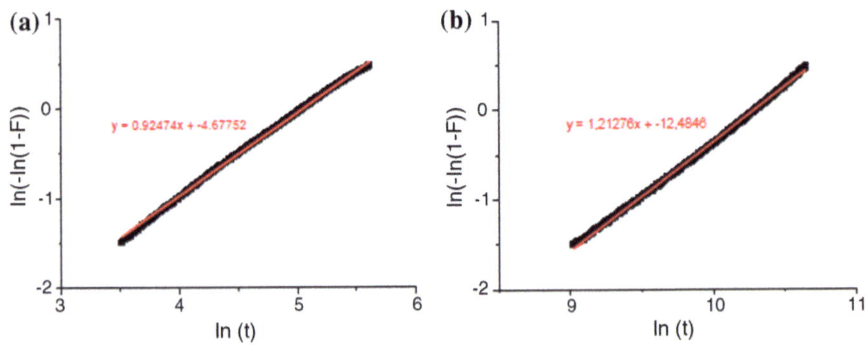

Fig. 11 Plots $\ln(-\ln(1 - F)) = \ln(t)$ on absorption for the "$LaCuMg_8$ activated powder" at **a** 300 °C and **b** 100 °C

Table 2 Rate constant (k) and n values, determined using the Erofeev-Avrami model, at different temperature, for absorption of the "$LaCuMg_8$ activated powder" in the area $0.2 \leq F \leq 0.8$

Température d'absorption (°C)	Constante n	Constante de réaction (k)
300	0.92	9.3×10^{-3}
250	1.08	1.2×10^{-3}
200	1.04	4.2×10^{-4}
150	1.06	7.1×10^{-5}
100	1.21	3.8×10^{-6}

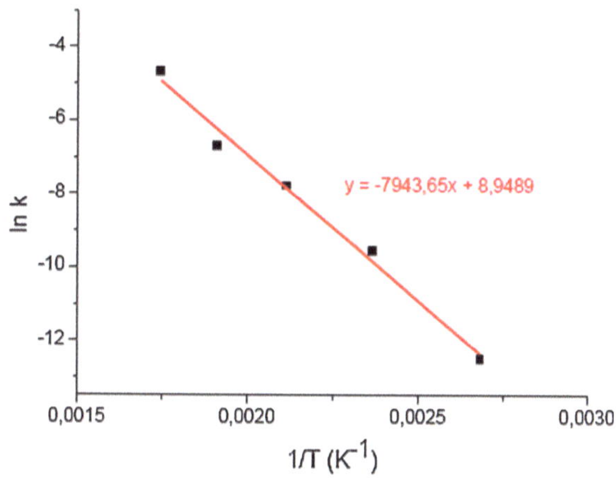

Fig. 12 Determination of the activation energy for the hydrogenation of the "$LaCuMg_8$ activated powder"

The activation energy obtained for $LaCuMg_8$ compound is greatly reduced compared to that of pure magnesium (i.e. 110–120 kJ/mol H_2) [96]. To date, the minimum activation energy values for magnesium reported in the literature are between 50 and 60 kJ/mol H_2 [95, 97]. Indeed, Bobet et al. [95] show that the doping with magnesium nano Cr_2O_3 leads to an activation energy of 51 kJ/mol H_2. Thus, for $LaCuMg_8$ system, an activation energy of the same order of magnitude is achieved but without any mechanical grinding.

4.3 *Optimization of the Sorption Properties—Role of Severe Plastic Deformation*

In 2001, Zhang et al. [98, 99] showed that the cold rolling of the compound Ti–22Al–27Nb (at.%) allows an improvement of the sorption kinetics. The improvement is explained by the appearance of dislocations and an optimum deformation rate.

These results have been confirmed by the team of Prof. Huot. They show that the cold rolling can improve: (i) the absorption kinetics, (ii) the activation, and (iii) the air resistance compared to samples obtained by grinding [100, 101].

The sorption properties of $LaCuMg_8$ cold rolled have been studied. It turns out that cold rolling improves significantly the activation process. Indeed, after 20 passes, activation (i.e. decomposition) is complete in only 3 h (i.e. nine times faster than the genuine sample). It is worth pointing out that after 150 passes, the activation is also improved but if the subsequent kinetics of absorption remains the same, the storage capacity is divided by three. This sharp decrease can be explained by the presence of oxides, in which rate increases progressively with rolling (cold rolling is done in air). The oxides amount becomes so important that it becomes easily detectable by XRD. Improved activation for cold rolled sample is explained by the increase of microstrain as well as by the decrease of the crystallites size.

However, it is important to note that the positive effect of cold rolling exists only for the first absorption (i.e. activation). The decomposition observed during the first cycle eliminates dislocations and defects induced by rolling. The gradual elimination of the effects of rolling highlighted by Zhang et al. [98, 99] is here exacerbated by the decomposition of the material (e.g. faster strain relaxation).

5 A Brief Presentation of Two Other New Ternary Alloys Based on Magnesium

5.1 The $Gd_{13}Ni_{9.5}Mg_{77.5}$ Compounds

This compound has a small domain of existence around the original composition. However, the solubility limits are not known precisely. It is also important to note that despite a complete optimization of the synthesis parameters, this compound is never perfectly pure. Indeed, a second phase with a composition close to the first one is often observed. A typical X-ray diffractogram can be indexed with a face-centered cubic lattice with a parameter of about 4.55 Å. Note that this experimental lattice parameter is consistent with the one determined using the geometric relationships of the cubic lattice and radii of atoms ($r_{average}$ = 1.59 Å = 13 * r_{Gd} + 9.5 r_{Ni} + 77.5 r_{Mg}). However, some peaks are not indexed with this structure. In addition, for close initial compositions, XRD is similar while EPMA analyzes present significant differences.

Due to the low crystallinity of the samples, it is necessary to use transmission electron microscopy to obtain information on the crystal structure. Some diffraction patterns are shown in Fig. 13.

A fourfold axis (Fig. 13a) and threefold axis (Fig. 13b) were observed indicating the existence of a cubic lattice. The most intense diffraction spots can be indexed considering a cubic lattice with a $\approx$ 4.55 Å. The similarity observed between

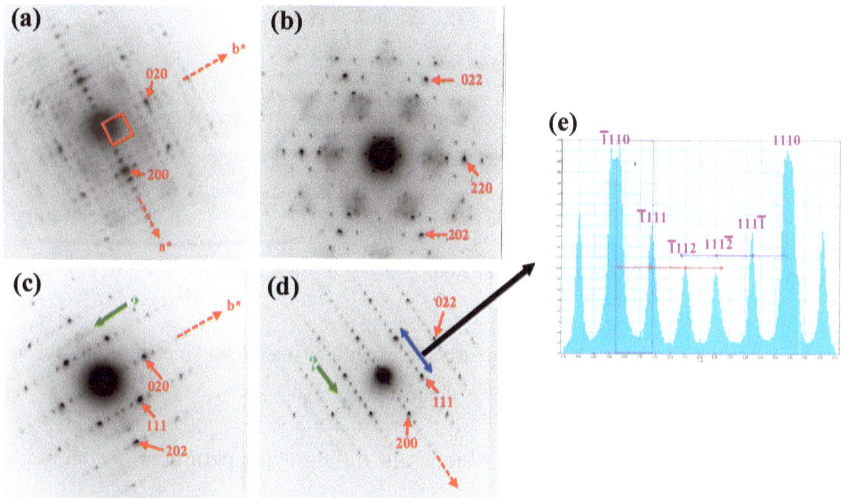

Fig. 13 Electron diffraction patterns of $Gd_{13}Ni_{9.5}Mg_{77.5}$. On figure **e**, evolution of the intensity of each diffraction spot along the *a** axis and indexation of diffraction spots, considering a modulated incommensurate structure vector of 0.42 *a**

Fig. 13c, d, respectively rotated along the axis *a** and *b** with a rotation angle in accordance with cubic symmetry, provides further evidence of the cubic symmetry.

However, many diffraction spots can not be indexed by considering the simple cubic lattice. The introduction of a five-order superstructure, in the (a, b) plane, allows to index all the diffraction spots but with some slight offsets. A more detailed analysis (i.e., the evolution of the intensity along one direction, Fig. 13e) shows that the superlattice spots are not equidistant. It is, therefore, necessary to introduce a modulation vector considering the average lattice. A modulation vector $q1 = 0.42$ a* allows a complete indexation (Fig. 13).

As for the compound $LaCuMg_8$, the compound $Gd_{13}Ni_{9.5}Mg_{77.5}$ slowly absorbs hydrogen (e.g. 3wt%.) at a temperature of 330 °C, and this absorption causes a decomposition according to the following reaction:

$$Gd_{13}N_{9.5}Mg_{77.5} + 90.5H_2 \rightarrow 13GdH_2 + 9.5Mg_2NiH_4 + 58.5MgH_2$$

The Rietveld refinement of the X-ray diffractogram of the activated sample confirms the reaction mechanism and proves that the reaction is complete (i.e. wt% of each phase determined by XRD is consistent with the theoretical ones). After decomposition, the mixture absorbs hydrogen reversibly according to the following reaction mechanism:

$$13GdH_2 + 9.5Mg_2NiH_4 + 58.5MgH_2 \leftrightarrow 13GdH_2 + 9.5Mg_2Ni + 58.5Mg + 77.5H_2$$

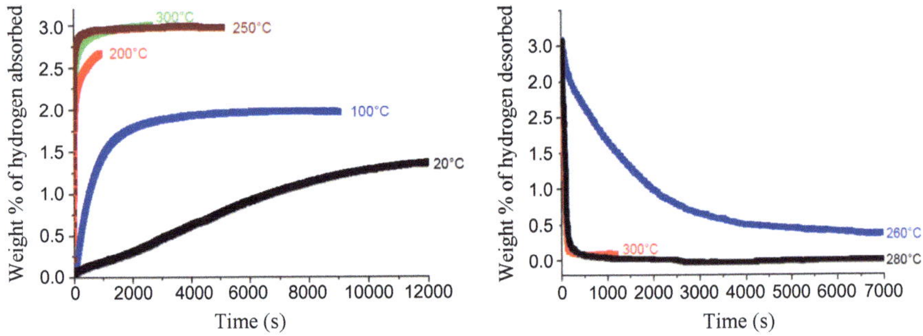

Fig. 14 Absorption at 10 bar (*left*) and desorption under 0.2 bars (*right*) kinetics of the activated $Gd_{13}Ni_{9.5}Mg_{77.5}$

The absorption and desorption kinetics at different temperatures are shown in Fig. 14.

The activated $Gd_{13}Ni_{9.5}Mg_{77.5}$ powder absorbs hydrogen at room temperature. However, the decrease in temperature causes a decrease of absorption capacity (i.e. partial reaction) and a decrease in absorption kinetics. For example, at 20 °C, a weight capacity of 1.3% is obtained in about 4 h. The desorption kinetics shows complete reversibility at a temperature as low as 260 °C, with good kinetics.

The mixture obtained by the decomposition of $Gd_{13}Ni_{9.5}Mg_{77.5}$ allows both, obtaining good sorption kinetics, and also a decrease of the desorption temperature of hydride MgH_2. The results obtained for this mixture are in agreement with those of Zaluska et al. [102] and Li et al. [103]. Indeed, for a mixture MgH_2 + Mg_2NiH_4 ground in the mass proportions 65–35%, Zaluska et al. get excellent kinetic and desorption temperature lowered of 100 °C compared to MgH_2. They explain this improvement by the presence of Mg_2NiH_4. Similarly, for the Mg + 20 wt% Ni–Y mixture milled 40 h, Li et al. get good kinetics attributed to the catalytic effect of both YH_3 and Mg_2NiH_4.

The behavior of the compound in hydrogen $Gd_{13}Ni_{9.5}Mg_{77.5}$ is interesting and it is the same for its physical properties. For example, the measurement of electrical resistivity is shown in Fig. 15. The values are compared to that of pure magnesium (to verify the influence of the addition of Ni and Gd) and to a powder mixture composed of (13 at.%), Gd (9.5 at.%) Ni and (77.5 at.%) Mg. A very similar electrical behavior for pure Mg and the physical mixture (13 at.%) Gd (9.5 at.%) Ni (77.5 at.%) is highlighted as expected. However, for the compound $Gd_{13}Ni_{9.5}Mg_{77.5}$, the behavior is very different. Its conductivity is greatly decreased compared to the other two samples. In addition, it is the only compound for which a change in behavior at 40 K is observed. Thus, the compound $Gd_{13}Ni_{9.5}Mg_{77.5}$ presents original electrical properties as above 40 K, it follows a semi-metallic behavior. This transition at 40 K is in agreement with measurements of magnetic properties (Fig. 15). This shows that this compound might have applications other than those related to hydrogen storage.

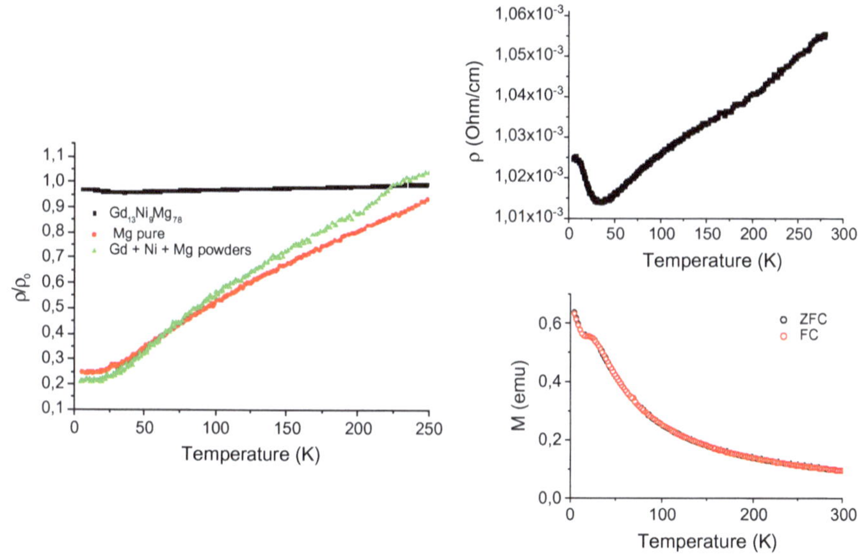

Fig. 15 Relative resistivity as a function of temperature for (i) $Gd_{13}Ni_{9.5}Mg_{77.5}$ (*black curve*), (ii) pure magnesium (*red curve*) and (iii) mixture of powder of Gd, Ni, Mg (*green curve*). On the *right*, a zoom of the relative resistivity of $Gd_{13}Ni_{9.5}Mg_{77.5}$ is shown as well as the magnetization as a function of temperature at 30 kOe

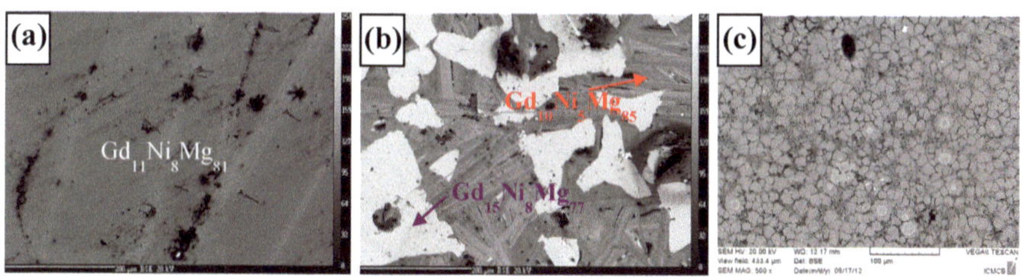

Fig. 16 Morphologies of the mixture $Gd_{13}Ni_9Mg_{78}$ after cooling **a** at 2 K/min, **b** at room temperature and **c** by copper mold casting

The mechanical properties of this compound can be also briefly discussed. First of all, it is important to remind that the hardness of the magnesium is 45 HV. As shown in Fig. 16, the same initial mixture leads to several morphologies depending on the cooling rate (Fig. 16).

For each sample, an important difference of the Vickers Hardness was observed. The sample cooled at 2 K/min generally shows a homogeneous microstructure. Nevertheless, the hardness ranges from 100 to 140 HV. The room temperature quenched compound exhibits two phases with a close composition (and the same crystal structure) but with a very important difference of the hardness. Indeed, the hardness is respectively 260 HV for the $Gd_{15}Ni_8Mg_{77}$ (white phase on the

Fig. 16b) phase and 100 HV for the $Gd_{10}Ni_5Mg_{85}$ (gray phase on the Fig. 16b) phase. This compound could be described as an in situ composite. The cooper mold casting sample shows little spherical grains (around 10 μm) with a grain boundary surrounding. Both compositions are really close to the one of the room temperature quenched compound. An average hardness of 250 HV was measured for this compound. Thus, changing the cooling rate it is possible to adjust and control the mechanical properties of the compound $Gd_{13}Ni_9Mg_{78}$.

5.2 *The $NdNiMg_5$ Compounds*

The structure of this compound is original because it is made of Mg blocks stacked along the c-axis (Fig. 17). These blocks are separated by NiNd layers. The Mg blocks are made of a close-packed array of magnesium atoms with Mg–Mg distances ranging from 3.108 to 3.223 Å. These distances are close to the average Mg–Mg distance of 3.203 Å in Mg metal [104]. The connectivity of the empty $[Mg_4]$ tetrahedra and $[Mg_6]$ octahedra corresponds to the one observed in fcc structure.

This structure presents similarities with metallic magnesium and offers a lot of free space (to be filled with H). The hydrogen absorption should be possible and relatively easy. It appears that under 10 bar of hydrogen, absorption occurs at 300 °C. As for the other two compounds presented previously, the hydrogen absorption leads to decomposition of the compound according to the reaction:

$$NdNiMg_5 + (12.2/2)H_2 \rightarrow NdH_{2.2} + Mg_2NiH_4 + 3MgH_2$$

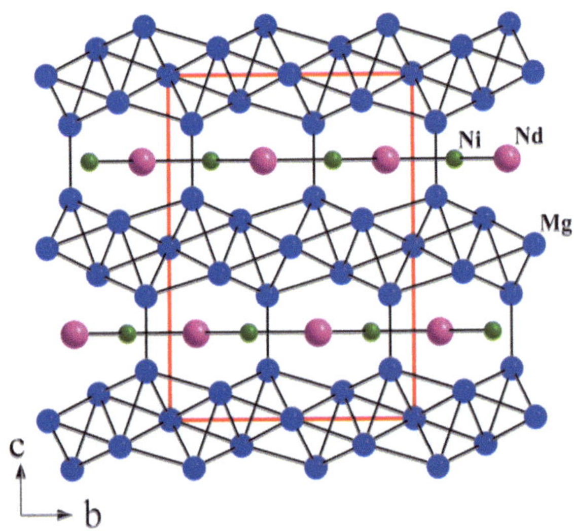

Fig. 17 View along the a axis of the structure of $NdNiMg_5$. The Mg–Mg and Ni–Nd bonds are drawn

The hydrogen sorption behavior of this compound should then be very close to the behavior of the compound $Gd_{13}Ni_{9.5}Mg_{77.5}$.

6 Conclusion

In this chapter, we introduced three new phases in the RE–TM–Mg system. The "fishing approach" used appears as a great way to discover new magnesium-rich phases. The $LaNiMg_8$ crystallizes in a structure derived from that of La_2Mg_{17} with a structural disorder. The compound $Gd_{13}Ni_{9.5}Mg_{77.5}$ crystallizes with a cubic structure, and we highlighted a modulation in this structure. Finally, the compound $NdNiMg_5$ crystallizes within a new structure in which we highlighted a 2D structure that will probably lead to original physical properties.

Under hydrogen, these three compounds decompose into (i) binary rare earth hydrides, (ii) ternary hydrides Mg–TM–H, and (iii) magnesium hydride. This decomposition is in good agreement with the model of Miedema. After decomposition, the intimate mixture thus obtained can be cycled. Sorption properties obtained are comparable or even higher than the one of ball milled magnesium. However, our intimate mixtures are obtained without lengthy and expensive mechanical action (both in time and energy). To change the stability of these compounds under hydrogen, it could be possible to make solid solutions by partial replacement Ni/Al, Mg/Ca or by using mixtures of rare earth. Such solid solution could help stabilized the quaternary hydrides at the expense of binary hydrides. Finally, it is important to note that if only three new compounds have been presented in this chapter, more than 10 new phases have been identified and are under study.

The semi-metallic behavior of $Gd_{13}Ni_{9.5}Mg_{77.5}$ also offers interesting prospects. More generally, it is clear that the study of the electrical and mechanical properties of these compounds is expected to highlight new areas of application other than the storage of hydrogen.

References

1. Wagemans RWP, Van Lenthe JH, De Jongh PE, Van Dillen AJ, De Jong KP (2005) Hydrogen storage in magnesium clusters: quantum chemical study. J Am Chem Soc 127 (47):16675–16680
2. Vajo JJ (2011) Influence of nano-confinement on the thermodynamics and dehydrogenation kinetics of metal hydrides. Curr Opin Solid State Mater Sci 15:52–61
3. Jeon K-J, Moon HR, Ruminski AM, Jiang B, Kisielowski C, Bardhan R, Jeffrey J (2011) Urban, air-stable magnesium nanocomposites provide rapid and high-capacity hydrogen storage without using heavy-metal catalysts. Nat Mater 10:286–290
4. Barcelo S, Rogers M, Grigoropoulos CP, Mao SS (2010) Hydrogen storage property of sandwiched magnesium hydride nanoparticle thin film. Int J Hydrog Energy 5:7232–7235

5. Zhang X, Yang R, Yang J, Zhao W, Zheng J, Tian W, Li X (2011) Synthesis of magnesium nanoparticles with superior hydrogen storage properties by acetylene plasma metal reaction. Int J Hydrog Energy 36:4967–4975
6. Adelhelm P, de Jongh PE (2011) The impact of carbon materials on the hydrogen storage properties of light metal hydrides. J Mater Chem 21:2417–2427
7. Fichtner M (2009) Properties of nanoscale metal hydrides. Nanotechnology 20:204–209
8. Wang CX, Yang GW (2005) Thermodynamics of metastable phase nucleation at the nanoscale. Mater Sci Eng R Rep 49:157–202
9. Reardon H, Hanlon JM, Hughes RW, Godula-Jopek A, Mandal TK, Gregory DH (2012) Emerging concepts in solid-state hydrogen storage: the role of nano-materials design. Energy Environ Sci 5:5951–5979
10. Blanco AAG, de Oliveira JCA, Lopez R, Moreno-Pirajan JC, Giraldo L, Zgrablich G, Sapag K (2010) A study of the pore size distribution for activated carbon monoliths and their relationship with the storage of methane and hydrogen. Colloids Surf A Physicochem Eng Aspects 357:74–83
11. Kim KC, Dai B, Johnson JK, Sholl DS (2009) Assessing nanoparticle size effects on metal hydride thermodynamics using the Wulff construction. Nanotechnology 20:204–211
12. Hwang YK, Hong D-Y, Chang J-S, Jhung SH, Seo Y-K, Kim J, Vimont A, Daturi M, Serre C, Férey G (2008) Amine grafting on coordinatively unsaturated metal centers of mofs: consequences for catalysis and metal encapsulation. Angew Chem Int Ed 47:4144–4148
13. Zlotea C, Latroche M (2012) Role of nanoconfinement on hydrogen sorption properties of metal nanoparticles hybrids. Colloids Surf A Physicochem Eng Aspects 439(December 20):117–130
14. Fierro V, Szczurek A, Zlotea C, Mareche JF, Izquierdo MT, Albiniak A, Latroche M, Furdin G, Celzard A (2010) Experimental evidence of an upper limit for hydrogen storage at 77 K on activated carbons. Carbon 48:1902–1911
15. Loidl A, Knorr K, Müllner M, Buschow KHJ (1981) Magnetic properties of some rare earth magnesium compounds RMg2. J Appl Phys 52(3):1433–1438
16. Buschow KHJ (1976) Magnetic properties of some rare earth magnesium compounds RMg3. J Less-Common Metals 44:301–306
17. Darriet B, Pezat M, Hbika A, Hagenmuller P (1980) Application of magnesium rich rare-earth alloys to hydrogen storage. Int J Hydrog Energy 5:173–178
18. Fornasini ML, Manfrinetti P (1986) GdMg5: a complex structure with a large cubic cell. Acta crystallographica C 42:138–141
19. Evdokimenko VI, Kripyakevich PI (1940) Ueber die loeslickeit von lanthan in aluminium, magnesium und den homogenen legierungen des magnesiums und aluminiums. Zeitschrift für angewandte chemie 46(6):357–364
20. Janot R, Cuevas F, Latroche M, Percheron-Guégan A (2006) Influence of crystallinity on the structural and hydrogenation properties of Mg2X phases (X = Ni, Si, Ge, Sn). Intermetallics 14:163–169
21. Buschow KHJ, Bouten PCP, Miedema AR (1982) Hydrides formed from intermetallic compounds of two transition metals: a special class of ternary alloys. Rep Prog Phys 45:937–1039
22. Latroche M (2004) Structural and thermodynamic properties of metallic hydrides used for energy storage. J Phys Chem Solids 65:517–522
23. Latroche M, Percheron-guégan A (2005) Hydrogen storage properties of metallic hydrides. Annales de chimie science des matériaux 30(5):471–482
24. Govind, Suseelan Nair K, Mittal MC, Lal K, Mahanti RK, Sivaramakrishnan CS (2001) Development of rapidly solidified (RS) magnesium–aluminium–zinc alloy. Mater Sci Eng A 304–306:520–523
25. Pettersen G, Westengen H, Hoier R, Lohne O (1996) Microstructure of a pressure die cast magnesium—4wt.% aluminium alloy modified with rare earth additions. Mater Sci Eng A207:115–120

26. Peng Q, Hou X, Wang L, Wu Y, Cao Z, Wang L (2009) Microstructure and mechanical properties of high performance Mg–Gd based alloys. Mater Des 30:292–296
27. Inoue A (1998) Amorphous, nanoquasicrystalline and nanocrystalline alloys in Al-based systems. Prog Mater Sci 43:365–520
28. Petricek V, Dusek M, Palatinus L (2006). Jana2006. The crystallographic computing system. Institute of Physics, Praha, Czech Republic
29. Couillaud S, Gaudin E, Bobet JL (2011) Rich magnesium ternary compound so-called $LaCuMg_8$ derived from La_2Mg_{17}. Structure and hydrogenation behavior. Intermetallics 19:336–341
30. Schulz R, Boily S, Huot J (1999) Apparatus for titration and circulation of gases and circulation of an absorbent or adsorbent substance. Patent 09/424,331
31. Gross K, Chartouni D, Leroy E, Zuttel A, Schlapbach L (1998) Mechanically milled Mg composites for hydrogen storage: the relationship between morphology and kinetics. J Alloy Compd 269:259–270
32. Schubert K, Anderko K (1951) Kristallstruktur von NiMg2 und AuMg2. Naturwissenschaften 38(11):259
33. Noreus D (1985) Structurally related phenomena in Mg2NiH4. Chemica Scripta 26A:103–106
34. Noreus D, Werner P-E (1982) Structural studies of hexagonal Mg2NiHx. Acta Chemical Scandinavica A36:847–851
35. Darnaudery JP, Pezat M, Darriet B (1983) Influence de la substitution du cuivre au nickel dans Mg2Ni sur le stockage de l'hydrogène. Journal Less-Common Metals 92:199–205
36. Zaluski L, Zaluska A, Ström-Olsen JO (1995) Hydrogen absorption in nanocrystalline Mg2Ni formed by mechanical alloying. J Alloy Compd 217:245–249
37. Abdellaoui M, Cracco D, Percheron-Guegan A (1998) Structural characterization and reversible hydrogen absorption properties of Mg Ni rich nanocomposite materials synthesized by mechanical alloying. J Alloy Compd 268:233–240
38. Selvam P, Viswanathan B, Swamy CS, Srinivasan V (1988) Studies on the thermal characteristics of hydrides of Mg, Mg2Ni, Mg2Cu and Mg2Ni1-xMx (M = Fe Co, Cu, or Zn; $0 < x < 1$) alloys. Int J Hydrog Energy 13(2):87–94
39. Latka K, Kmiec R, Pacyna AW, Mishra R, Pöttgen R (2001) Magnetism and hyperfine interactions in Gd2Ni2Mg. Solid State Sci 3:545–558
40. Pöttgen R, Fugmann A, Rodewald UC, Niepmann D (2000) Intermetallic cerium compounds with ordered U3Si2 type structure. Zeitschrift naturforschung 55b:155–161
41. Mishra R, Hoffmann RD, Pöttgen R (2001) New magnesium compounds RE2Cu2Mg (RE = Y, La-Nd, Sm, Gd-Tm, Lu) with Mo2FeB2 type structure. Zeitschrift naturforschung 56b:239–244
42. Hoffmann R-D, Fugmann A, Rodewald UC, Pöttgen R (2000) New intermetallic compounds Ln2Ni2Mg (Ln = Y, La-Nd, Sm, Gd-Tm) with Mo2FeB2 structure. Z Anorg Allg Chem 626:1733–1738
43. Rieger W, Nowotny H, Benesovsky F (1964) Die Kristallstruktur von Mo2FeB2 – Kurze Mitteilung. Monatshefte fur chemie und verwandte teile anderer wissenschaften 95:1502–1503
44. Couillaud S, Gaudin E, Andrieux J, Gorsse S, Gayot M, Bobet JL (2012) Study of the hydrogenation mechanism of LaCuMg8 ternary phase: the decomposition induces kinetics improvement. Int J Hydrogen Energy 37:11824–11834
45. Rodewald UC, Chevalier B, Pöttgen R (2007) Rare earth-transition metal-magnesium compounds—an overview. J Solid State Chem 180:1720–1736
46. Chotard JN, Filinchuk Y, Revaz B, Yvon K (2006) Isolated [Ni2H7]7- and [Ni4H12]12- ions in La2MgNi2H8. Angewandte chemie international edition 45:7770–7773
47. Chevalier B, Krolak AA, Bobet J-L, Gaudin E, Weill F, Hermes W, Pöttgen R (2008) On the strongly correlated electron hydride Ce2Ni2MgH7.7. Inorg Chem 47(22):10419–10424

48. Renaudin G, Guénée L, Yvon K (2003) LaMgNiH7, a novel quaternary metal hydride containing tetrahedral [NiH4]4- complexes and hydride anions. J Alloy Compd 350:145–150
49. Kadir K, Sakai T, Uehara I (1997) Synthesis and structure determination of a new series of hydrogen storage alloys; RMg2Ni9 (R = La, Ce, Pr, Nd, Sm and Gd) built from MgNi2 Laves type layers alternating with AB5 layers. J Al Compds 257:115–121
50. Kadir K, Sakai T, Uehara I (2000) Structural investigation and hydrogen storage capacity of LaMg2Ni9 and (La0.65Ca0.35) (Mg1.32Ca0.68)Ni9 of the AB2C9 type structure. J Al Compds 302:112–117
51. Kadir K, Kuriyama N, Sakai T, Uehara I, Eriksson L (1999) Structural investigation and hydrogen capacity of CaMg2Ni9: a new phase in the AB2C9 system isostructural with LaMg2Ni9. J Al Compds 284:145–154
52. Geibel C, Klinger U, Weiden M, Buschinger B, Steglich F (1997) Magnetic properties of new Ce-T-Mg compounds (T = Ni, Pd). Phys B 237–238:202–204
53. Guénée L, Favre-Nicolin V, Yvon K (2003) Synthesis, crystal structure and hydrogenation properties of the ternary compounds LaNi4Mg and NdNi4Mg. J Alloy Compd 348:129–137
54. Bobet J-L, Lesportes P, Roquefere J-G, Chevalier B, Asano K, Sakaki K, Akiba E (2007) A preliminary study of some "pseudo-AB2" compounds: RENi4Mg with RE = La, Ce and Gd. Structural and hydrogen sorption properties. Int J Hydrog Energy 32:2422–2428
55. Aonol K, Orimo S, Fujii H (2000) Structural and hydriding properties of MgYNi4: a new intermetallic compound with C15*b*-type Laves phase structure. J Alloy Compd 309:L1–L4
56. Kadir K, Noreus D, Yamashita I (2002) Structural determination of AMgNi4 (where A = Ca, La, Ce, Pr, Nd and Y) in the AuBe5 type structure. J Alloy Compd 345:140–143
57. Osamura K, Murakami Y (1978) Crystal-structures of CuSnMg and Cu4SnMg ternary compounds. J Less-Common Metals 60:311–313
58. Stan C, Andronescu E, Asano K, Sakaki K, Bobet J-L (2008) In situ X-ray diffraction under H2 of the pseudo-AB2 compounds: YNi3.5Al0.5Mg. Int J Hydrog Energy 33:2053–2058
59. Stan C, Andronescu E, Predoi D, Bobet J-L (2008) Structural and hydrogen absorption/desorption properties of YNi4−*x*Al*x*Mg compounds (with $0 \leq x \leq 1.5$). J Alloy Compd 461:228–234
60. Stan C, Asano K, Sakaki K, Akiba E, Couillaud S, Bobet J-L (2009) In situ XRD for pseudo Laves phases hydrides highlighting the remained cubic structure. Int J Hydrog Energy 34:3038–3043
61. Luo ZP, Zhang SQ (2000) High-resolution electron microscopy on the X-Mg12ZnY phase in a high strength Mg-Zn-Zr-Y magnesium alloy. J Mater Sci Lett 19:813–815
62. Park ES, Chang HJ, Kim DH (2007) Mg-rich Mg–Ni–Gd ternary bulk metallic glasses with high compressive specific strength and ductility. J Mater Res 22(2):334–338
63. Solokha P, De Negri S, Pavlyuk V, Saccone A, Marciniak B (2007) Crystallochemistry of the novel two-layer RECuMg4 (RE = La, Tb) ternary compounds. J Solid State Chem 180:3066–3075
64. Teresiak A, Gebert A, Savyak M, Uhlemann M, Mickel Ch, Mattern N (2005) In situ high temperature XRD studies of the thermal behaviour of the rapidly quenched Mg77Ni18Y5 alloy under hydrogen. J Alloy Compd 398:156–164
65. Kalinichenka S, Röntzsch L, Baehtz C, Kieback B (2010) Hydrogen desorption kinetics of melt-spun and hydrogenated Mg90Ni10 and Mg80Ni10Y10 using in situ synchrotron. X-ray diffraction and thermogravimetry. J Alloy Compd 496(1–2):608–613
66. Gebert A, Khorkounov B, Wolff U, Mickel Ch, Uhlemann M, Schultz L (2006) Stability of rapidly quenched and hydrogenated Mg–Ni–Y and Mg–Cu–Y alloys in extreme alkaline medium. J Alloy Compd 419:319–327
67. Hagihara K, Yokotani N, Umakoshi Y (2010) Plastic deformation behavior of Mg12YZn with 18R long-period stacking ordered structure. Intermetallics 18(2):267–276
68. De Negri S, Giovannini M, Saccone A (2007) Constitutional properties of the La–Cu–Mg system at 400 °C. J Alloy Compd 427:134–141

69. Opainich IM, Pavlyuk VV, Bodak OI (1996) Crystal structure of a Ce2Fe2Mg15 compound. Crystallogr Rep 41(5):813–816
70. Florio JV, Baenziger NC, Rundle RE (1956) Compounds of thorium with transition metals. II. Systems with iron, cobalt and nickel. Acta Crystallogr A 9:367–372
71. Givord D, Givord F, Lemaire R, James WJ, Shah JS (1972) Evidence of disordered substitutions in the "Th2Ni17-type" structure. Exact determination of the Th-Ni, Y-Ni and Er- Co compounds. J Less-Common Metals 29:389–396
72. Johnson Q, Smith GS (1967) Refinement of the Th2Ni17-Type structure: CeMg10.3. Acta Crystallogr A 23:327–329
73. Isnard O, Miraglia S, Soubeyroux JL, Fruchart D, Stergiou A (1990) Neutron diffraction study of the structural and magnetic properties of the R2Fe17Hx(Dx) ternary compounds (R = Ce, Nd and Ho). J Less-Common Metals 162:273–284
74. Tereshina I, Nikitin S, Suski W, Stepien-Damm J, Iwasieczko W, Drulis H, Skokov K (2005) Structural and magnetic properties of Dy2Fe17Hx (x = 0 and 3) single crystals. J Alloy Compd 404–406:172–175
75. Block G, Jeitschko W (1987) Tb2Mn17C3-x with filled Th2Ni17-type structure and some structural and magnetic properties of related compounds. J Solid State Chem 70:271–280
76. Fischer P, Halg W, Schlapbach L, Yvon K (1978) Neutron and X-ray diffraction investigation of deuterium storage. J Less-Common Metals 60(1):1–9
77. Zachariasen WH, Holley CE, Stamper Jnr JF (1963) Neutron diffraction study of magnesium deuteride. Acta Crystallogr A 16:352–353
78. Yajima S, Kayano H, Toma H (1977) Hydrogen sorption in La2Mg17. J Less-Common Metals 55:139–141
79. Slattery DK (1995) The hydriding-dehydriding characteristics of La2Mg17. Int J Hydrog Energy 20(12):971–973
80. Khrussanova M, Pezat M, Darriet B, Hagenmuller P (1982) Le stockage de l'hydrogène par les alliages La2Mg17 et La2Mg16Ni. J Less-Common Metals 86:153–160
81. Khrussanova M, Terzieva M, Peshev P (1986) On the hydriding kinetics of the alloys La2Mg17 and La2−*x*Ca*x*Mg17. Int J Hydrog Energy 1(5):331–334
82. Sun D, Gingl F, Nakamura Y, Enoki H, Bououdina M, Akiba E (2002) In situ X-Ray diffraction study of hydrogen-induced phase decomposition in LaMg12 and La2Mg17. J Alloy Compd 333:103–108
83. Wang L, Wang X, Chen L, Gao L, Xiao X, Chen C (2006) Effects of surface modification on the electrode behavior of ball-milled La2 Mg17 + 200 wt% Ni composite in alkaline solution. J Alloy Compd 420:306–311
84. Gao XP, Lu ZW, Wang Y, Wu F, Song DY, Shen PW (2004) Electrochemical hydrogen storage of nanocrystalline La2Mg17 alloy ball-milled with Ni Powders. Electrochem Solid-State Lett 7(5):A102–A104
85. Reilly JJ, Wiswall RH (1967) The reaction of hydrogen with alloys of magnesium and copper. Inorg Chem 6(12):2220–2223
86. Huot J, Liang G, Boily S, Van Neste A, Schulz R (1999) Structural study and hydrogen sorption kinetics of ball-milled magnesium hydride. J Alloy Compd 293–295:495–500
87. Palumbo M, Torres FJ, Ares JR, Pisani C, Fernandez JF, Baricco M (2007) Thermodynimic and ab initio investigation of the Al-H-Mg system. Comput Coupling Phase Diagrams Thermochem 31:457–467
88. Gorsse S, Shiflet GJ (2002) A thermodynamic assessment of the Cu-Mg-Ni ternary system. Comput Coupling Phase Diagrams Thermochem 26(1):63–83
89. Andreasen A, Sørensen MB, Burkarl R, Møller B, Molenbroek AM, Pedersen AS, Vegge T, Jensen JN (2006) Dehydrogenation kinetics of air-exposed MgH2/Mg2Cu and MgH2/MgCu2 studied with in situ X-ray powder diffraction. Appl Phys A 82:515–521
90. Shao H, Wang Y, Xu H, Li X (2005) Preparation and hydrogen storage properties of nanostructured Mg2Cu alloy. J Solid State Chem 178:2211–2217
91. Karty A, Grunzweig-Genossar J, Rudman PS (1979) Hydriding and dehydriding kinetics of Mg in a Mg/Mg Cu eutectic alloy: pressure sweep method. J Appl Phys 50(11):7200–7210

92 Au M, Wu J, Wang Q (1995) The hydrogen storage properties and the mechanism of the hydriding process of some multi-component magnesium- base hydrogen storage. Int J Hydrogen Energy 20(2):141–150
93. De Bruijn TJW, De Jong WA, Van den berg PJ (1981) Kinetic parameters in Avrami-Erofeev type reactions from isothermal and non-isothermal experiments. Thermochimica Acta 45:315–325
94. Barkhordarian G, Klassen T, Bormann R (2004) Effect of Nb2O5 content on hydrogen reaction kinetics of Mg. J Alloy Compd 364:242–246
95. Bobet JL, Kandavel M, Ramaprabhu S (2006) Effects of ball milling condition and additives on the hydrogen sorption properties of Mg + 5wt% Cr2O3 mixtures. J Mater Res 21 (7):1747–1752
96. Wu CZ, Wang P, Yao X, Liu C, Chen DM, Lu GQ, Cheng HM (2006) Hydrogen storage properties of MgH2/SWNT composite prepared by ball milling. J Alloy Compd 420:278–282
97. Liang G, Huot J, Boily S, Van Neste A, Schulz R (1999) Catalytic effect of transition metals on hydrogen sorption in nanocrystalline ball milled MgH2–Tm (Tm = Ti, V, Mn, Fe and Ni) systems. J Alloy Compd 292:247–252
98. Zhang LT, Ito K, Vasudevan VK, Yamaguchi M (2002) Effects of cold-rolling on the hydrogen absorption/desorption behaviour of Ti–22Al–27Nb alloys. Mater Sci Eng A329–331:362–366
99. Zhang LT, Ito K, Vasudevan VK, Yamaguchi M (2001) Hydrogen absorption and desorption in a B2 single-phase Ti-22Al-27Nb alloy before and after deformation. Acta Materalia 49:751–758
100. Couillaud S, Enoki H, Amira S, Bobet JL, Akiba E, Huot J (2009) Effect of ball milling and cold rolling on hydrogen storage properties of nanocrystalline TiV1.6Mn0.4 alloy. J Alloy Compd 484:154–158
101. Dufour J, Huot J (2007) Rapid activation, enhanced hydrogen sorption kinetics and air resistance in laminated Mg–Pd 2.5at.%. J Alloy Compd 439:L5–L7
102. Zaluska A, Zaluski L, Ström-Olsen JO (1999) Synergy of hydrogen sorption in ball-milled hydrides of Mg and Mg2Ni. J Alloy Compd 289:197–206
103. Li Z, Liu L, Jiang L, Wang S (2007) Characterization of Mg-20wt% Ni-Y hydrogen storage composite prepared by reactive mechanical alloying. Int J Hydrog Energy 32:1869–1874
104. Swanson HE, Tatge E (1959) structure of Mg. J Res Nat Bur Stand 46:318–327

Erratum to: Nanoenergy

Flavio L Souza and Edson R Leite

Erratum to: F.L. Souza and E.R. Leite (eds.), *Nanoenergy*, https://doi.org/10.1007/978-3-319-62800-4

The original version of the book was inadvertently published without updating the editor names "Flavio Leandro de Souza and Edson Leite" as "Flavio L Souza and Edson R Leite", respectively, in Frontmatter and Cover page. The erratum book has been updated with the change.

The updated online version of this book can be found at https://doi.org/10.1007/978-3-319-62800-4

F.L. Souza and E.R. Leite (eds.), *Nanoenergy*,
https://doi.org/10.1007/978-3-319-62800-4_11

Index

F.L. de Souza and E. Leite (eds.), *Nanoenergy*,
DOI 10.1007/978-3-319-62800-4

MIX
Papier aus verantwortungsvollen Quellen
Paper from responsible sources
FSC® C105338

If you have any concerns about our products,
you can contact us on
ProductSafety@springernature.com

In case Publisher is established outside the EU,
the EU authorized representative is:
Springer Nature Customer Service Center GmbH
Europaplatz 3, 69115 Heidelberg, Germany

Printed by Libri Plureos GmbH
in Hamburg, Germany